BARRON'S

CDL

Commercial Driver's License
Truck Driver's Test

Mike Byrnes and Associates, Inc.

Dedication

This book is dedicated to the late Mike Byrnes, who left a legacy of enthusiasm and dedication to excellence in truck driver education. We continue on in that tradition, guided by his principles and concern for you, the student.
Mike Byrnes & Associates, Inc.
Devorah Fox, President

Acknowledgements

Special recognition is due John M. Rojas, Director, Del Mar College Transportation Training Services, Corpus Christi, Texas, and Joyce Walters. You are deeply appreciated.

Published by Kaplan North America, LLC d/b/a Barron's Educational Series
1515 West Cypress Creek Road
Fort Lauderdale, FL 33309
www.barronseduc.com

ISBN: 978-1-5062-8763-8

10 9 8 7 6 5 4 3 2 1

Kaplan North America, LLC d/b/a Barron's Educational Series print books are available at special quantity discounts to use for sales promotions, employee premiums, or educational purposes. For more information or to purchase books, please call the Simon & Schuster special sales department at 866-506-1949.

Table of Contents

How to Use This Book

If you want to haul cargo as a professional truck driver, this book will help you to pass the Commercial Driver's License Tests. You must pass these tests to get your Commercial Driver's License (CDL).

In this book, you will learn what the CDL is. You'll learn who must have one. You'll find out how to apply for a CDL. You'll get the information that our government says you must have to be a commercial driver. Plus, you'll find comprehensive review, extensive practice, and helpful tips on taking the CDL tests.

We hope that you'll find the process painless, maybe even enjoyable.

This book is divided into three main sections:

Part One—In this part, you'll learn general information about the CDL and the tests that you must take to get one. Review the various question types you may encounter, and study the important test-taking tips and tools for success. Also be sure to take the diagnostic test to find out your strengths and areas that need improvement.

Part Two—Within these chapters, you will review important topics and facts about commercial truck driving that you must have at your command to pass the CDL tests. For each chapter, first attempt the PASS Pre-trip questions that open the chapter to see what you know. Then, after reviewing the contents of the chapter, try the PASS Post-trip questions at the end of the chapter to gauge how much you've learned.

Part Three—After completing your review of Parts One and Two, it's time to take the sample CDL tests in Part Three. Answer every question, and then consult the Answer Keys in Appendix A to see how well you did. Within this part, you'll also find some final tips on making your test day worry-free!

We, the authors, are honored that you have chosen this book to prepare for the CDL. We know how important this license is to you. Our pledge is to give you every bit of help we can. Your time is valuable, so we won't waste it with information you don't need. We want you to pass the CDL. Your success is our success.

We have a long history of experience in preparing people for successful careers as commercial vehicle drivers. Now, with this book, we can help with that final step toward that career: getting a CDL. This book will help you with that one important extra step.

Congratulations on your decision to prepare yourself for taking the CDL tests. Many drivers skip this step. They take the tests unprepared. They may feel that their driving experience or training is enough. Their test results show that this isn't true. Even experienced drivers with good driving records fail the CDL tests. The failure is not because the applicants don't know how to drive. Many times, they simply don't do well on tests.

Many people don't do well on tests, even when they know their stuff. They don't recognize the correct answer, even when they know what it is, because of the way the question is worded. Perhaps that's been your experience. If so, preparing for the CDL tests with this book is an especially smart move. You will not only get the facts you need for the tests, you will also learn how to be a good test-taker. You will learn skills you will be able to use on any type of test, now and in the future.

So let's get to it!

A BRIEF OVERVIEW OF THE CDL

What is the Commercial Driver's License anyway? Very simply, the Commercial Driver's License grants a license to certain drivers to drive certain vehicles. Let's examine that piece by piece to see what that means to you.

License

When you receive a driver's license, any kind of driver's license, a body of government has granted you permission to drive. In this case, it's a state government. That makes driving a regulated activity.

Why should the government regulate driving? Shouldn't we all be allowed to drive if we wish? Take a closer look at that question.

What if you could not hear or see well? What if you did not know the rules of the road, or refused to obey them? You would be an unsafe driver. You would be a danger to others on the road. To ensure everyone's safety, the state government judges your fitness to drive before granting you a license. You are given permission to drive. It's not a right granted to everyone.

What is a commercial driver's license? A commercial driver's license is different from a car or motorcycle driver's license. It gives **certain drivers** the license to drive **certain vehicles**. These are commercial vehicles.

Certain Vehicles

What vehicles are commercial vehicles? Briefly, these are:

- vehicles with gross vehicle weights of 26,001 pounds or more
- trailers with gross vehicle weights of 10,001 pounds or more
- vehicles that transport hazardous materials that require placards
- buses designed to carry 16 or more people (including the driver)

Later in this book, you'll find "commercial vehicle" defined in greater detail. We will explain terms like "gross vehicle weight" and "placards."

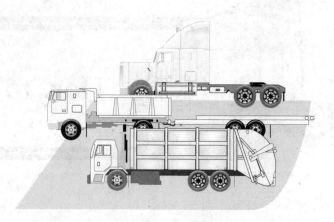

Figure 1. Drivers of vehicles like these need Commercial Driver's Licenses.

Certain Drivers

Who are those "certain drivers?" Only those drivers who pass the Commercial Driver's License tests may hold a CDL. There are different CDLs for different driving jobs, and tests that go with them. You may take all the tests, or only a few. **All** commercial vehicle drivers must pass at least the basic knowledge test.

There are many reasons for a Commercial Driver's License law. Here are three:

- public safety
- driver quality
- national standards

Driver Quality

The Commercial Motor Vehicle Safety Act seeks to remove unqualified drivers from the road. All drivers of commercial vehicles must show they have a minimum of knowledge and skill. This is shown through testing.

National Standards

You just read, however, that it is the state that gives out licenses, not the federal government. This is still true. What the Commercial Vehicle Safety Act of 1986 did was set standards that all states must follow in granting licenses.

In the past, states did not always use the same standards when giving out driver's licenses. As of July 1988, only 32 states issued some form of a classified driver's license. (A classified license is one that states what type of vehicle the license holder could drive.) Of these 32 states, only 12 required a behind-the-wheel test to obtain the license. Some of the remaining states accepted proof of training, either on the job or through a school, instead of a license test. Some states required neither testing nor training.

So you see, there were no "standards." It was easier to get a license in some states than in others.

The federal government has offered states a model test to follow. Most states do follow this model. This book follows the same model. Later, you'll learn how to find out if your state follows this model, and what to do if it doesn't.

You will still get your license from your state. When you do, you'll have to have the same basic knowledge and skills that drivers in all the other states must have. The Commercial Driver's License Test will test those skills.

Think of it this way. In trucking's early days, driving a heavy vehicle used to be a simple job. Now, there are a great many laws covering driving. The equipment is more complex. Traffic conditions leave no room for error. Driving is no longer simple. The driver of a heavy vehicle must be better informed and more skilled than ever before. Other drivers want to feel safe and confident when they are on the road.

It's much like other forms of professional licensing. Doctors, lawyers, and accountants all must pass licensing exams before they can go into practice. The license assures consumers that these professionals have at least the basic skills and knowledge required to do their job. Now, the motoring public can have the same confidence about the heavy vehicle drivers with whom they share the road.

With your CDL in hand, you will not only drive like a professional, you will have the license to prove it.

You may not be delighted about having to take this test. You may resent it. You may be fearful about it. Put those bad feelings aside. Resentments and fears won't help you. The whole process will go much better if you approach it positively. Once you have completed this book, you can tackle those tests with confidence.

PART ONE

Chapter 1

Introduction

Were you surprised by the size of this book? Are you wondering if there really is so much to learn before taking the CDL tests? Are you thinking, "I hope I don't have to read all of it?"

Perhaps you won't. We'll show you how to determine just how much of this book you do need. This is one feature of the **PASS** system you'll use to prepare for your CDL tests. In this chapter, you will learn:

- how to get the most information from reading
- success strategies for taking tests
- what other resources will help you prepare for the CDL test
- how much you already know about subjects tested by the CDL
- how to measure the progress of your preparation
- what agencies exist to help you get your CDL
- how to apply for a CDL

To complete this chapter you will need:

- scrap paper
- a pencil and an eraser
- a calendar with room to write on
- colored pens, markers, pencils, or crayons

PASS is a technique that you'll apply to getting through the rest of the book. Perhaps you don't have much spare time in which to prepare for the CDL tests. You will appreciate the way **PASS** helps you make the most of that time. **PASS** will help you stay on track, and stay motivated.

PASS includes tips on how to read. Perhaps you don't think there's any particular trick to this. You may be in for a surprise. You'll also learn how to take a test, any test.

USING PASS

The letters **PASS** stand for two things. One is the word "PASS"—your final goal. Your reason for reading this book in the first place is that you want to "**PASS**" the CDL test.

"**PASS**" also stands for "**P**ersonal **A**ction **S**trategy for **S**uccess." It's the system we suggest you use to prepare for taking the test.

PASS IS PERSONAL

This means that you can tailor it to your needs and preferences. Each person preparing for the CDL tests is different. Your skills, knowledge, and experience are different from your co-driver's, your friend's, or your spouse's.

Your preferences are different, too. Some people like to read. Some people don't. Some people like to work alone. Others do better in a group. Some people can chug along quite nicely under their own power. Other people need a little pat on the back now and then to keep going.

Which type are you? Maybe you're not so sure. To find out, take stock of your needs and make note of your preferences.

Needs

First, how much do you already know about driving? Have you just finished a training program? Your knowledge may be fresh and quite complete. You may feel confident about your knowledge.

Maybe you're trying to upgrade your license. Have you been driving for some time? You may not be sure of what you do know, or wonder if it's going to be enough. The quickest way to get a reading on this would be—you guessed it—take a test.

Throughout this book you will find many tests—tests before you read a chapter, and after. If you simply don't like tests, think of them as inspections. In fact, we'll call them **PASS** Pre-trip and **PASS** Post-trip Inspections.

A vehicle inspection is a routine part of the driver's job. It's required by law, and it's important for safety. You will likely inspect your vehicle both before your run and after. In this book, you'll "inspect" your knowledge both before and after you read a chapter.

Figure 1-1. Testing your knowledge "pre-trip" is like inspecting your vehicle pre-trip.

When you pre-trip your vehicle, you take stock of your equipment before going to work. You are looking for defective or missing equipment. When you "pre-trip" yourself before starting a chapter, you will be looking for defective or missing areas in your knowledge about the chapter's subject.

Then, when you post-trip your vehicle, you check to see what, if anything, has changed since you last looked. It's the same with the **PASS** Post-trip test. You'll look for changes. You'll check for any change in your knowledge about the subject covered by the chapter you just read.

Before you go any further, then, pre-trip your knowledge about truck driving.

DIAGNOSTIC CDL TEST

The following is a sample CDL test that contains questions similar to the ones on an actual CDL Knowledge Test. The questions are similar in terms of the question type presented and the information covered. You may take all the time you need to complete the test. However, do complete the test in one sitting, much as you would under true test conditions. Time yourself to see how long it takes.

You may not use any notes or reference materials while you are taking this sample test. You may not ask for help from another person. You can't do any of these things when taking the CDL Knowledge Test.

This sample test is made up of one type of test question. This is the multiple-choice question. First read the numbered question or statement. Then read the three choices—A, B, and C. If the numbered item is a question, pick the lettered item that answers the question correctly. If the numbered item is a statement, pick the lettered item that completes the statement and makes it true.

Write your answers on the answer sheet provided, not on the test itself. That will make it easier to score.

When you have finished marking your answers on this answer sheet, turn to page 8. That's where you will find the answers so you can score your own test. You can also find out what different scores mean starting on page 8.

Mark your answer sheet with the date on which you first take the diagnostic test. Fill in the time you spend taking the test, and record your score. Store it away because you will come back to it later.

DIAGNOSTIC TEST ANSWER SHEET

Test Item	Answer Choices				Self - Scoring Correct	Incorrect
1.	A	B	C	D	☐	☐
2.	A	B	C	D	☐	☐
3.	A	B	C	D	☐	☐
4.	A	B	C	D	☐	☐
5.	A	B	C	D	☐	☐
6.	A	B	C	D	☐	☐
7.	A	B	C	D	☐	☐
8.	A	B	C	D	☐	☐
9.	A	B	C	D	☐	☐
10.	A	B	C	D	☐	☐
11.	A	B	C	D	☐	☐
12.	A	B	C	D	☐	☐
13.	A	B	C	D	☐	☐
14.	A	B	C	D	☐	☐
15.	A	B	C	D	☐	☐
16.	A	B	C	D	☐	☐
17.	A	B	C	D	☐	☐
18.	A	B	C	D	☐	☐
19.	A	B	C	D	☐	☐
20.	A	B	C	D	☐	☐

Cut Here

1. If you do not have a Hazardous Materials Endorsement on your Commercial Driver's License, you can _____.
   A. never haul hazardous materials
   B. haul hazardous materials when the load does not require placards
   C. haul hazardous materials when the load will not cross state lines

2. You are driving a 40-foot vehicle at 45 mph on dry pavement in clear daylight weather. To be safe, the least amount of space you should keep in front of your vehicle is the distance you travel in _____.
   A. three seconds
   B. four seconds
   C. five seconds

3. The best way to handle a tailgater is to _____.
   A. increase speed to put more space between you and the tailgater
   B. increase the space in front of your vehicle
   C. signal the tailgater when it is safe to pass you

4. When steering in order to avoid a crash, _____.
   A. apply the brakes while turning
   B. don't turn any more than needed to clear whatever is in your way
   C. avoid countersteering

5. A vehicle carrying a load that extends four feet or more beyond the rear of the body must have _____ at the extreme end of the load.
   A. one white light
   B. two amber reflectors
   C. one red flag

6. If the road you are driving on becomes very slick with ice, you should _____.
   A. downshift to stop
   B. stop driving as soon as it is safe to do so
   C. drive at a varying speed

7. You wish to turn right from a two-lane, two-way street to another. Your vehicle is so long that you must swing wide to make the turn. The proper way to make this turn is shown in _____.
   A. Figure A   B. Figure B   C. Figure C

8. When the roads are slippery, you should _____.
   A. drive alongside other vehicles
   B. decrease the distance that you look ahead of your vehicle
   C. make turns as carefully as possible

9. A load's center of gravity _____.
   A. is only a problem if the vehicle is overloaded
   B. should be kept as high as possible
   C. can make a vehicle more likely to roll over on a curve.

10. Heading down a long, steep hill, your brakes begin to fade, then fail completely. You should _____.
    A. downshift
    B. pump the brake pedal
    C. look for an escape ramp or escape route

11. Medical certificates must be renewed every _____.
    A. year
    B. two years or as specified by a physician
    C. four years

12. To correct a rear-wheel acceleration skid, _____.
    A. apply the brake
    B. increase acceleration to the wheels
    C. stop accelerating and push in the clutch

13. To avoid a crash, you drove onto the right shoulder. You are driving on the shoulder at 40 mph. The safest way to return to the pavement is to _____.
    A. make sure the pavement is clear. Come to a complete stop. Then steer back onto the pavement
    B. brake hard to slow the vehicle. Then steer sharply onto the pavement
    C. keep moving at 40 mph, steering very gently back onto the pavement

14. Which of these statements about overhead clearance is true?
    A. If the road surface causes the vehicle to tilt, you should drive close to the shoulder.
    B. You should assume posted clearance signs are correct.
    C. A vehicle's clearance can change with the load carried.

15. Which of these statements about tires is true?
    A. Tires of mismatched sizes should not be used on the same vehicle.
    B. Radial and bias-ply tires can be used together on the same vehicle.
    C. A tread depth of 2/32 inch is safe for the front tires.

16. You can put out _____ fire with water.
    A. a tire
    B. a gasoline
    C. an electrical

17. While driving at 50 mph on a straight, level highway, a tire blows out. The first thing you should do is _____.
    A. begin light braking
    B. stay off the brake until the vehicle slows on its own
    C. quickly steer onto the shoulder

18. The purpose of retarders is to _____.
    A. provide emergency brakes
    B. apply extra braking power to the non-drive axles
    C. help slow the vehicle while driving and reduce brake wear

19. Hydroplaning _____.
    A. only occurs when there is a lot of water
    B. only occurs at speeds above 55 mph
    C. is more likely if tire pressure is low and if there is a lot of water

20. What is the proper way to test for leaks in a hydraulic brake system?
    A. Step on the brake pedal and the accelerator at the same time and see if the vehicle moves.
    B. Move the vehicle slowly and see if it stops when the brake is applied.
    C. With the vehicle stopped, pump the pedal three times, apply pressure, then hold for five seconds and see if the pedal moves.

Scoring the Diagnostic Test

Count the number of questions you answered incorrectly. Include in the count of wrong answers any questions you did not answer (left blank).

ANSWERS

1. B	**6.** B	**11.** B	**16.** A
2. C	**7.** B	**12.** C	**17.** B
3. B	**8.** C	**13.** A	**18.** C
4. B	**9.** C	**14.** C	**19.** C
5. C	**10.** C	**15.** A	**20.** C

Scoring

Five or more wrong answers: less than 80 percent
Four wrong answers: 80 percent
Three wrong answers: 85 percent
Two wrong answers: 90 percent
One wrong answer: 95 percent
No wrong answers: 100 percent!

Using Your Diagnostic CDL Knowledge Test Score

How did you do on the Diagnostic CDL Knowledge Test? Did you really take it? You should.

The diagnostic test is very much like the CDL Knowledge Test offered by many states. All states must ask at least 30 questions. Many will ask more. Most will have multiple-choice questions like the one on the Diagnostic CDL Knowledge Test. Taking the diagnostic test will do at least two things for you:

- it gives you a feeling for what it's like to take a long, multiple-choice question test
- it gives you an idea of how well prepared you already are for an actual CDL Knowledge Test

Here's how to interpret your score:

Less than 80 percent—If you scored less than 80 percent, you did not pass the diagnostic test. You wouldn't pass the real CDL Knowledge Test, either. You must have a score of 80 percent or more to pass. Don't go crazy! You took this test with very little preparation. Remember, many experienced and skilled drivers have failed the CDL Knowledge Test because they took it without preparation.

California was the first state to start commercial driver licensing. In the first few months of testing, failure rates were around 57 percent. Word quickly got out that people needed to prepare for this test. As more and more people did, failure rates dropped. Later reports stated that failure rates were down to 15 percent in some areas.

Your score may not surprise you. You may have already known you would need some preparation for the CDL Knowledge Test. Give yourself plenty of time to go through this book thoroughly. Plan to do every exercise, take every practice test, and use the other suggested resource materials until you get perfect scores. Allow yourself time to give extra effort to the subject areas that you missed on the diagnostic test. Then you can tackle the actual CDL tests with confidence.

Between 80 and 100 percent—You passed the test. You can feel good, but don't get too cocky. What about the questions you missed? Do you want to risk missing them on the real CDL Knowledge Test? Pinpoint those subject areas and plan to give them special attention.

If you scored 100 percent, congratulations! You are already well prepared to take the CDL Knowledge Test. Just the same, you should work through the rest of this book. After all, this was just a 20-question sample test. There will be at least 30 questions on the actual CDL Knowledge Test. You want to be prepared for them all.

You might use a "scan" or "review" technique described later. This technique will allow you to target those areas where your knowledge is the weakest. Or, you may wish to do a more thorough preparation, just to be sure.

Time Needed

One factor that will influence your test preparation plan is "time." How much time will you need?

How do you learn new things? Think back to the last time you had to learn something new. Say it was a work-out program. How did you go about learning the exercises? Did you sign up for a class, or get a trainer? Did you get a how-to book from a bookstore, and create a program for yourself?

These methods are good. Neither is better than the other. If you tend to learn new things by reading about them, using this book will feel quite natural. You will like being able to use it anytime, anywhere. You won't need any special equipment, or have to wait for an instructor to be available.

If you would rather be "shown how" than "read about how," you won't have the advantage of using your preferred learning method. You'll have to cut yourself some slack for that.

Preparing for the CDL tests will take anywhere between 28 and 90 hours. How many hours you will need depends on your requirements and your level of commitment.

If you just finished a formal training program, much of the information in this book will sound familiar. Your preparation for the CDL tests may be more like a review. Otherwise, you may find that many of the terms and ideas in this book sound new. You may need to go over the material more than once.

Your commitment to give this material your full attention will help shorten the time you need to spend. If you allow yourself to be distracted, you will have to go over the material more than once to absorb it.

You may need anywhere between 28 and 90 hours to prepare adequately for the CDL, but these are merely estimates. Your experience will be unique. There is no "right" answer. The real amount of time you will need to prepare for the CDL test is however much it takes for you to feel confident. Take all the time you need, and all the time you can afford to spend.

Preferences

When you were in school, did you feel you could do a whole lot better if only you could set the schedule?

Now you can. Perhaps you're a "night person." If you think the best time to pick up this book would be 2:00 in the morning, then by all means, that's when you should do it. Your spouse may think you're nuts at first. You'll just have to explain that this is what works best for you.

Find it hard to work alone? Many people do. Try teaming up with someone else who must prepare for the CDL tests. Having a "study buddy" can help you stay focused and on track. It can make the whole process just a little more pleasant. Your buddy's skills and experience will likely be different from yours. This different viewpoint can be valuable.

You can choose where you want to work. It can be at home, in the public library, in the park—wherever suits you. In the section on how PASS is a system, you'll find suggestions for a work space that will aid the preparation process. You can pick the tip that most appeals to you.

PASS IS ACTION

Action is a very important part of PASS. You can't just put this book under your pillow at night and expect the knowledge to seep into your head while you sleep. You must go after it. Tell yourself you are going to PASS the CDL tests, come hell or high water.

Set a date on which you plan to take the tests. Then state your goal: "I will PASS the CDL tests on (date)." Write it down. Post it someplace where you will see it every day. Try the bathroom mirror, or the refrigerator door. Writing your goal statement will fix it in your mind. Looking at it every day will keep you on track.

Collect the resource materials and tools you will need. You may need other books besides this one. When you do, they are listed at the beginning of each chapter. Have them all in place before you get to work.

You'll be taking notes as you read, so you'll need blank paper, blank index cards, a digital notepad, or a notebook. (Of course, you will also need a pen or pencil, too!) You will also find a color highlighter, colored pens, pencils, markers, or crayons helpful.

PASS IS A SYSTEM

Now that you have a goal, you need a plan to reach that goal. You need a way to measure your progress toward that goal, and a way to tell when you have reached the goal.

Find Time to Work

In the section on the personal part of PASS, you figured out how much time you will need to complete the program. How much time do you have?

As an adult, you can't devote the entire day to learning and studying. You have other jobs. Other responsibilities, such as family or work, take up your day.

Wondering where you are going to find the time to prepare for the CDL tests? If so, you may find it helpful to identify the free time that you do have. One suggestion is to create a "Free Time Locator," on which you draw yourself a weekly calendar and "x" out all the time periods that are taken up by work, chores, eating, and sleeping. Basically, just block out any time periods that are not "free time."

Then, highlight the time periods that remain. You might identify some free early morning hours after the family has gone off to work or school. Maybe there's some time available in the evening, just before dinner, or late at night, when your surroundings are quiet. No matter when these hours occur, you will commit them to preparing for the CDL tests.

If you have an irregular work schedule, you may find this exercise tough to complete. If you don't know when your duty shifts will be, you may not be able to identify when you will have free time. In that case, approach this exercise from another angle. Think back over the past few weeks. When were you not working? You may need to map a few weeks to see any pattern of free time.

Even after doing the exercise this way, you may not find any pattern of free time. You know it's there, though. You know you have times when you are not working, sleeping, or doing necessary chores. Make a commitment to use these hours to prepare for the CDL tests.

Set Mini-Goals

Your final goal is to pass the CDL tests and get your license. If achieving this goal is far in the future, you may need something more immediate. Setting mini-goals will help you to move along. It will break up a large task into smaller, more manageable units.

You could aim to finish a certain number of pages in each session. You could aim to continually improve your score on the **PASS** Post-trips. Set mini-goals that are well within your ability to achieve. That way, you will end each session with a feeling of achievement. Then you will look forward to the next one.

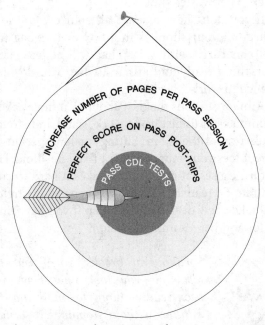

Figure 1-2. Aim to achieve your goals.

Set a Schedule

Earlier, you figured out when you will have time available to prepare for the CDL tests. Now you will combine that information with your mini-goals and make up a schedule. Making a schedule will help you to make the best use of your time. It will help you to stay on track.

For this exercise, you will need a calendar. It should be large enough so there is writing space for each day. If you don't have a large calendar, sketch one out on scrap paper, or use any calendar you can find.

Pencil in your mini-goals. If, for example, Test Day, the final goal, is 15 days away, you would write "Pass CDL Test" on the calendar day 15 days away from today. Perhaps you set a mini-goal of six hours of CDL preparation each day. You would write "CDL prep = 6 hours" on each calendar day until Test Day.

Fill in the calendar with your personal Test Day and mini-goals.

Now compare the calendar with your Free Time Locator (if you made one). You may find that you do not have an equal amount of spare time each day. Adjust your goals to suit reality. If you have no free time on one day, plan to make up the time somewhere else.

Use your calendar, your needs and mini-goals, and the Free Time Locator to schedule appointments with yourself.

Chart Your Progress

Now that you've made up your schedule, be sure to use it. Put it someplace where you will see it often. It will help keep you on track. You should avoid falling behind. Falling behind will only make you feel anxious. Being nervous will not enhance your learning ability.

Yes, emergencies do happen. In spite of your best intentions, you may have to miss a session. Take immediate steps to catch up. Don't allow yourself to fall too far off the schedule.

When you do complete a session, cross it off your calendar with a big, colorful "X." It may sound silly, but you will be surprised at how good this simple act feels. It's part of ending each session on a positive note. It's part of recognizing your achievements.

From time to time, review the progress you have made. You will likely notice that your skills are improving. You may be getting higher and higher scores on the **PASS** Post-trips. You may find that you can cover more pages in one session than you could at the beginning. These improved skills will help you **PASS** the CDL tests.

Your Work Space

The space in which you work can help you or hurt you. You'll base where you will do your preparation in part on your needs and preferences.

Most people do their best work when they have a minimum of distractions. That means don't try to read this book and watch TV at the same time. You'll probably remember more about what you watched than what you read.

On the other hand, you might not want to be in a room that's completely silent. When you first begin this program, you may find yourself easily distracted by little noises around you, such as traffic outside, or people talking in the next room.

A radio or stereo can be helpful. Find something to listen to that is neutral. That is, you don't want music you like so well that you end up dancing instead of reading. What you are looking for is something that will block out the other, more disturbing noises. This masking sound is called "white noise." "Easy listening" music is often a good choice. Have it on at a low volume.

As you improve your powers of concentration, you will be able to do without the "white noise." You will be able to tune out the little distracting noises that used to disturb you, all by yourself.

Find a space where you can work without interruption. If you are working at home, try the bedroom or kitchen. Ask those around you not to disturb you. You may have to "leave home" if you cannot find the privacy you need. Try going to a park or a public library.

Mostly, you will want to work by yourself. When someone else is with you, it's too easy to distract each other. However, if you have a hard time working in a vacuum, find a buddy. It would be best if this person is also preparing for the CDL tests. Find someone you can meet with from time to time, or call up on the phone.

The two of you can compare notes. You may have understood something that was a problem for your partner (or the other way around). Talking it out could clear up the matter.

If you are the competitive type, set up a little challenge. It could be just the extra spark you need to get cracking when you are tempted to goof off. Just knowing your buddy will ask you about your progress will keep you from skipping a session.

Get in the Mood

Here are some tips that help you focus on the task at hand and make it more pleasant. Like making big, red "X's" on your calendar, these tips may seem silly at first. They do help you focus on your work.

Pick out something special that you will use only when you work on this program. It could be a colorful calendar that you will devote to this project. Maybe it's a special coffee cup. Put on your "lucky shirt." Whatever it is, it will remind you to concentrate on your CDL test preparation, and only that.

If possible, try to work at the same time every day. This helps form a habit. When you start having the feeling that it's time to crack the books, you start work already warmed up mentally.

Have all the supplies you need with you before you begin. If you have a space all to yourself that won't be disturbed, that's great. Make sure all your books, paper, markers, and snacks are in place. Then you won't have to get up in the middle of a sentence to get a cup of coffee or a pencil. These interruptions will only break your train of thought. When you resume, you will have to backtrack. That wastes time.

If you don't have a space all your own, at least collect all your supplies in a box or tote bag that you can carry with you. That will at least keep everything ready and at hand when you do see a chance to work.

Aim for sessions that last about 45 minutes. After 45 minutes of undisturbed work, stand and stretch. Walk around the room. Sitting for long periods of time forces your muscles into a fixed position. This in turn causes fatigue, which makes it hard to concentrate. So read for 45 minutes, then get up, stretch, and take a few deep breaths. Then if you have time, go for another 45-minute session.

 Speaking of breaks, when was the last time you took one? Have you been reading along for more than 45 minutes without stopping? If so, now would be a good time to get up and stretch. Then come back to this page.

When working for 45 minutes at a time gets easier, increase the number of 45-minute sessions you string together. The goal is to get better at focusing your attention for long stretches of time. It will probably take longer than 45 minutes to complete the CDL tests.

As you get closer to your test date, make your **PASS** sessions more like actual test-taking. If you have been using "white noise" to help you concentrate, practice doing without it. Go through several sessions without coffee, soda, or snacks.

End your session on a positive note. Part of the trick is knowing when to quit. Once you have met your goal for the session, stop there. If you have gone through the number of pages you set out to do, and have passed the **PASS** Post-trip, call it a day.

You may be tempted to try "getting ahead." If you try to do too much, though, you may trip yourself up. If you get over-tired and don't do as well as you wanted to, you'll just end up feeling defeated. Again, if you have achieved what you have set out to do, that's enough. Pat yourself on the back, "X" out that mini-goal on your calendar, and put the work away until the next session.

Be in Good Physical Condition

You will have an easier time of preparing for and taking the CDL tests if you are in good physical condition. Does that surprise you? Did you say, "What am I training for, the Olympics?" It's true, though. If your health is generally good, you will be able to concentrate for longer periods. You will remember more of what you read. You will be more alert, sharper, better able to make decisions. So think of it as training for the Olympics, if you like. Do what you can to get, and stay, in good health and good physical condition.

PASS IS SUCCESS

The main goal of this program is, of course, to **PASS** the CDL tests. The ultimate reward will be getting your Commercial Driver's License. On the way to achieving your final goal, you have set mini-goals. Just as you have mini-goals, you will have mini-successes.

Celebrate those successes. Crossing off a completed session is one way of doing that. Treat yourself to a favorite food, or movie that you wanted to see. Make a list of small items that please you. Whenever you achieve a mini-goal, reward yourself with an item from that list.

Remember that "success" is achieving a goal, whatever the goal was. Your goal is to pass the CDL tests. You don't have to get 100 percent, you simply have to get 80 percent to pass. So if you pass with 80 percent, feel good about it! That was all you had to do.

You don't even have to pass the tests on the first try. Of course you would prefer to pass the first time. It's time-consuming and costly to have to take the tests more than once. Yet people do fail the tests on the first try, and pass on the second. Why? Often it's because taking tests like these was such a new experience for them. But on the second try, they had enough familiarity with the testing process to pass easily.

Plan to pass the CDL tests the first time. But if you don't, don't give up and look for some other career. You'll probably pass the second time. It's nothing to feel bad about. The important thing is, you did pass.

To sum up, **PASS** is a system to help you tackle learning something new. The letters **PASS** stand for Personal Action Strategy for Success. It's personal, because you make a judgment about how much you need to learn in order to pass the CDL tests. You decide how hard you have to work to learn it. You take into account your own preferences, and style.

It's action, because you don't just sit back and wait for the knowledge to come to you. You go after it aggressively.

It's a strategy, because you don't take aimless stabs at preparing for the CDL tests. You make a plan, and you follow that plan.

It's success, because you are going to pass the CDL tests. You are going to get your license.

So much for **PASS**. Next, you'll learn about **SEE**. Before you go on to Chapter 2, take a little break.

Chapter 2

Using SEE to Prepare for the CDL Tests

In this chapter, you'll learn about:

- reading techniques that result in maximum recall
- the difference between a dictionary and a glossary
- how to use a dictionary and a glossary
- what a "**PASS** Billboard" is

To complete this chapter, you will need:

- a dictionary
- a pencil or pen
- blank paper, a notebook, or an electronic notepad
- colored pencils, pens, markers, or highlighters

Are you here at Chapter 2 already? We thought we told you to take a little break after Chapter 1.

The letters in **SEE** stand for **S**can, **E**xamine, and **E**xtract. It's the technique we suggest you use to get the most out of this book, or any book you use to learn something new. You may think there's not much to know about a particular subject. Or, you may feel that you already know everything you need to know. We think you'll be surprised.

If, on the other hand, you have had a good grounding in "how to read," these techniques may sound familiar. If so, feel free to move rapidly through the material.

BE RELAXED AND ALERT

You will get the most from your **PASS** sessions if you are relaxed and alert when you begin. Avoid working when you are tired, or tense from the day's stresses.

Warm Ups

If you are tired, try some light exercise. If you've ever been involved in a work-out program, you may already know a warm-up routine. That would work. Any one of the following would also help you to feel more awake. Take a cool shower. Go for a brisk, 20-minute walk, jog, or bicycle ride. Jump rope or do "jumping jacks." Get just enough exercise to get your circulation going again, but not so much that you tire yourself out!

Stretches

To relieve stress, try some simple stretching exercises.

Stand and raise your arms straight up over your head. Really reach for the ceiling. Then let go. Bending from the waist, reach straight out in front of you. Reach to the left, then to the right. Bend over from the waist and reach toward your toes. Don't bounce or strain to touch your toes. Just hang there for a few minutes. Then come up slowly and take a deep breath.

Neck Rolls

"Neck rolls" is a stress-reliever you can do from a seated position. Here's how to do them. Slowly, allow your head to droop forward, as if you were going to rest your chin on your chest. Then, just as slowly, raise your head. Slowly, tilt your head to the right, as if you could touch your right ear to your right shoulder. Raise your head. Slowly, drop your head back and look up at the ceiling. Bring your head back to the face-forward position. Last, tilt your head to the left.

Repeat the whole series five times.

Next, keeping your neck straight, swivel just your head to the right and look along the line of your right shoulder. Swivel slowly back and to the left. Return to the face-forward position. Then repeat the movement, beginning at the left this time. Do this five times also.

As you do the neck rolls, breathe deeply. Breathe in slowly and deeply through your nose. Breathe out through your mouth.

Clear Your Mind

Imagine your mind is a blackboard, and you're erasing it so you can start your **PASS** session with a clean slate. Or, picture yourself gathering up your worries and preoccupations and putting them in a box. See yourself putting the box on a shelf. You know you can get the box down and work on those problems later. For now, though, you are going to be 100 percent involved in preparing for the CDL tests.

Get Ready

Make a small ritual of getting down to work. Think of it as suiting up for a ball game. This little ritual puts you in a frame of mind to focus on one thing only.

Gather all your supplies. Make sure you have everything we discussed in Chapter 1. To review, that would be:

- this book
- pencils, pens, and markers
- paper or index cards for taking notes
- your **PASS** calendar and schedule

You should also have a dictionary. A small, paperback dictionary is all you need.

Your supplies should also include any manuals you may have gotten from your state Department of Motor Vehicles, or from your employer, or both.

If you own a calculator, have that on hand, too. You can manage without one, but if you happen to own one, you will find it handy.

Settle down in the work area you have set aside for your **PASS** sessions. If you are in the cab of your truck or in some public place, clear the area immediately around you. Make some room to spread out your books and papers. Remove as many distractions from view as possible.

PASS Billboard

Everything You Ever Wanted to Know About Dictionaries

All the reading strategies in the world won't help you if you don't understand the meaning of the words you read. Fortunately, there's help. It's called a dictionary.

Many people reject the use of a dictionary because they don't understand what it's for or how to use it. A dictionary has words, and their meanings, called definitions. It contains other information, such as how to pronounce words and use them in sentences. In many dictionaries, you can also find out something about the history of the word—where it came from.

For the most part, you will be interested in definitions and pronunciations.

Dictionaries are set up in alphabetical order. That means words that begin with the letter "A" are at the beginning, and words that begin with "Z" are at the end. The word's second letter further organizes the order of the words. So, a word like "aardvark," whose second letter is "a," would come before a word like "abalone," whose second letter is "b." But "azure" (which begins with "a") still comes before "bath," (which begins with "b").

You can estimate where in the dictionary you will find a word, based on its spelling. Then you use the words at the top of the pages to guide you further. The guide words tell you the first and last words you will find on that particular page. When the word you are looking for falls between the two guide words (in alphabetical order), you know you are on the right page.

For example, you may be reading this sentence: "Position your tractor in front of the trailer so the fifth wheel is lined up with the kingpin." You want to know the meaning of the word "kingpin." First, find the section of the dictionary that has words beginning with "k." Next, use the guide words to find the pages with words beginning "ki . . . ." Say you find a page with the guide words "kindred/kinship." This tells you the first word on this page is "kindred," and the last word is "kinship." "Kingpin" comes after "kindred" and before "kinship" alphabetically, so you know you have found the correct page.

The first piece of information you usually find is how the word is broken up into segments called syllables. This information is most useful in writing. For the most part, you will only be interested in syllables as far as pronunciation is concerned. To pronounce a word correctly, you must know which segment, or syllable, gets the emphasis.

Take the word "present" as a very simple example. There are two syllables to this word: "pre" and "sent." If you give the first syllable the emphasis (PREsent), you have the word "present" as in "The governor is present at today's State Truck Roadeo." If you stress the second syllable (preSENT), you have the word "present" as in "I now present to you the winner of the State Truck Roadeo."

The next type of information the dictionary usually gives you is more details on how to pronounce the word. You'll see some symbol to tell you which syllable gets the emphasis. Other symbols stand for different sounds. An explanation of what the symbols mean is usually at the bottom of the page. You'll often find example words that give you an idea of the sound a symbol stands for.

Using our example of "kingpin," you may find the symbol for the sound of the first "i" is an "i" with a half-circle over it. The example at the bottom of the page shows this is the sound of "i" as in the word "hit." "Kingpin" is a fairly easy word, and you probably already know how to say it. But you will find the pronunciation guide helpful with a word like "synchromesh."

Next, you'll often find information about what "part of speech" the word is. In other words, how is it used in a sentence? For our purposes here, you probably already know how the word is used in a sentence. You want to know what it means. Fortunately, that's usually the next piece of information.

Following the definition, you may find some more symbols. These have to do with the history of the word. Most of our English words have "roots" in a foreign language.

Back to the definition. Many words have more than one meaning, or different shades of meaning. The most common definition is given first. The least common one is given last. How can you tell which definition is the one you want? Try each definition in your sentence to see which one makes sense.

Let's say your dictionary gives you three definitions for "kingpin." The first is "the central pin in an arrangement of bowling pins." The next definition is "the leader or central figure in an organization." The last definition is "the part of a coupling device found on a semi-trailer."

Use each of these definitions in your sentence. Remember our example sentence? It was: "Position your tractor in front of the trailer so the fifth wheel is lined up with the kingpin." Try definition number one in the sentence. Clearly, we are not talking about bowling here, so this definition is not the one you want. Try definition number two. Since we're not talking about organized crime, this definition is no help, either. Definition number three is therefore the one we're after. Now you know that "kingpin" is part of the coupling device.

What if your dictionary has definitions one and two, but not three? A small dictionary probably doesn't. That's because "kingpin" as we're using it is a technical term. It's not used this way in common speech. How then would you find out what it means?

This is when a glossary comes in handy. A glossary is a mini-dictionary of special or technical words, or words with meanings that are unique to a particular use. Many words used in the trucking industry fit this description. Take, for instance, words like "fifth wheel," "pancake," "brownie," or "yard goat." These words have special meanings when used to talk about trucking. You're not likely to find these meanings in an ordinary dictionary.

There is a glossary included in this book. It's organized much like a dictionary. It defines technical terms and common words that have special meanings when used in the trucking industry.

You need both a dictionary and a glossary to get the full meaning of what you read here. Make sure you include both in your list of CDL test preparation supplies.

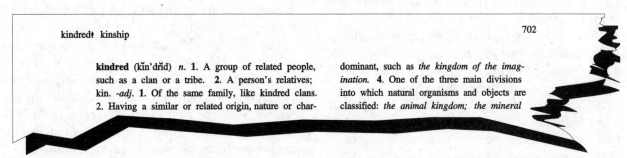

kindred‡ kinship   702

kindred (kĭn'drĭd) *n.* **1.** A group of related people, such as a clan or a tribe. **2.** A person's relatives; kin. -*adj.* **1.** Of the same family, like kindred clans. **2.** Having a similar or related origin, nature or char-

dominant, such as *the kingdom of the imagination.* **4.** One of the three main divisions into which natural organisms and objects are classified: *the animal kingdom; the mineral*

Figure 2-1. Words at the top of the page guide you through the dictionary.

If you have access to the Internet, you will find online dictionaries even easier to use. Simply type the word for which you need a definition in the space provided, and click on "Search."

BE SYSTEMATIC

Now you have everything in place and are ready to get to work. So you start with the first word and read until your eyes fall out, right? Wrong. We suggest you be more systematic than that.

First, take the **PASS** Pre-trip. Take stock of what you already know about the subject, and what you don't. As you read the chapter, be on the lookout for answers to the Pre-trip questions you missed.

Next, instead of simply reading the chapter, we suggest you use **SEE**: Scan, Examine and Extract.

Scan

First, scan the material you have chosen to read. Perhaps your goal for one **PASS** session is to read the chapter on vehicle inspections. Read the chapter title. Get the main topic firmly in your mind.

What are the sections that make up the chapter on vehicle inspections? They are "walk-around" and "in-cab" inspections. Right away, this tells you something about the subject this chapter covers. Even before you begin to read the chapter, you know there are two types of inspections, walk-around and in-cab.

Read the objectives at the start of the chapter. They tell you what you can expect to know when you are done. They are mini-goals for the chapter. Keep these objectives, these mini-goals, in mind as you read. Mentally cross them off as you achieve them.

You may also find a list of supplies you'll need to complete the chapter. Gather the supplies together before you sit down. That way you won't have to interrupt your train of thought to go hunting for a pencil.

Now scan the chapter pages. Note the main topic headings and the subtopic headings. Take a brief look at any charts, tables, or illustrations. Start getting warmed up to the subject.

As you scan, do any questions occur to you? Write them down. You'll look for the answers later when you read the chapter. That will help you to read more closely and keep you from just skimming over the material.

Examine

Now that you have some familiarity with the topic, get down to serious business. This is the time for close reading. Use the list of questions you formed when you scanned. Keep the objectives in mind. Look for the answers to these questions.

Use your highlighter or marker as you read. If this is your book, you should feel free to mark it up in any way that helps you. Find an answer to one of your questions? Highlight it. Read something that you find surprising or new? Highlight it. If you find a section you don't quite understand, circle it. Come back to it later. It may make more sense then. If it's very difficult or technical, you may have to read it several times before you understand it.

Use your dictionary or glossary when you come to an unfamiliar word. Highlight the word, and write the definition in the book's margin. Or, write the new words and their definitions on your notepaper or in your notebook.

Extract

When you have finished the number of pages you set out to cover, take the time to review in your mind what you have read. Write a little summary in your notebook. Get a hold of your "study buddy," your spouse, even one of your kids. Make a little oral report to them on what you have read. If you can explain it to someone else, it's a good sign you understand it yourself.

Now take the **PASS** Post-trip. Pretend this is an actual test, and you can't look up the answers. Then score yourself. Did you miss any questions? If so, turn to the table of contents or the index. Find the part of the chapter that deals with the subject of the questions you missed.

Reread that information. Try to figure out why you missed it. Did you simply not read it with enough attention? Maybe you didn't understand it completely the first time. Highlight that section of the chapter with a different color. You may need to come back to those sections more than once before you feel you have grasped the information.

Before you conclude a **PASS** session, do one last review. Scan the pages all over again. Include in your scan all the material you highlighted. Review the definition of the words you looked up.

You're done reading when you have:

- met all the objectives stated at the top of the chapter
- answered all the questions that occurred to you during the scan
- answered the **PASS** Pre-trip questions correctly
- answered all the **PASS** Post-trip questions correctly

TOOLS FOR SUCCESS

This book has been written to do more than just provide information. The information is presented in a way that's easy to read. We hope you will find the book organized in a way that makes sense.

This book includes several other helpful tools. You've already been introduced to some of these tools. They are the table of contents, the glossary, the objectives, and the list of supplies. Here are some others.

Index

We've mentioned the index, which you can find at the back of this book. The index is like a table of contents. It lists the subjects covered by this book and on which pages you can find them discussed. The index is more detailed than the table of contents. Also, the subjects are listed in alphabetical order. The table of contents, on the other hand,

PASS BILLBOARD
How to Learn from a Picture

Along with words, you'll find pictures in this book. Some of them are charts or tables that offer you a different way to look at facts or figures. Many are technical illustrations. That is, they are drawings of equipment parts, sometimes even entire systems. Once you have examined the picture, you should be able to locate or recognize the same parts or systems on an actual truck.

You could stare at the picture for an hour or two, until the image is burned into your brain. But there's a better, faster way to learn from a picture. Refer to Figure 2-2 as you read about this method.

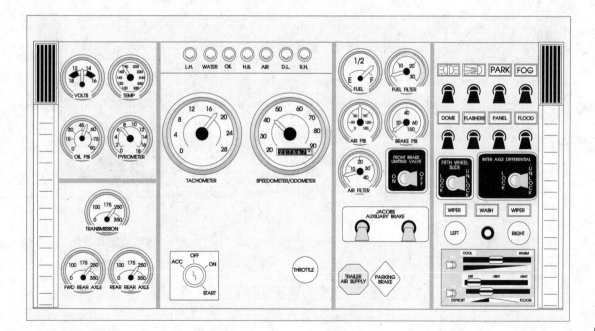

Figure 2-2. Typical CMV dashboard.

Pick a system for scanning the picture. You could use vertical (up-and-down) or horizontal (side-to-side) rows, whichever you like. Here's how the horizontal pattern works.

Start at the top left. Work your way across the top, toward the right edge of the picture. Follow the arrows and labels that name each part. Say the name of the part out loud (or, if you are not alone, at least say it to yourself). Move your eyes down a little and scan back towards the left edge of the picture. Continue moving your eyes back and forth across the picture until you arrive at the bottom right, having looked at all the parts.

Using Figure 2-3 as an example, start with the voltmeter at the top left of the picture. Move your eyes to the right along the instrument cluster at the top, to the fog light switch. Now look at the controls down the right side of the picture. Move your eyes along the controls across the middle and to the left, where the oil pressure gauge is. Now look back to the right, across the tachometer and speedometer to the flood light switch. Moving back toward the left, study the controls leading to the transmission temperature gauge. Finish by looking at the bottom row of gauges and controls, ending with the air-conditioning controls at the bottom right.

As you look at the illustration, say the name of each part. Describe it, aloud or to yourself. You might say something like this:

"Inter-axle differential control. Rectangular. Toggle switch with two positions. Lock is to the left. Unlock is to the right."

Scan the picture several times, until you think you have it down. Then cover up the picture. Get a sheet of paper and a pencil. Sketch the picture as you remember it. You don't have to be an artist. Just get all the parts with their labels, in their approximate sizes and shapes, and in their approximate positions.

Now compare the picture, and your sketch. How much did you recall? If it was a complex system, you may have missed something. Make a note to return to the picture and repeat the exercise from time to time until you can recall every part perfectly.

It doesn't matter which pattern you choose. All the pattern does is give you a system so you see every part of the picture. It keeps you from missing anything. Pick a pattern, and use the same pattern each time. Use the same pattern to sketch or recall the picture that you used to study it in the first place.

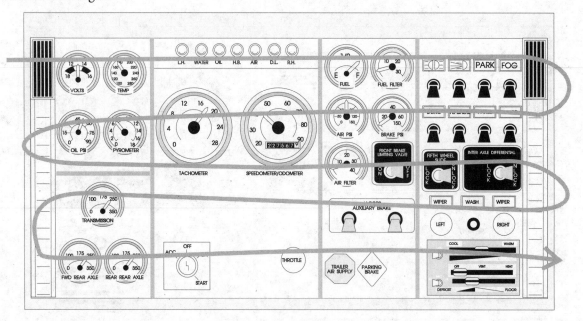

Figure 2-3. Use a scanning pattern to "learn" from a picture.

follows the order in which the subjects appear in the book, from first page to last.

Illustrations

You'll find many charts, tables, diagrams, and other illustrations in this book. You know the old saying, "A picture is worth a thousand words." You'll find illustrations that help explain ideas or procedures, or describe equipment. Examine these closely.

The pictures have a dual numbering system similar to the pages. Each picture, or figure, has a number, followed by a dash, followed by another number. Like the pages, the first number is the chapter number. The second number is the order in which the picture appears in that chapter. So, Figure 3-1 is in Chapter 3, and it's the first picture you will find there. Figure 3-3 is the third picture in Chapter 3. Figure 8-3 is the third picture in Chapter 8, and so on. From time to time, you will be told to "see . . ." or "refer . . ." to a figure. The numbering system will help you find it quickly.

Some pictures are included simply to give you a mental break from all this reading!

PASS BILLBOARDS

You may have been wondering what those **PASS** Billboards are you've been seeing. **PASS** Billboards are much like the billboards you see along the roadside. Sometimes they have very important, useful information. Coming into a strange town late at night, you may have welcomed billboards that told you about nearby motels or truck-stops. Sometimes billboards are just silly stuff you barely notice before turning your eyes back to the road.

Our **PASS** Billboards contain information that's a little off the beaten track of the main subject. Sometimes they are helpful hints or alternative techniques. Other times they contain extra information that will broaden your understanding of the subject.

If it doesn't disrupt your train of thought too much, read them as you come to them. Otherwise, come back and read them when you finish the chapter.

Once in a while, the **PASS** Billboards offer a story or joke that, like some of the cartoons you see, give you a break from the more serious reading. Read them when you feel yourself becoming men-

tally tired. They will refresh you and help you to feel more alert.

PASS Billboards are set off from the main text with a special border.

PASS Tips

From time to time, you will see some special messages. These are **PASS** tips. Like **PASS** Billboards, they are extra material or helpful hints. Unlike **PASS** Billboards, they are shorter, and more directly related to what you are reading at the time.

Tests

You may not have thought of a test as a helpful tool. Tests do help you learn, however, by measuring your progress.

There are several ways to measure what you have retained from your reading. You could try writing a copy of the chapter from memory. Then you could compare it to the original to see how you did. You could recite the chapter from memory to someone who is reading along in the book. That person could tell you how close your recall comes to the real thing. These methods would take a lot of time and could even be thought of as overkill. After all, you're not concerned with whether you remember every word. All you want to know is whether you recall the important ideas and facts.

A simpler way to do this is to answer a few questions. These questions address the major points and items of information. Yes, that is a test. But there is no "pass" or "fail" to the tests given in this book. They are truly measuring tools, yardsticks of what you remember. If you answer all or most of the questions correctly, you know you have retained much of what you read. If you can only come up with a few correct answers, you know you did not retain much.

Why didn't you? Only you can answer that. Perhaps you were not reading with your full attention. Maybe you were distracted. Perhaps you read too quickly, or skipped over sections that didn't look important. Whatever the reason, you have learned something useful. The next time you read something, if you want to remember more of it, you will have to go about reading a different way.

All these tests do many things for you:

- they give you practice taking tests.
- they give you examples of different kinds of tests.
- they help you measure your progress in your own **PASS**.

The more you practice taking tests, the more relaxed and confident you will be about taking the CDL tests, or any test for that matter. Remember, one of our goals is simply to reduce your anxiety about taking tests. Practice—or rehearsals, if you prefer—will help achieve that goal.

The **PASS** Pre-trips and Post-trips are offered in different forms. There are "essay" questions, true/false, multiple-choice, and matching questions. These are not the only kinds of test questions, but they are the most common. The CDL test you will take could use any of these types. It may combine several types. It could even use them all.

You will use different methods for answering each type of test question. One of the things we will do in Chapter 3 is demonstrate those methods.

Last, these tests help you measure your progress. The **PASS** Pre-trip is a special type of test. While the **PASS** Post-trip tells you how far you have come, the Pre-trip tells you where you were when you started. The **PASS** Pre-trip will help you take stock of how much you already know about the subject you are about to tackle.

We promised you we would not waste your time. Why spend time and energy on what you already know? If you are pressed for time, the **PASS** Pre-trip will help you decide what parts of this book you can go over lightly. Then you can spend the rest of your valuable time where it is needed the most.

The **PASS** Pre-trip will ask three questions about the chapter's subject. You may be able to answer all three questions easily. If so, you can expect the chapter to seem familiar to you. You should be able to sail right through it.

Don't assume, however, that you can skip the entire chapter. After all, these are only three out of many possible questions.

You may feel confident about only one or two of your answers to the **PASS** Pre-trip. Make special note of the questions you could not answer. Pay particular attention to the parts of the chapter that cover those subjects.

Most likely, you will not be able to answer any of the questions. That's to be expected. If you were already well-prepared for the CDL tests, you wouldn't need this book in the first place. If you can't answer any of the **PASS** Pre-trip questions, resolve to pay close attention to the chapter that follows. You may find that through experience or training, you already know the material. You may simply need to learn different terms, or slightly different procedures.

The **PASS** Post-trip will measure whether you have achieved your goal. That goal is usually to learn something or some things about the subject in the chapter. The **PASS** Post-trip will include the three questions that made up the **PASS** Pre-trip. Once you have finished the chapter, you should be able to answer those questions correctly. If you miss the questions a second time, that calls for a close review of the chapter.

There is no "good" or "poor" in your results on either test. Your scores are for your eyes, and for your use only. They are simply to guide you in your CDL test preparation efforts.

You're just about ready to tackle the real business of preparing for the CDL tests. You have at least one system for getting the most out of the time you spend reading. That's the **S**can, **E**xamine, and **E**xtract system. You have a strategy for learning from illustrations. That's the "scan pattern" system we described in this chapter. You know you have several tools available to help you, such as a dictionary.

You need only one more piece before you dive in. That's the "secret" of taking and passing tests. We will reveal that "secret" in Chapter 3.

Chapter 3

About Tests

In this chapter, you'll learn about:

- the purpose of tests
- the different types of tests
- the different types of test questions and answers
- the way tests are scored
- common test-taking mistakes and how to avoid them
- smart guessing
- what makes the CDL tests special
- how to use the sample tests in this book

To complete this chapter, you will need:

- a dictionary
- a pencil or pen and an eraser
- blank paper, a notebook, or an electronic notepad
- colored pencils, pens, markers, or highlighters

After reading Chapter 2, you may have a different view of tests. Now you see how they can be tools to help you in your learning process.

Thinking about tests you have taken in the past, you may feel some were a real challenge. Hard. Even impossible. You may not have felt very positive about them at all. Some tests are harder than others. Why is that so?

There are several reasons. One is that the questions test very fine details of the subject you have studied. You may not have studied with an eye for such fine detail. Another reason is the test questions may be complex. One single question might test several points of information. You have to be able to cover each point to answer the question correctly.

Some tests are hard because they are simply bad tests. They were badly written. Writing a test is harder than you might think. It's not uncommon to find poor tests where answer choices don't really fit the question. Or, the test may ask questions about material that wasn't covered. Another thing that test writers do is write "trick" questions.

They have the mistaken idea that this makes a more challenging test. All it does is make an unfair test.

You shouldn't have either of these problems with the sample tests in this book. They were not written to trick you. Every effort was made to make them fair and appropriate questions. You can feel confident about using them to measure your progress.

Nor should you have these problems with your CDL tests. You may find the CDL test questions hard, especially if you are not fully prepared. They may test your knowledge of certain details, and they may be complex questions. You would be unlikely to find the test questions tricky or unfair, though.

Even when asked good, fair, test questions, people still manage to give the wrong answers. It may surprise them to find they failed the test. They may claim they worked very hard to prepare for it. When you ask them more about it, you find that they did prepare well. So why did they fail?

It's because there is more to passing a test than simple preparation. There is a wrong way and a right way to take a test. In this chapter, you'll learn some of the common mistakes test-takers make. That will help you avoid making those mistakes.

You'll see how different types of test questions are put together. Knowing how the question was put together will make finding the right answer that much easier.

There are many forms of tests and types of test questions. We'll focus on those you are likely to find on CDL tests.

TYPES OF TESTS

To get your CDL license, you will take two types of tests. These are knowledge tests and performance (skills) tests. Knowledge tests test what you know. Performance tests test what you can do.

"Knowing" and "doing" are not the same. What if you are a new, inexperienced driver? You may "know" how to double-clutch. That is, you can list all the steps involved in double-clutching. Knowing how may not be enough to keep you from grinding the gears when you try to put your knowledge into practice, however!

If you are an experienced driver, you may be able to double-clutch quite smoothly. Try to explain what you are doing to someone else, without showing them. It might be harder to explain double-clutching than it is to do it. You may find you don't "know" all the steps.

The CDL law requires drivers to have certain knowledge and certain skills. It's not enough to "know" how to back a truck. You'll have to show you can do this maneuver, and do it safely. On the other hand, knowing the details of the Safety Act is enough. There's no performance test for this.

Some aspects of being a licensed commercial driver will be tested by both knowledge and performance tests. One example is vehicle inspection. You will be asked knowledge questions about vehicle inspections. You will also be asked to do a vehicle inspection.

The Knowledge Test is the first CDL test you will take. You must pass the Knowledge Test before you can take any performance tests. So let's look at knowledge tests first.

Knowledge Tests

There are many ways to test what people know. You could ask them to give a report, for example. The method the CDL Knowledge Test uses is to ask you questions and check to see if your answers are correct.

The most common way to present such questions is in a written test. On the CDL Knowledge Test, you will be given a form with the questions.

You may put your answers right on the same form, or you may be given a separate form on which to put your answers.

You will record your answers in one or more ways, based on the type of questions asked. You may have to write out a sentence or two. You may be asked to look at a selection of possible answers and pick the one or ones that you feel answer the question correctly. You may be told to check off or circle the right answer.

The examiner may give you an answer form that will be scored by machine. This will be a form with many numbered or lettered boxes. Using a pencil, you fill in the box that has the same number or letter as your answer choice. A machine compares the pencil marks you make with a form that has been filled out correctly. The places where your form does not match are scored as incorrect answers.

You may take your CDL Knowledge Test on a computer. Don't let this concern you. It doesn't make the CDL test harder or special. It's just a different way of presenting the questions to you and recording your answers. You will not be expected to have any experience with computers to do this. You will be given any information you need about taking a test on a computer at the time you take the test. You'll probably find you enjoy taking a test this way.

Some states make an oral test available. Instead of you reading the questions from a form, an examiner will read the questions to you. You will tell the examiner what your answer is instead of writing it down.

You may have to make special arrangements to take an oral exam in your state. Check with your Department of Motor Vehicles in advance if this option interests you.

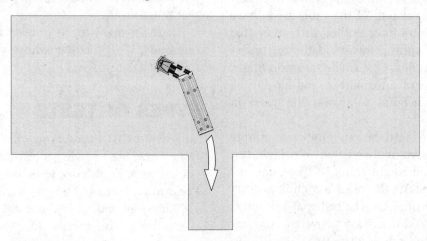

Figure 3-1. It's not enough to "know" how to back a truck. You'll have to show you can do it.

TYPES OF TEST QUESTIONS

There are many different types of test questions. In your schooling, you have likely run into them all at one time or another. Here are some of the most common types:

- essay
- matching
- fill-in-the-blank
- true/false
- multiple-choice

Your CDL Knowledge Test will probably have multiple-choice and true/false type questions. But we'll discuss the other types briefly in case you run into those types, too. For now, we'll just introduce you to these test question types. Later in the chapter, you'll learn strategies you can use to answer them.

Essay

An essay question requires you to write or recite your answer, usually at some length. This gives you the advantage of being able to express your answer in your own words. A disadvantage is you don't get any suggested answer choices to jog your memory.

Questions you may be asked during the Skills Tests are like essay questions. That's because you will recite your answers in your own words. No one will prompt you or give you hints about what the right answer is.

Matching

Matching tests offer you two lists. One list contains the "questions." These can be words, phrases, or sentences. The other list contains the answers. These can also be words, phrases, or sentences. You are supposed to match one item from the first list with one item from the second. Usually, it's a one-to-one match. That is, both lists have the same number of items. You should not have any unmatched items left over when you are done.

Matching questions are a good way to test your understanding of terms. You match the term with its definition. You may also be asked to match a regulation to a description of what that regulation requires or when it passed into law. You may even be asked to pick an item from one list that completes a sentence in the other list, making it a true

sentence. This is like a fill-in-the-blank question, which you'll learn about next.

In matching questions, you must find your answer among the list of suggested answer choices. You cannot supply an answer that isn't on the list, even if you think it's a better answer. On the plus side, having a list of answer choices in front of you may aid your recall.

Fill-in-the-blank

In this type of test question, you are given a statement that is missing a word, or several words. You write out the word or words that is missing. Often, you are given a list of words or phrases to choose from. One of the words or groups of words on the list correctly completes the sentence. You must choose the item that completes the statement and makes it true.

True/False

This type of test question usually presents you with a statement. You must decide if the statement is true or false. It sounds easy, but don't be fooled. The statement can be quite complex. It may have several parts and conditions. All the parts of the statement must be true for the entire statement to be true. If any part of the statement is false, the entire statement is false.

Multiple-Choice

This is a very common type of question on a knowledge test. It's one you're likely to find on many CDL Knowledge Tests. The multiple-choice question begins with a part called a stem. This can be a question or a statement. It can be complete or incomplete. It is followed by two or more answer choices. You choose the answer that completes the statement and makes it true. If the "stem" is a question, you pick the correct answer from the list of choices.

TYPES OF TEST ANSWER CHOICES

So much for types of test questions. Let's look at answer choices. There's a little more to this than just "the right answer" and "the wrong answer."

As we said, on an essay question, you can phrase the answer in your own words. How you phrase your answer usually doesn't matter, as long as you give the examiner enough evidence that you have the knowledge being tested.

On the other types of test questions, you have to choose an answer from a list of choices. These answers are written in someone else's words. They may not be your words. You may not find the answer phrased quite the way you would have phrased it. Nevertheless, you have to choose one of the items on the list.

More than other types of test questions, multiple-choice questions can have different types of test answer choices. You may run into them all on your CDL Knowledge Tests. They are:

- only one answer is correct
- one answer is the best answer
- all of the answers are correct
- none of the answers are correct

Only One Answer Is Correct

On most multiple-choice tests, only one answer choice is correct. All the others are wrong. There will usually be a note or instruction if this isn't the case.

One Answer Is the Best Answer

You may be instructed to choose "the best answer" or something similar. Don't make the mistake of thinking this is the same as "the right answer." More than one answer may be correct. But one answer will clearly be the best response. This is best described by example. Try this multiple-choice question:

When you approach an intersection marked by a red octagonal stop sign, you should _____ before you proceed through the intersection.
A. come to a full stop
B. come to a full stop and check for other traffic
C. come to a full stop, check for other traffic, and wait for the intersection to clear

As you can see, choice A isn't wrong. But it's not the correct choice for this question. If you were to choose choice A, you would be marked "wrong."

Choice B is a better answer. At a stop sign, you should stop and check for other traffic. This still isn't the correct answer choice for this test question, though.

The correct answer is choice C. Of the three answer choices, it has the most complete and correct description of what you should do at a stop sign. You should not only stop and check for other traffic, but you should only proceed if the intersection is clear.

These types of questions are often used to test your knowledge of a procedure or system. Sometimes it's simply not enough for you to know only one step of the procedure or one piece of the system. You must know the entire procedure or system for your knowledge to be any good. A test question that asks you to recall all the steps or parts truly tests the completeness of your knowledge.

All the Answers Are Correct

Sometimes, when all the answer choices seem to be correct, they are! Some test questions offer an answer choice that reads "all of the answers are correct" or "all of the above." When you see "all of the above," it's the last of a series of answer choices. If the answer choices that came before it in the series are all correct, you would choose "all of the above."

Here's an example of "all of the above":

Which of the following terms describes a type of wheel?
A. disc
B. spoke
C. Budd
D. all of the above

The correct answer is D. All three terms, disc, spoke, and Budd, describe types of wheels. This is a type of "best answer" question. The best answer is "all of the above." It is the only answer that would be correct. Yes, it's true that "spoke" is a type of wheel. You may feel it's correct to choose B as an answer choice. While B is not wrong, it is

PASS BILLBOARD
How To Calculate Percentages

You'll be keeping track of your progress by scoring your **PASS** Post-trips. It would be handy to convert your scores to percentages. If you don't remember how to calculate a percentage, use this chart to estimate your score.

Number of right answers	Number of questions									
	100	90	80	70	60	50	40	30	20	10
100	100%	—	—	—	—	—	—	—	—	—
90	90	100	—	—	—	—	—	—	—	—
80	80	89	100	—	—	—	—	—	—	—
70	70	78	88	100	—	—	—	—	—	—
60	60	67	75	86	100	—	—	—	—	—
50	50	56	63	71	83	100	—	—	—	—
40	40	44	50	57	67	80	100	—	—	—
30	30	33	38	43	50	60	75	100	—	—
20	20	22	25	29	33	40	50	67	100	—
10	10	11	12	14	15	20	25	33	50	100
9	9	10	11	13	15	18	22	30	45	90
8	8	9	10	11	13	16	20	27	40	80
7	7	8	9	10	13	14	18	23	35	70
6	6	7	8	9	10	12	15	20	30	60
5	5	6	6	7	8	10	13	17	25	50

also not "the best answer." You would be marked wrong for choosing B as your answer. Again, the "best" answer, the only right answer in this case, is "all of the above."

None of the Answers Are Correct

Here is another variation on the "best" answer test question. It's the opposite of "all of the above." This is the test question where none of the answer choices are correct. Be very careful with this type of test question. You must be absolutely sure that no answer choice is correct before you choose "none of the above."

SCORING KNOWLEDGE TESTS

Most knowledge tests are scored by counting the number of questions that were answered correctly. These are compared with the total number of questions on the test. This comparison is expressed as a percentage. If you took a 100-question test, and answered 85 of the questions correctly, you get a score of 85 percent. A score of 70 percent on a 50-question test means you answered 35 of the 50 questions correctly.

You may have received a letter score, like A, B, or C, on tests you took in school. Here, the letter represents a group of percentages. For instance, the letter A might represent the group of scores from 90 percent to 100 percent. Someone scoring 91 percent would receive an "A," and so would someone scoring 99 percent. However, your score on the CDL Knowledge Test will probably be expressed as a percentage.

On most tests, you must answer a certain number of questions correctly to pass the test. Federal law states that you must score 80 percent or better to pass the CDL Knowledge Test. Your state may require CDL applicants to achieve even higher scores in order to pass the test.

SKILLS TESTS

Beside Knowledge Tests, you will take Skills Tests to get your CDL. You will be tested on your ability to do certain maneuvers, such as backing and turning. You will also be tested on your skill in driving in traffic.

Sometimes, a knowledge test will be combined with a skills test. As you do an inspection, for example, the examiner may ask you questions

Figure 3-2. You'll be tested on your skill in driving in traffic.

about the systems you are inspecting. In a sense, you will be taking two tests at the same time. You will not only show you can do a proper inspection, you will also show you have good knowledge of your equipment.

SCORING SKILLS TESTS

Skills tests are a little harder to score than knowledge tests. There are many ways to do a maneuver. Which is the "correct" way? Take the "skill" of making right turns. Is it enough simply to get your rig around the corner? Is part of the skill how long it takes to turn the corner? Does it count if you scrape the curb or go out into the middle of the intersection?

What about the examiner who is measuring your skill? Will two people score the same skill the same way?

Yes, it's hard to score skills tests—but not impossible. The answer to all the questions we just asked about scoring performance tests is "standards." So that skills tests can be scored fairly, standards have been set for what is to be considered a "good" performance.

Let's look at right turns again. Most people would agree that it's not good to scrape the curb. So this condition is made part of the test. The test would then not simply be, "Make a right turn"; it

would be, "Make a right turn, without scraping the curb." To pass the test, you must meet both conditions. You must make the turn, and you must avoid scraping the curb.

You can feel confident that your CDL Skills Tests have been very carefully designed. They will have very specific and detailed descriptions of what skills are expected. In making a right-hand turn, for example, your vehicle's rear wheels should come as close to the curb as possible without scraping it. You should also not turn too widely. If you scrape the curb, or turn wider than five feet, your level of skill is not considered "passing."

Conditions like these ensure a fair test and fair scoring. They mean everyone who takes the test is judged by the same standards. They make it possible to measure your level of skill, rather than your looks, your jokes, the attractiveness of your vehicle, or some other factor.

These conditions keep the examiner's personal feelings out of the picture. It doesn't matter how the examiner would have made that turn. It doesn't matter if the examiner likes you or likes the make of truck you drive. All the examiner has to decide is if your performance met all the conditions set in the test.

You will often be told what the conditions are. For the right-hand turn test, the examiner might say, "You must make a right-hand turn without

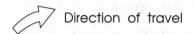 Direction of travel

 Traffic cone or marker

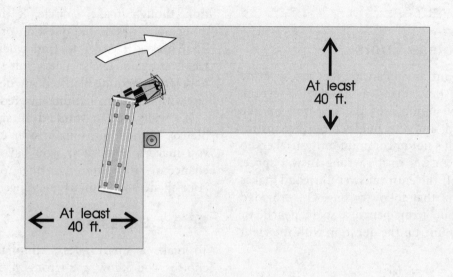

At least
40 ft.

At least
40 ft.

Figure 3-3. The CDL Skills Tests are scored to specific standards.

scraping the curb and without turning wider than five feet away from the curb."

You will also be told if you must meet all the conditions, or only most of them. Sometimes, it only takes one error to fail a test. This is the case on some of the CDL Skills Tests. Breaking a traffic law or causing an accident during the Skills Test results in automatic failure.

TEST-TAKING TIPS

The best test-taking tip we could give would be "prepare well." There's nothing like having a good grasp of the material that will be tested. Along with that, these tips will help you not to make careless errors.

Understand the Instructions

Before you begin any test, make sure you understand the instructions. On a written test, read any information that comes before the questions. It's there you will find instructions to choose "the right answer" or "the best answer."

If no instructions have been given, ask the examiner. Ask if you should use a pencil or pen to take the test. Find out if you may use scrap paper. Will there be a time limit? Don't make a pest of yourself. Do ask enough questions so you feel you understand the test conditions.

Don't Rush

If there is no time limit, relax and take all the time you need. Rushing will cause you to make careless errors. Take each question or skill test item one at a time. Clear your mind of the other parts of the test.

Read Carefully

Read knowledge test questions carefully. Before you answer, be certain you understand clearly what's being asked. Read any suggested answers just as carefully. When two or more answers seem alike, describe to yourself the difference between them.

Answer the Easy Ones First

Unless you are told to do otherwise, scan through written tests quickly. Answer all the questions you can answer off the top of your head. Then work on the trickier ones. Finally, before you hand in the test, review the easy ones just to make sure you didn't misread them, or write down "A" when you meant to write "B."

Avoid Careless Errors

When filling out answer forms that are separate from the question forms, avoid careless errors. Make sure the number of the question on the question form and the number of the answer blank match. It's not uncommon for test-takers to put the right answer in the wrong answer space. That throws off the entire answer form and makes all the answers that follow wrong. On provided forms, use your scrap paper, a spare pencil, or your finger to line up the question with the right answer blank.

Write Neatly

Write neatly on printed forms. Mark your answer choices with firm, bold, marks. This is especially important when the answer form will be "read" by a machine. The machine can't read marks that are too light, or that fall outside the fill-in area. Stray marks will be scored as wrong answers.

Don't Change Your Answer

It's recommended that you don't change an answer choice once you have made it. Usually, your first decision about the answer is the best one. More often than not, when test-takers change their answer, they change it to the wrong one!

If you do decide to change your answer, erase the old answer completely. Make sure it's clear that your new answer choice is the one you mean.

Again, with machine-scored tests, this is vital. If you do not erase completely, the machine may see two answer choices marked. This will be scored as a wrong answer, even if one of the answers happens to be the right choice.

Don't Leave Blanks

Don't leave any blanks on written tests. If you don't know the answer, plan to come back to the question later. Often, other questions on the test will refresh your memory of the subject. Even if that doesn't happen, you can make a smart guess. We'll cover that next. Before you leave the question, though, make a circle, X, arrow, or some other mark beside the question or answer blank. That makes it easier to find when you want to tackle it again. It also helps you to see that you have left something blank so you don't hand in the test with questions left unanswered.

It's really not to your advantage to leave any blanks. If there are four answer choices, even if you make a wild guess, you have a one-in-four chance of hitting the right one. You can improve your chances with smart guessing.

Smart Guessing

To make a smart guess, eliminate any answer choices you know are wrong. Does one of the remaining answer choices have words like "all," "every," "always," or "never"? Consider that answer carefully. It's rare that something is always so or never the case. You may be able to eliminate this answer choice after all.

That narrows down your choices. You may be left with only two choices, instead of four. A wild guess will then give you a 50-50 chance of getting the right answer. By now, one of the remaining two answers may look pretty good to you. You may feel quite certain it's the right one after all.

Penalties for Guessing

The only time you want to leave questions unanswered is if there is a penalty for guessing. Some tests are scored as the percentage of right answers compared with the number of total questions answered. This is different from tests where the right answers are compared with the total number of questions asked. On such a test, it's better to leave a question blank if you don't know the answer.

There will usually be some information on the test that tells you when there is a penalty for guessing. Your CDL Knowledge Tests will not likely have such penalties. Unless you are told otherwise, make smart guesses on the questions that have you stumped.

TIPS FOR DIFFERENT TEST QUESTION TYPES

Matching

On a matching test, look for any instructions first. You may be told that the first list is a list of terms and you are to match terms to their definition. Or, each list may have a heading that describes the items in the list. Look for clues about what relationship the two lists have to each other.

If there are an equal number of items in each list, you probably have to make a one-to-one match. If there are more items offered on the answer list than there are on the question list, this isn't the case. You will clearly have some left over. Or, you may have fewer answers than there are questions. In that case, some answers are used more than once. You guessed it—these variations make the test more challenging.

Scan the lists quickly. Make any matches that jump out at you right away. If you may have scrap paper, you may want to use pieces of it to cover up any answer choices you have eliminated in this first pass. Or, draw a line through them (unless you're told not to write on the form). Since you have already "used" them, you don't need to consider them as answers for any of the other questions. If you remove them from "view," you reduce the chance of mistakenly choosing them a second time.

Next, work on any remaining unmatched "question" items. Taking each one in turn, try to answer it without looking at the answer choices. If the question item is a term, define it. If it's a regulation, what does it regulate? If it's an item of equipment, what does it do?

Look over the list of matching choices. Do you see your definition there? If so, match the two. Draw a line through or cover up the answer you have chosen.

Continue to make all the matches you can. Skip over any you can't answer at all. Come back to those when you have done all the "easy" ones. By then you will have eliminated some more answer choices. You'll have a short list of answers to choose from, instead of the entire range.

Consider each of the answer choices with the remaining questions. Examine the questions and answers closely. You may notice some things right away that help you to make the correct matches.

Take these three sample matching question and answer choices.

Term	Definition
duals	riding on the surface of the water
hydroplaning	two tires on one axle
check valve	a part of the braking system

The first item, "duals" seems to be asking about a group of things. Only one of your remaining answers refers to a group. That is "two tires on one axle." That's the correct match for "duals."

Notice also that "hydroplaning" seems to be asking about an action, rather than a thing. "Riding on the surface of the water" is the only answer choice left that is also an action rather than a thing. That's the correct match for "hydroplaning."

Now you're left with one question item, "check valve," and one answer item, "a part of the braking system." Match those two up and you are done.

Use this logic if you are simply flat out stuck on some test questions. Stop racking your brain for definitions you don't know. Take a mental step back and simply look at the words. Try to find any way in which they relate to each other or seem alike. That may lead you to the match the test is asking you to make.

True/False

Usually true/false questions are worded as statements. You decide if it is a true statement or not. Sometimes, true/false questions are worded as questions. An example would be, "Are on-duty drivers who take narcotics prescribed to them by a doctor breaking the law?" If you would answer the question "yes," your answer choice should be "true." If you would answer the question "no," your answer choice should be "false." (In this case, the answer is "yes." The "question" is "true.")

Remember that all parts of the statement or question must be true for you to say it's a true statement. If any part is false, then "false" is your answer choice. Take this true/false question, for example:

It is safe for an on-duty driver to take over-the-counter cold remedies without a prescription.
A. True
B. False

The answer to this question is "B. False." Yes, it's true that over-the-counter cold remedies can be taken without a prescription. But it's not true that on-duty drivers can take them safely. Many of these medications cause drowsiness. It would not be safe for a driver to take them while driving. If the driver in the question were "off-duty," the answer to the question would be "A. True."

Be on the lookout for words like "always," "ever," "never," "all," and "complete." These words should make you read the statement very carefully. Be certain that your answer, true or false, covers every possible case, since a question like this does not allow for any exceptions.

True/false questions look easy. That's why it's especially important to read them closely.

Multiple-Choice

Multiple-choice questions are a little easier to answer if you follow a system. We'll assume we're dealing with the "only one answer is correct" type of question.

First, read the question "stem" to yourself. Then read the stem with the first answer choice. Does the combination make a true statement? If so, this answer choice is correct. If not, go on to the next answer choice.

Read the question stem with each answer choice in turn. Note which combinations make a true statement. Most times, only one answer choice will make a true statement of the question stem. If more than one answer seems to be correct, use the elimination and smart guessing techniques already described to narrow down the choices.

Read "best answer" test questions and their answer choices, very carefully. Describe to yourself the difference between the answer choices. Does one include more details than the others? If all the details are true, this may be the "best" answer. It's often true that an answer choice that is a lot shorter, or a lot longer, than all the rest is the correct answer.

All the Answers Are Correct

Be careful when one of the answer choices is "all of the above." Just because "all of the above" is an answer choice doesn't mean it's the correct one. Select it only if you have convinced yourself that all the answer choices are correct. It may be that one or more of the answer choices is indeed wrong.

Try each answer choice with the stem. Make certain that each one completes the question stem and makes it a true statement or answers the question correctly. If all the answer choices meet this test, then the answer choice to select is "all of the above."

PASS PRACTICE TESTS

Throughout the rest of this book, you will find many sample tests to practice with. You'll have a chance to work with true/false, matching, fill-in-the-blank, and multiple-choice questions. Try all the techniques described in this chapter.

After you have scored your own tests, examine the questions you missed. Did you make careless errors, or were you simply not prepared enough? Keep track of your results. See if practice helps you make fewer careless errors.

You'll also find some practice performance tests. They will include specific conditions your performance must meet to be considered "passing." Keep these conditions in mind when you self-score your performance.

This concludes Part One. You've covered test preparation and test-taking in general. In Part Two, you'll get into preparation specifically for taking the CDL tests. Before leaving Chapter 3, though, there's one last item to cover.

In the following chapters, you may read about facts or procedures that differ from what you already know or do. The terms used for equipment or maneuvers may not be the ones you are familiar with or use. That doesn't mean you are wrong. Your way may be just fine. You may even be better informed. But to get a CDL, you'll have to do things "by the book." You must "speak the language" used by the CDL tests and the examiners.

When you find a way of doing something or describing something that's different from what you are used to, make a special note of it. Review it and get it firmly in your mind. You don't want to be surprised by it on the test. If you're looking only for answers phrased "your way," you won't recognize the right answers when you see them.

PART TWO

PART TWO

Chapter 4

Commercial Driver Licensing

In this chapter, you will learn about:

- what the Commercial Driver's License is
- who must have a Commercial Driver's License
- how to get a Commercial Driver's License
- penalties for not having a Commercial Driver's License
- areas of knowledge and skill covered by the Commercial Driver's License
- Commercial Driver's License Tests
- other Safety Act rules regarding commercial driver licensing

To complete this chapter, you will need:

- a dictionary
- a pencil or pen
- blank paper, a notebook, or an electronic notepad
- colored pencils, pens, markers, or highlighters
- a CDL preparation manual from your state Department of Motor Vehicles, if one is offered
- a driver license manual from your state Department of Motor Vehicles
- the Federal Motor Carrier Safety Regulations pocketbook (or access to U.S. Department of Transportation regulations, Parts 383, 387, and 390-399 of Subchapter B, Chapter 3, Title 49, Code of Federal Regulations)

PASS PRE-TRIP

Instructions: Read the three statements. Decide whether each statement is true or false. If it is true, circle the letter A. If it is false, circle the letter B.

1. If you do not have a Hazardous Materials Endorsement on your Commercial Driver's License, you can never haul hazardous materials.
   A. True
   B. False

2. It is legal to have a Commercial Driver's License from more than one state if you lived in that state when you got that license.
   A. True
   B. False

3. There is a national Commercial Driver's License required by federal law, but your home state may decide not to adopt the CDL.
   A. True
   B. False

How did you do with these first **PASS** Pre-trip questions? Did you feel confident about your answers? If you felt that you don't know the first thing about commercial driver licensing, don't worry. When you finish this chapter, you will know the answers to the **PASS** Pre-trip questions and more.

CDL BACKGROUND

The Commercial Driver's License is a special type of driving license. Before we explore this special license, it would be helpful for you to have a clear understanding of just what a driver's license is—any type of driver's license.

Driver Licensing

When you receive a driver's license, you have been granted permission to drive by a body of government. That makes driving a regulated activity.

To get your license to drive any vehicle, you go to the Motor Vehicles Department of your home state. This is because it is your state that grants you the license to drive. It's not the city, county, or federal government.

Commercial Driver Licensing

A Commercial Driver's License (CDL) grants permission to people who have passed one or more tests. By passing, they have shown that they have the knowledge and skill to drive certain types of vehicles in certain conditions. The CDL law is quite specific about the types of vehicles, conditions, and drivers concerned. You'll learn about these later in this chapter.

It takes special skill and knowledge to do many important jobs in today's society. Airline pilots, police officers, and other professionals prepare for and take tests to earn special licenses. They take pride in that special testing and licensing. The licensed commercial driver can also take pride in being a member of a recognized skilled profession.

FMCSR PART 383: COMMERCIAL DRIVER'S LICENSE STANDARDS

The CDL is part of a larger body of regulations. These are the regulations set by the U.S. Department of Transportation in Subchapter B, Chapter 3, Title 49 of the Code of Federal Regulations (CFR).

You should, at least once, read these regulations in their "official" language. If you would like your own copy of the Federal Motor Carrier Safety Regulations (FMCSR), you can order one from any number of companies that supply materials to the trucking industry.

An "appendix" in a book is extra material. Like our **PASS** *Billboards, it's material that you may find helpful or may not need at all. That's why you will usually find appendices (that's the plural form of "appendix") at the back of a book, after the main material. This book has several appendices.*

The FMCSR is a compact book and usually not very expensive. If you can't afford your own personal copy of the FMCSR, check with the reference librarian at your local public library. The library will either have an FMCSR or a copy of Title 49 CFR, if not both. You can't take either out of the library, however. You can only look at them there.

You can even find the FMCSR on the Internet at:

http://www.fmcsa.dot.gov

CMVSA 1986

In 1986, the Congress of the United States of America passed the Commercial Motor Vehicle Safety Act (CMVSA). The goal of this act was to improve driver quality and remove unsafe and

---

PASS BILLBOARD
Abbreviations and Acronyms

You will see many abbreviations and acronyms in this book. An abbreviation is a shortened form of a word. For example, "Dept." is an abbreviation. It is a shortened form of the word "Department." You can tell it's an abbreviation because it ends with a period. This is true whether it appears at the beginning, in the middle, or at the end of a sentence.

An acronym is a word formed from the first letters of several words in a group. Taking the first letters of the words in the group "Department of Transportation" forms the acronym "DOT." You can tell it's an acronym,

and not a word by itself, because it appears in all capital letters.

In this book, you will always be properly introduced to the abbreviation or acronym. First you will see the whole word, name, or group of words spelled out. That will be followed by the abbreviation or acronym enclosed in curved brackets called parentheses. After the abbreviation or acronym has been introduced in this way, it will be used throughout the book with no further explanation.

However, if you ever forget what an acronym stands for, you can find it in the glossary.

unqualified commercial motor vehicle (CMV) drivers from the highway.

The Federal Highway Administration (FHWA) of the U.S. Department of Transportation (DOT) responded to this Act by changing some regulations and adding Part 383 to the FMCSR. These changes fall into two main areas of interest to commercial drivers:

- the single license law
- commercial driver testing and licensing

These changes occurred at the federal level of government. It's still state governments that issue licenses to CMV drivers. Each state's licensing standards must comply with the federal regulations, however. Within certain limits, states may develop their own tests and licenses for commercial drivers. These tests and licenses must, however, at least meet the federal requirements. State standards can be higher or more demanding than the federal standards, but they can't be lower or less demanding.

The bottom line is, although state governments issue driving licenses, there is a national Commercial Driver's License required by federal law. Effective April 1, 1992, your home state must have a CDL program that meets the federal requirements.

These requirements are set out in Part 383 of the FMCSR. If you have a copy of the FMCSR, follow along as we discuss each section of that Part.

Purpose and Scope

The purpose and scope of the CDL program is described in FMCSR Part 383, Subpart A (General), Section 383.1. As you know by now, the purpose has two targets. They are:

- to help reduce or prevent truck and bus accidents, deaths, and injuries by requiring drivers to have a single commercial motor vehicle driver's license
- to disqualify drivers who operate commercial motor vehicles in an unsafe manner

How does the CDL program achieve this? We'll answer that question briefly, then address each point in greater detail later in the chapter.

First, a CMV driver may have only one CMV driver's license. This is known as the single license requirement.

Next, drivers must notify their current employers and their home state of certain convictions.

Drivers must provide previous employment information when they apply for jobs as CMV drivers.

Employers must not hire anyone with a suspended license to operate a commercial motor vehicle.

Persons convicted of certain criminal acts or traffic violations are subject to fines, suspensions, loss of license, and other penalties.

Commercial motor vehicle drivers must pass tests and receive a special license.

States must give knowledge and skills tests to all qualified applicants for commercial driver's licenses.

There are different classes or types of licenses. To each class or license type, there are additional items that can be added that allow the person to operate specific categories of vehicles or haul certain types of cargo within a particular license class. These categories are called **endorsements**.

Drivers must meet specific knowledge and skills requirements to be licensed to drive certain vehicles and haul certain cargo.

Commercial driver licensing tests must meet specific standards. Commercial driver license applicants must achieve a stated minimum score on these tests.

States must keep records on licensed commercial drivers.

With the exception of some military drivers, and a few other types of CMV drivers under specific conditions, Section 383.3 states that the rules in FMCSR Part 383 apply to all who operate commercial motor vehicles in interstate, foreign, or intrastate commerce. These rules also apply to the employers of these drivers.

Definitions

Section 383.5 defines many of the terms used to describe the CDL program. For the most part, these are terms that won't be found in a dictionary, or terms that have special meanings. For instance, what exactly is a commercial vehicle? A commercial vehicle is one that is used in commerce. The CDL law defines this as trade, traffic, or transportation in the United States (or any area controlled by the United States). The trade can be between a

PASS BILLBOARD

There are three words in the English language that have common, everyday meanings. These are "may," "should," and "must." In common speech, you may hear these words used as if they all had the same meaning, or nearly the same. They don't. It usually doesn't matter much.

When laws are written, however, the meaning of words must be very clear. "May," "should," and "must" have unique and specific meanings when used in a legal sense. They refer to three different levels of "permission."

"May" refers to a relaxed level of permission. When a law says you "may" do something, it means you are allowed. If there is more than one action possible, the choice is yours to make. This is usually used when some conditions have been set. If the situation meets the condition, then you "may" choose to act or not to act.

Take, for instance, a law that says you "may" make a right turn at a red light. This means there is nothing that would stop you, legally, from making that turn. Of course, there may be reasons besides whether it is legal or not for why you wouldn't want to make that right turn. Perhaps you're concerned that oncoming traffic turning left will turn into your lane. So you stop at the red light and wait for a green light before you turn. As far as the law is concerned, though, a right turn at a red light is permitted.

Sometimes, you might see the word "can" used instead of "may." "Can" refers to ability rather than permission. That is, you might have permission to do something, but not the ability. You might have permission to make a right-hand turn, but if your steering system breaks down, you might not have the ability to make it. This is splitting hairs, though. It's all right to assume that if a law says you "can" do something, it means the same as you "may" do something.

"Should" gives you a little less permission than "may." It carries recommendations with it. It's usually used with some conditions. The

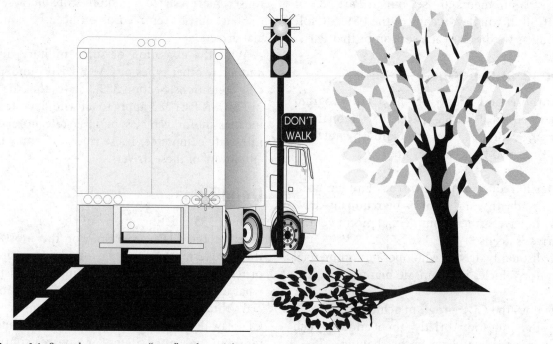

Figure 4-1. Some laws state you "may" make a right turn at a red light.

"right on red" law would more likely be stated "You should stop at a red light. If it is safe to do so, you may make a right turn at a red light." The recommendation is that you stop at the red light. The condition is that it must be safe to proceed. Once the situation has met that condition, you have permission to choose whether to turn or not.

The recommendation in "should" laws is more than an idle suggestion. It's usually best to do what is recommended. Although you have permission to make your own choice, you will rarely get into trouble if you just do what the law would prefer you to do.

The last level of permission is "must." A law that states you "must" do something (or "must not" do something) gives you no choice

at all. Don't even bother to think about it, to second-guess it, or make excuses. Just do it (or don't do it, if the law says you must not). If the law says "You must stop at a red light," then stop. Don't think, "Well, I'm the only one on the road, I may as well go ahead." It doesn't matter if you are the only one who has been on that road in 20 years! The law says "You must stop." So stop.

Keep the differences between these words in mind when you read about regulations or try to decide their meaning.

There's one last word you'll see often in the FMCSR but rarely hear in everyday speech. That word is "shall." As used in the FMCSR, it gives the same level of permission as "must."

place in a state and a place outside the state, even outside the United States.

"Interstate" means between states. "Intrastate" means within the borders of a single state. "Foreign," of course, would mean trade with a foreign country. Canada and Mexico are considered foreign countries. See Figure 4-3 on page 40.

That defines "commercial vehicle." The CDL law is not concerned with every commercial vehi-

cle, however. For instance, a small taxicab or limousine that transports fewer than 16 people would not be covered by the CDL. This section further defines commercial vehicles as vehicles that:

- weigh 26,001 pounds or more
- carry 16 or more people, including the driver
- haul hazardous materials that require placards or are listed as a select agent or toxin

The "weight" referred to is the gross combination weight rating (GCWR). Gross combination weight rating is defined as the value stated by

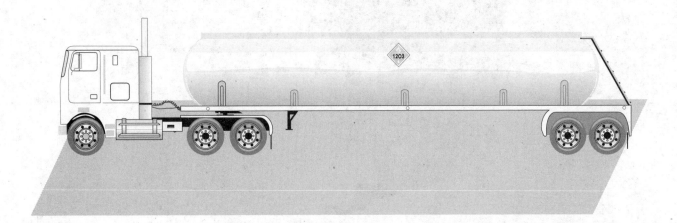

Figure 4-2. Drivers must meet specific knowledge and skills requirements to be licensed to drive certain vehicles and haul certain cargo.

the manufacturer as being the maximum loaded weight of a combination vehicle. A combination vehicle is one that articulates. That means it has joined sections that can move independently of each other.

The GCWR in this case is 26,001 or more pounds, including a towed unit with a gross vehicle weight rating of more than 10,000 pounds. Gross vehicle weight rating (GVWR) is the value stated by the manufacturer as the maximum loaded weight of a single vehicle. In other words, this is the most that a vehicle and its load can weigh without exceeding the rating intended by the manufacturer.

The weight definition in the CDL law also applies to any vehicle with a GVWR of 26,001 or more pounds.

A "commercial motor vehicle" is also a vehicle that is designed to carry 16 or more people, including the driver. "Designed" means the manufacturer gave it a "seat rating." What if you drive your church's Little League team to a game in the church bus van that is designed to carry more than

16 people? Would you need a CDL? Yes, you would. As long as the bus has a 16-seat rating, you would need a CDL to do this. And, you would need a passenger endorsement. You'll learn about endorsements in greater detail later in this chapter.

Certain cargo is considered "hazardous material" when it's being transported. These are defined by the Hazardous Materials Transportation Act. They are listed in Parts 171–180. (Some of these parts concern the transport of hazardous material by boat or plane.) They are also discussed in the FMCSR. Some materials are only hazardous when hauled in large enough quantities. At that point they are termed hazardous, and the vehicle hauling them must bear hazardous materials placards. This is described in further detail in Title 49 CFR Part 172, Subpart F and 42 CFR Part 73.

A vehicle of any size that is hauling materials termed hazardous by the Hazardous Materials Transportation Act is a "commercial motor vehicle." So is a vehicle that is placarded for hazardous cargo or hauling a select agent or toxin.

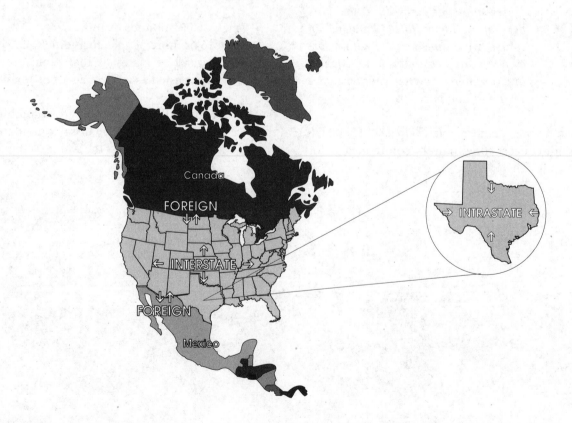

Figure 4-3. "Interstate" means between states. "Intrastate" means within the borders of a single state. Canada and Mexico are considered foreign countries.

Here are some other important terms defined in this section:

"Alien" means any person who is not a citizen or a national of the United States.

When referred to by CDL laws, "alcohol or alcoholic beverage" means beer and wine as well as distilled spirits (liquor).

"Alcohol concentration" (AC) is a measure of how much alcohol is in a person's blood or breath. It's usually expressed as a percentage. You may also see the term "blood alcohol concentration" (BAC) used.

The terms "alcoholic beverage" and "AC" become important in discussions of drinking and driving.

"Controlled substance" refers to the illegal use of certain drugs. The CDL sets penalties for using drugs illegally.

"Disqualification" can mean a suspension, revocation, or complete loss of driving privileges. Whether it's suspension or loss depends on the crime or violation and can be for lack of qualification as a CMV driver.

"Driving a commercial motor vehicle while under the influence of alcohol" is clearly defined. You are driving under the influence (DUI) if your AC is 0.04 percent or more. Your state may set the percentage of AC even lower. There are also circumstances when you are automatically DUI if you refuse to be tested for drunkenness.

An "employee" is anyone who operates a CMV either directly for or under lease to an employer. An "employer" is anyone who owns or leases a CMV or assigns employees to operate one.

"Endorsement" is a term you'll run into a lot. It is a special allowance added to your CDL that permits you to drive certain vehicles or haul certain cargo.

A "felony" is a crime that is punished by death or a prison term of more than one year.

An "imminent hazard" is a condition that could likely mean death, serious illness, severe personal injury, or a serious threat to health, property, or the environment, which could occur before formal action is taken to lessen that risk.

A "representative vehicle" is one that is like the type of motor vehicle that a CDL applicant plans to drive.

Your home state is your state of domicile. This is the place you plan to return to when you are absent. It's where you have your main, permanent home.

A "tank vehicle" is one that transports gas or liquids in bulk. Portable tanks that carry less than 1,000 gallons are not included in this definition.

Single License Requirement

Section 383.21 states that you shall not have a license from more than one state. This is known as the Single License Requirement.

Testing and Licensing

Section 383.23 of the FMCSR states that effective April 1, 1992, all drivers of commercial vehicles must have taken and passed the CDL tests. No one may drive a CMV without a Commercial Driver's License.

There are certain drivers who are exempt from the CDL requirement. Drivers whose "state of domicile" is in a foreign country that does not have a CDL program don't get a regular CDL. These drivers get a "non-resident CDL." This non-resident CDL must meet the requirements of the regular CDL.

(There's an exception to this exception. Canada's and Mexico's CDL meets the requirements of the U.S. CDL. So Canadian and Mexican drivers must have a CDL from their country. They then may not have a U.S. CDL or a non-resident CDL. This would cause them to be in violation of the single license requirement.)

People who are learning to drive a CMV obviously don't have a CDL. Yet, they must be able to drive a commercial vehicle to train properly. These people should get a learner's permit from their state. (Different states have different names for this permit, such as "Instruction Permit.") A learner's permit allows learners to drive a CMV. When they are driving the CMV, they must be accompanied by someone who does hold a valid CDL for the same group-type CMV. Also, these learners must have a valid automobile driver's license, or at least pass all the vision, sign/symbol, and knowledge tests they would take to get an automobile driver's license. Your state may have additional requirements.

Notification Requirements and Employer Responsibilities

Once you have a CDL and are operating a CMV, you are subject to the notification requirements. These apply to traffic violations other than parking offenses. They apply no matter what type of motor vehicle you were driving.

Drivers who are convicted of a traffic violation in a state other than the one that issued their license must notify their home state. That is, they must notify the state that issues the CDL of a conviction they received for a traffic violation in another state. This notification must be made within 30 days from the date of the conviction.

CMV drivers who hold a CDL and are convicted of violating a state or local motor vehicle traffic control law must also notify their current employer about the conviction. This is true of violations in any type of motor vehicle, except parking violations. This notification must also be done within 30 days of the conviction.

The notification must be in writing, must be signed by the driver, and include:

- the driver's full name
- the driver's license number
- the date of conviction
- the specific offense or violation
- any suspension or loss of driving privileges that resulted from the violation
- whether the violation was in a CMV
- where the offense took place

If you have your driver's license suspended, revoked, or canceled, or lose your right to operate a CMV, you must notify your employer. This must be done before the end of the business day that follows the day that you lost your license. This notification must be made if the license is lost due to suspension, revocation, cancellation, or disqualification.

When applying for a job as a CMV driver, an applicant must provide the future employer with certain background information. Employers must require this information at the time the applicant submits the employment application. The background information must cover the past 10 years of employment. The information that must be provided consists of:

- a list of the names and addresses of the applicant's previous employers during the past 3 years, plus the names and addresses of the employers during the 7 years before that for whom the applicant operated a CMV.
- the dates the applicant was employed by these employers
- the reason for leaving such employment
- certification by the applicant that the information is true and complete
- additional information that the employer requires

Before the application is turned in, the employer must tell the applicant that these references will be checked. The future employer may contact former employers to confirm this work history.

Employers may not knowingly hire or require anyone to drive a CMV whose CDL has been suspended, revoked, or canceled and not yet restored, or who has been disqualified from operating a CMV. An employer may not knowingly hire or require anyone to drive a CMV who has more than one valid CDL. This prohibition also applies to out-of-service CMV drivers, their vehicles, or motor carriers. Drivers who are in violation of railroad-highway grade crossing laws are also prohibited.

Disqualifications and Penalties

As you might expect, there are serious penalties for violating these CDL laws. These are described in FMCSR Part 383.51.

Very simply, a driver who is disqualified may not drive a CMV, even a driver with a CDL. An employer may not allow a disqualified driver to operate a CMV.

What causes a driver to be disqualified? You can be disqualified when driving a motor vehicle if you are:

- driving while under the influence of alcohol

 Recall the definition of DUI? Quick, write down what you remember. Then review the earlier material on Section 383.5. Or, review the material in your copy of the FMCSR.

- driving while under the influence of a controlled substance
- operating a CMV with a BAC of 0.04 or greater
- refusing to take an alcohol test
- leaving the scene of an accident
- using the vehicle to commit a felony
- driving a CMV even though your CDL is revoked, suspended, or canceled
- causing a fatality through the negligent operation of a CMV
- using the vehicle to commit a felony involving manufacturing, distributing, or dispensing a controlled substance

Figure 4-4. Driving while under the influence of alcohol is grounds for disqualification.

The length of the disqualification depends on the offense. First-time offenders are disqualified for one year. This is unless hazardous materials requiring placards were involved. In that case, a first-time offender is disqualified for three years.

There is also a special penalty for drivers involved in certain felonies. This is the case of the driver who uses a CMV to commit a felony that involves the manufacture, distribution, or issue of controlled substances. This driver is disqualified for life. That's right. Such a driver can never legally drive a CMV ever again, even if this was the first conviction.

Some people never learn. So there are harsher penalties for repeat offenders. Repeat offenders are disqualified for life if they have already been convicted of DUI (alcohol or drugs), leaving the scene of an accident, or committing a felony. The law does provide for a second chance, though. A driver who has been disqualified for life can reduce the lifetime disqualification after 10 years. First, the driver must enroll in and successfully complete a rehabilitation program. This rehabilitation program must meet the state licensing agency's standards. After completing the program, the driver must apply to restore the CDL. Last, the driver must meet any requirements the state has for reinstating CDLs.

Having gotten the CDL restored, if the driver commits one of these offenses a third time, the driver is again disqualified for life. There are no more second chances this time. Three strikes and you're out!

There are also no second chances for drivers disqualified for using a CMV to manufacture, distribute, or issue controlled substances. On conviction, that offense earns a lifetime disqualification the first time out. This disqualification is permanent.

There are separate penalties for CMV drivers who commit serious traffic violations. Certain traffic violations are considered "serious traffic violations." These are:

- going more than 15 miles over the posted speed limit
- reckless driving
- improper or erratic lane changes
- following the vehicle ahead too closely
- traffic violations that result in fatal accidents
- driving a CMV without obtaining a CDL
- driving a CMV without the CDL in the driver's possession
- driving a CMV without the proper class of CDL or endorsements
- texting while driving

Texting includes, but is not limited to, emailing, instant messaging, entering a command or request to access a web page, pressing more than a single button to initiate or terminate a voice communication using a cell phone, or engaging in any other form of electronic text retrieval or entry, for present or future communication. It does not include inputting, selecting, or reading information on a global positioning system or navigation system; or pressing a single button to initiate or terminate a voice communication using a cell phone; or using a device capable of performing multiple functions (e.g., fleet management systems, dispatching devices, smart phones, citizens band radios, music players, etc.) for a purpose that is not otherwise prohibited.

For the first offense there is a 60-day disqualification. (There may also be other types of penalties for breaking traffic laws, such as fines or sentences.) If two offenses are committed within three years, the second conviction earns the driver a 60-day disqualification. Three violations in three years result in disqualification for 120 days.

CMV drivers will be disqualified for violating railroad-highway grade crossing laws. These violations occur when:

- the driver is not required always to stop, but fails to slow down and check that tracks are clear of an approaching train
- the driver is not always required to stop, but fails to stop before reaching the crossing, if the tracks are not clear
- the driver is always required to stop, but fails to do so before driving onto the crossing
- the driver fails to have enough space to drive completely through the crossing without stopping
- the driver fails to obey a traffic control device or an enforcement official's directions at a crossing
- the driver fails to negotiate a crossing because of insufficient undercarriage clearance.

The penalties for railroad-highway grade offenses are CMV disqualification for no less than 60 days for the first offense. A second conviction brings a penalty of no less than 120 days. Convictions after that bring a penalty of no less than one year.

Drivers who operate CMVs in violation of out-of-service orders are disqualified for no less than 180 days for the first conviction. The second conviction brings a disqualification penalty of no less than two years. Further convictions bring a disqualification of no less than three years. If the driver was transporting hazardous materials or passengers, the penalties are similar. On first conviction, the driver is disqualified for no less than 180 days. The second offense brings a disqualification of no less than three years.

Last, some drivers may be disqualified because they are determined to be an imminent hazard.

Remember, states may have additional penalties for these crimes. Disqualification as a commercial driver is simply the penalty set forth by federal law.

Other Testing and Licensing Requirements

The CDL law adds testing and licensing standards to qualification requirements that already exist. Applicants for a CDL who plan to drive interstate must meet the qualifications set forth in FMCSR Part 391. Persons who will drive intrastate must certify that they are not subject to Part 391 and must meet any state driver qualification requirements.

CDL applicants must pass a Knowledge Test and a Driving Test. (You'll learn about these tests in greater detail later in this section.) The driving test must be taken in a vehicle that is representative of the one that the CDL applicant plans to drive.

CDL applicants must supply the information required to fill out a CDL document and application. This will become the "Commercial Driver's License" or "CDL." The license itself is described in Subpart J of FMCSR 383.

The application asks for such information as:

- the full name, signature, and mailing address of the person getting the license
- physical and other information describing the licensed person
- a color photograph of the driver
- the driver's state license number
- the name of the state issuing the license
- the date the license was issued and when it expires
- the CMV or CMVs the driver is licensed to operate
- any endorsements
- an air brake restriction, if it applies
- the driver's social security number, unless this is a non-resident CDL

CDL applicants must then certify that they are not disqualified from driving. They must state that they do not have any other driving licenses from any other state. If they have any non-CDL driver's licenses, they must give them over to the state. They must provide the names of all the states in which they were previously licensed to drive any

PASS BILLBOARD

FMCSR Part 391 deals with the qualifications of drivers. Even with the CDL program, drivers still have to meet these qualifications or show they are exempt from them.

Briefly, drivers who meet the qualifications:

- are at least 21 years old
- can read and speak English enough to talk with the general public, understand highway signs, respond to officials, and fill out reports and records
- can safely operate the vehicle
- meet physical qualifications
- have a single valid commercial motor vehicle operator's license
- have notified their employer of any violations, or certified that there are none to report
- have not been disqualified to drive a motor vehicle
- have completed a road test or its equal

- can tell whether cargo is distributed properly
- can secure cargo

Figure 4-5. Drivers must be able to secure cargo.

This section also provides for drug testing. Applicants must take and pass a drug test before they can be hired.

---

kind of motor vehicle in the last 10 years. Last, applicants for the hazardous materials endorsement must comply with Transportation Security Administration requirements. They must provide proof of citizenship.

This section of the FMCSR (Section 383.71) also tells what to do if you already have a CDL and want to transfer it, renew it, or upgrade it.

If you move your permanent residence to another state, you must transfer your license. That way your CDL will be issued by your new state of domicile. You must apply for a new CDL from your new home state no more than 30 days after you have established your new domicile.

You provide your new state with the same type of information you gave to get your original CDL. You give up your CDL from your old home state. Then you get a new CDL from your new home state. As long as your old CDL was still valid, you do not need to retake the tests.

If you had a hazardous materials endorsement, you must show that within the last two years, you

passed the test to earn this endorsement. If this isn't the case, you can take and pass a hazardous materials test or pass a training course that the state accepts as a substitute. You must provide the proof of citizenship.

Commercial driver's licenses do expire. If you want to go on driving, you must renew it. The process is similar to transferring your license. You certify that you are or are not covered by FMCSR Part 391. You update the information you gave to get your CDL the first time. Renewing your hazardous materials endorsement involves the same process as transferring it.

As you progress in your driving career, you may want to upgrade your license. You may want to change the type of vehicle you are licensed to drive. You may want to add endorsements.

To upgrade your CDL, you again certify that you are or are not covered by FMCSR Part 391. You pass all the tests you took to get your original CDL. You take and pass any additional tests to earn the new endorsement.

Remember reading about the "non-resident CDL?" This license is for drivers whose "home state" is a foreign country without a CDL program. These drivers get one, and only one, non-resident CDL from any state with a CDL program. To get one, applicants must complete the same requirements for getting a regular CDL. They also must report any action that their home government, or any government, may make against their driving privileges. This report must be made to the state that gave them the non-resident CDL. This report must be made within the same time limits as the regular notification requirements discussed earlier.

Implied Consent

Getting a CDL carries with it an "implied consent to alcohol testing." This is stated in FMCSR 383.72. If you are stopped on suspicion of drunkenness, you may not refuse to take a breath test or other test for drunkenness. By accepting your license, you state, in advance, that you agree to take such tests. If you refuse anyway, you may be disqualified from driving. This was stated in FMCSR 383.51. Reread the section on Disqualifications and Penalties to see how these two sections relate to each other.

State Procedures

Section 383.73 of the FMCSR outlines the state's role in granting CDLs. Your home state takes your application. It gives you the appropriate tests and decides if you passed them. The state makes sure you take the test in the type of vehicle that you plan to drive. It conducts the process of transferring, renewing, or updating your CDL.

The state checks your driving record. In making this check, the state uses the Commercial Driver's License Information System (CDLIS). Through CDLIS, all the states can share the records they have on drivers they have licensed. Because of CDLIS, bad drivers can no longer "bury" their driving violations in one state and get a license from another. The state also checks with the National Driver Register to see if the applicant has been disqualified or has had a license revoked, suspended, or canceled in the three years prior to application.

It is the state that hands out penalties to applicants who provide false information on their applications.

Your CDL comes from your home state. Still, other states that have CDL programs honor your CDL. This is known as reciprocity.

The CDL law permits states to contract with a "third party" to give the CDL tests. This explains why you may actually take one or all your CDL tests at a truck driving school, motor carrier, or other private company, instead of on state grounds.

Last, this section permits states to excuse some applicants from having to take the Skills Test. If you have a current, valid license, and have a good driving record, your state may excuse you from taking the Skills Test. A "good driving record" means that, in the past two years, you have not had your license suspended, revoked, or canceled. You must not have been disqualified from driving in the past two years. Your driving record for the past two years must be free of traffic violations connected with an accident (other than parking violations). You must also prove that you are regularly employed as a CMV driver. You must show evidence that in the past, you took and passed a skills test and received a classified license.

 TIP *A classified license is one that states or limits the type of vehicle you may drive.*

You must prove that the skills test was behind the wheel of a vehicle representative of the one you plan to drive with your CDL. Last, you must show evidence that you have experience driving this type of vehicle. You must have had this experience in the two years that just passed before you applied for a CDL.

If you can prove all these things, your state may "waive," or excuse you from taking, the CDL Performance Test. Most CDL applicants will, however, have to take the test.

Vehicle Groups and Endorsements

You have now seen two terms, "representative vehicle" and "endorsements," used many times. FMCSR 383 Subpart F defines these two important terms.

The CDL law takes the view that different levels of skill are required to drive different types of commercial vehicles. The driver of a tractor hauling a liquid cargo tanker needs different skills from those that the driver of a straight truck does. The CDL law puts vehicles into three groups. These groups are described in FMCSR Part 393.91. They are as follows:

Group A is the **combination vehicle** group. A combination vehicle is any combination of vehicles with a GCWR of 26,001 or more pounds, as long as the GVWR of the vehicle or vehicles being towed is more than 10,000 pounds.

Group B vehicles are the **heavy straight vehicles**. This group includes any single vehicle with a GVWR of 26,001 or more pounds, or any such vehicle towing a vehicle of no more than 10,000 GVWR.

Group C is the **small vehicle** group. A small vehicle is any one vehicle, or combination of vehicles, that isn't a Group A vehicle or a Group B vehicle. Group C vehicles are further defined as designed to carry 16 or more passengers (including the driver) or used to haul hazardous materials (hazmat).

"Hazardous materials" includes not only substances that are hazardous by themselves. It also includes materials that are hazardous when hauled in large enough amounts that placards are required. The Hazardous Materials Regulations (Title 49 CFR Part 172 Subpart F) go into more detail about when placards are required.

You must take your CDL tests in a vehicle that is representative of the one you plan to drive. This means that your test vehicle must belong to the same group, A, B, or C, as your work vehicle. It doesn't mean that the two vehicles must be the same make or model. They simply must belong to the same group.

You may get your CDL to drive Group C vehicles and later want to be licensed to drive Group A vehicles. At that time, you must upgrade your CDL. Review the earlier material on testing and licensing where upgrading your CDL is discussed. Exceptions to these requirements are as follows:

A driver who is licensed for Group A may also drive Group B and C vehicles. This driver must simply have the added endorsements needed to drive these vehicles. For example, this driver may need to have a hazmat endorsement to drive Group C vehicles as well as Group A.

Also, drivers who have passed the knowledge and skills tests for Group B vehicles may also drive Group C vehicles. These drivers simply must have the added endorsements required for Group C vehicles, such as the hazmat endorsement.

Now to the endorsements. These are described in FMCSR 383.93. You recall that "endorsement" means authorization. Adding different endorsements to your CDL is like adding options to a car. If you don't live in an area where having air conditioning is a must, why go to the extra trouble and expense to put AC in your car? You choose which endorsements to add to your CDL based on the type of vehicle you want to drive.

The endorsements cover the following vehicle types:

- double/triple trailers
- passenger vehicles
- tank vehicles
- vehicles that require hazmat placards
- school buses

For some endorsements, you need only take another knowledge test. For others, you need a knowledge and a skill test. Here are the endorsements again, with the type of test you must take to earn them:

- double/triple trailers: knowledge test
- passenger vehicles: knowledge and skills test
- tank vehicles: knowledge test
- vehicles that require hazmat placards: knowledge test
- school buses: knowledge and skills test

Your CDL will be coded according to the group type of vehicle you may operate and any endorsements or restrictions you have. These codes are:

- A for Combination Vehicle Group
- B for Heavy Straight Vehicle
- C for Small Vehicle
- T for Double/Triple Trailers
- P for Passenger
- N for Tank Vehicles

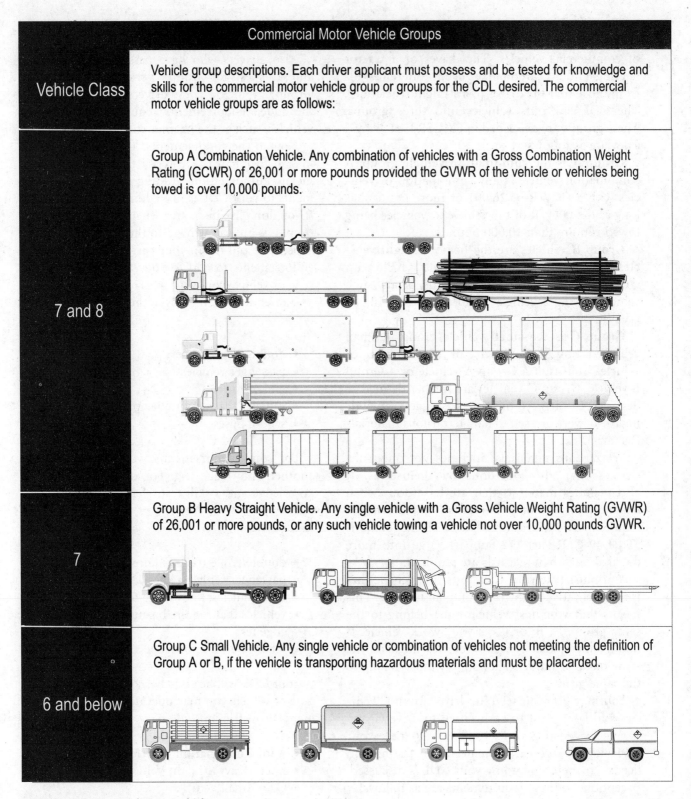

Commercial Motor Vehicle Groups	
Vehicle Class	Vehicle group descriptions. Each driver applicant must possess and be tested for knowledge and skills for the commercial motor vehicle group or groups for the CDL desired. The commercial motor vehicle groups are as follows:
7 and 8	**Group A Combination Vehicle.** Any combination of vehicles with a Gross Combination Weight Rating (GCWR) of 26,001 or more pounds provided the GVWR of the vehicle or vehicles being towed is over 10,000 pounds.
7	**Group B Heavy Straight Vehicle.** Any single vehicle with a Gross Vehicle Weight Rating (GVWR) of 26,001 or more pounds, or any such vehicle towing a vehicle not over 10,000 pounds GVWR.
6 and below	**Group C Small Vehicle.** Any single vehicle or combination of vehicles not meeting the definition of Group A or B, if the vehicle is transporting hazardous materials and must be placarded.

Figure 4-6. Commercial motor vehicle groups.

Commercial Driver's License Tests	Endorsements
Knowledge Tests. Each driver applicant will have to take one or more knowledge tests, depending on what group of license and what endorsements are needed. The CDL tests include:	
The General Knowledge Tests, taken by all applicants	A
The Air Brakes Test, which you must take if your vehicle has air brakes. The Combination Vehicles Test, which is required if you want to drive combination vehicles. The Hazardous Materials Test, which is required if you want to haul hazardous materials. The Tanker Test, which is required if you want to haul liquids in bulk. The Doubles/Triples Test, which is required if you want to pull double or triple trailers.	H N T
The General Knowledge Tests, taken by all applicants.	B
The Air Brakes Test, which you must take if your vehicle has air brakes. The Hazardous Materials Test, which is required if you want to haul hazardous materials. The Tanker Test, which is required if you want to haul liquids in bulk.	H N
The General Knowledge Test, taken by all applicants.	C
The Air Brakes Test, which you must take if your vehicle has air brakes. The Hazardous Materials Test, which is required if you want to haul hazardous materials. The Tanker Test, which is required if you want to haul liquids in bulk.	H N

Figure 4-7. Commerical driver's license tests.

- H for Hazardous Materials
- X for Tank Vehicle and Hazardous Materials combined
- S for School Bus

Your state may have other codes for other combinations.

An endorsement gives authorization. A restriction does the opposite. It limits what you can do. The CDL program includes one restriction. This is the air brake restriction described in FMCSR 383.95. If you wish to drive a CMV with an air brake system, you must take and pass a Knowledge Test on air brakes. You must take your skills tests in a vehicle with air brakes. If you fail the Knowledge Test, or take the Skills Test in a vehicle without air brakes, an air brake restriction is put on your CDL. This means you are not allowed to drive a CMV with an air brake system. This includes braking systems that rely in whole or in part on air brakes.

Last, applicants for Group A CDLs may be asked to take an additional knowledge test on combination vehicles. This is to test how much they know about coupling and uncoupling and inspecting combination vehicles. Of course, the skills test must be taken in a representative vehicle.

You're nearly done with this chapter. What's left to cover are the knowledge and skills required to earn a CDL. You'll also get a brief preview of what may be included on the tests. You'll get a look at the test procedure.

First, quickly recap what you've covered this far:

- CDL background
- driver licensing
- commercial driver licensing
- why a CDL is needed
- how the CDL laws came to be and where they may be found
- definition of terms
- single license requirement
- notification requirements
- disqualification and penalties
- vehicle groups
- endorsement and restriction

It wouldn't hurt to feel well acquainted with the material that has been presented up to this point. You've covered a lot. You may wish to stop now and review the first part of this chapter before going on.

Required Knowledge and Skills

Subpart G of FMCSR 383 lists all the areas of knowledge and skills a CMV driver must have to drive safely. The main knowledge areas are as follows:

- safe operations regulations
- commercial motor vehicle safety control systems
- safe vehicle control
- basic control
- shifting
- backing
- visual search
- communication
- speed management
- space management
- night operation
- extreme driving conditions
- hazard perceptions
- emergency maneuvers
- skid control and recovery
- relationship of cargo to vehicle control
- vehicle inspections
- hazardous materials knowledge
- air brake knowledge

Drivers of combination vehicles must know about coupling and uncoupling. They must also know how to inspect combination vehicles.

The main skill areas are:

- basic vehicle control
- safe driving
- air brakes
- pre-trip inspection
- driving

At least some of the skills must be tested in "on-street" conditions. Some of them may be tested off the street or with a truck simulator.

Additional areas of knowledge and skills are listed for drivers who want the different endorsements.

Does that sound like the table of contents for this book? That's with good reason. The chapters that follow will provide detailed information in all these areas.

Sections 383.111 and 383.113 of the FMCSR list the specific items of knowledge and skills required in each area. Sections 383.115, 383.117, 383.119, and 383.121 list the knowledge and skill areas required for the endorsements. The appendix to Subpart G goes into yet more detail.

We won't list those details here. Instead, we'll list them in the chapters that relate to the main knowledge and skill areas. However, if you have a current copy of the FMCSR, you can preview them now.

FMCSR 383.131 directs states to provide a study manual for CDL applicants. These driver information manuals explain how to obtain a CDL. They describe the vehicle groups and endorsements. They contain the facts that relate to the required areas of knowledge and skills.

The state driver information manual also gives details of testing procedures that apply in that state. This would include directions for taking the test and time limits, if any.

You should definitely get your state's driver information manual. You will find much of it contains the same information that this book does. However, your state's manual will also cover any special laws or requirements beyond the federal standards that your state has for CDL applicants. Read the state driver information manual thoroughly. If there are practice tests or exercises, do them.

Also, make sure to get any additional manuals or booklets your state Department of Motor Vehicles (DMV) may have on laws and requirements for commercial vehicle drivers in that state.

On the same subject, here's another state manual you should have. Read your state's driver's manual for a regular automobile license. This is especially important for new drivers. The CDL manual, and even this book, will assume you have read this manual. You must be familiar with general rules of the road and basic safe driving practices. These are covered in the automobile driver's manual. The automobile driver's license book will have information about road markings, right-of-way, and other subjects that all drivers must know.

FMCSR 383.133 covers testing methods. It gives states guidelines on how to construct and score

Figure 4-8. All drivers must know the rules of the road, including the meanings of all signs.

their tests. The guidelines in this part help ensure that tests will be fair.

Last, FMCSR 383.135 states the minimum passing scores. CDL applicants must get at least 80 percent on the Knowledge Test. "Passing" on the skills tests depends on what's being tested. In all cases, an applicant who breaks a traffic law automatically fails the test. So does any applicant who causes an accident during the test.

That's it. That's all the particulars on the CDL laws. Step-by-step, you've gone through each one as it appears in the FMCSR. You should become familiar with the laws just as they are written in the FMCSR. This chapter aimed to explain them in more detail. The following chapters will give you the facts you need to pass the knowledge and skills tests and earn the doubles/triples or tank vehicles or hazardous materials endorsements if you want them. Air brakes will be covered in detail so you can avoid the air brake restriction if that is your goal.

PASS BILLBOARD

In most states, you will be able to walk into any DMV office during normal working hours and take the CDL test. This may not be the case in all areas, however.

In some states, only certain DMV offices will be prepared to give the entire set of CDL knowledge and skill tests. Sometimes, you may have to go somewhere else to get the skill tests. You may even be required to make an appointment to take the test. In other words, you may not be able to just walk into any DMV office on the spur of the moment and test for your CDL.

When you first apply for your CDL, the state must check your record. The state will check with CDLIS and the NDR. There may be a delay while this check is being made. Further, applicants seeking the hazardous materials endorsement, or a renewal or transfer, will have to get Transportation Security Administration approval. You may not be allowed to take your tests until the record checks are complete. There may also be a delay before you can take skills tests. Some states may not have enough examiners or sites to test all the people who want to be tested.

These delays can throw a monkey-wrench into your timetable for getting your CDL. Find out about them now.

Most applicants will be expected to bring the following items:

- photo identification or driver's license
- Social Security card

- medical card (certification of medical exam)

Know, before you go, if there is anything else your state requires. You may need some lead time to collect any special documents the state would like to see. If you don't have a medical card, or if it has expired, get your DOT physical before you apply for your CDL. That way you will have a valid medical card.

Now's a good time to find out what the process is in your state. Start by contacting your state DMV office. Refer to Appendix B at the back of this book. You may already know the location of a DMV office near you.

Find out:

- where you may take the Knowledge Test
- where you may take the Skills Test
- where you may test for air brakes or endorsements
- if you must make an appointment to take any part of the test
- how much the test costs and whether there are separate fees for the endorsement tests
- how long the tests take, or whether there are time limits
- what, if anything, you should bring to the test

This book does not contain details on the passenger endorsement.

Before you start Chapter 5, test your knowledge of the CDL laws with this **PASS** Post-trip. Use the answer sheet provided on page 55.

PASS POST-TRIP

Instructions: Read the statements. Decide whether each statement is true or false. If it is true, circle the letter A. If it is false, circle the letter B.

1. Federal law requires that the CDL Knowledge Test must be identical in every state.
   A. True
   B. False

2. The CDL Knowledge Test must be taken every two years.
   A. True
   B. False

3. If you do not have a Hazardous Materials Endorsement on your Commercial Driver's License, you can never haul hazardous materials.
   A. True
   B. False

4. It is legal to have a Commercial Driver's License from more than one state if you lived in that state when you got that license.
   A. True
   B. False

5. There is a national Commercial Driver's License required by federal law, but your home state may decide not to adopt the CDL.
   A. True
   B. False

6. A driver who fails the air brake test receives an air brake endorsement.
   A. True
   B. False

7. A driver with a Group A CDL may drive Group B or Group C vehicles with no further testing at all.
   A. True
   B. False

8. You must have a brake endorsement on your CDL to drive a commercial vehicle if the vehicle has hydraulic brakes.
   A. True
   B. False

9. No one from any state can drive a commercial motor vehicle without a CDL effective April 1, 1986.
   A. True
   B. False

10. A CDL applicant who wants a doubles/triples endorsement must take the regular knowledge and skills test, plus an extra knowledge and an extra skills test.
    A. True
    B. False

CHAPTER 4 ANSWER SHEET

Test Item	Answer Choices		Self - Scoring
			Correct Incorrect
1.	A	B	☐ ☐
2.	A	B	☐ ☐
3.	A	B	☐ ☐
4.	A	B	☐ ☐
5.	A	B	☐ ☐
6.	A	B	☐ ☐
7.	A	B	☐ ☐
8.	A	B	☐ ☐
9.	A	B	☐ ☐
10.	A	B	☐ ☐

Cut Here

Chapter 5

Federal Motor Carrier Safety Regulations

In this chapter, you will learn about:

- the qualifications of motor vehicle drivers
- procedures and equipment for safe vehicle operations
- the effects of fatigue, poor vision, hearing, and general health on safe motor vehicle operation
- the hours of service of drivers
- commercial motor vehicle inspection, repair, and maintenance
- the transportation of hazardous materials

To complete this chapter you will need:

- a dictionary
- a pencil or pen
- blank paper, a notebook, or an electronic notepad
- colored pencils, pens, markers, or highlighters
- a CDL preparation manual from your state Department of Motor Vehicles, if one is offered
- the Federal Motor Carrier Safety Regulations pocketbook (or access to U.S. Department of Transportation regulations, Parts 383, 387, and 390-399 of Subchapter B, Chapter 3, Title 49, Code of Federal Regulations)

PASS PRE-TRIP

Instructions: Read the statements. Decide whether each statement is true or false. If it is true, circle the letter A. If it is false, circle the letter B.

1. If a drug or medicine you are taking was prescribed for you by your doctor, it is both safe and legal to take it while on-duty.
   A. True
   B. False

2. The main reason for the post-trip inspection is to let the vehicle owner know about vehicle problems that may need repair.
   A. True
   B. False

3. No driver may drive a vehicle hauling any amount of hazardous material without a hazardous materials endorsement.
   A. True
   B. False

If you were able to answer all these questions easily, you are very familiar with FMCSR Parts 391, 392, 393, 395, 396, and 397. You'll need to be! They are regulations CMV drivers live with every working day.

If you were not able to answer the **PASS** Pre-trip questions easily, you have some homework to do. You've got to bone up on these sections of the Safety Regulations.

Why, you ask? It's because the first item in the list of required knowledge and skills that all CMV drivers must have is knowledge of "safe operations regulations." This is stated in FMCSR Part 383. The "safe operations regulations" happen to be the FMCSR sections that were just listed.

FMCSR Part 383.110 states as a **general requirement** that all drivers of commercial motor vehicles shall have knowledge and skills necessary to operate a commercial motor vehicle safely as contained in this subpart.

FMCSR Part 383.111 states as **required knowledge** that all CMV drivers must have knowledge of the following general areas:

Safe operations regulations Driver-related elements of the regulations contained in 49 CFR Parts 391, 392, 393, 395, 396, and 397, such as:

- motor vehicle inspection, repair, and maintenance requirements
- procedures for safe vehicle operations
- the effects of fatigue, poor vision, hearing, and general health on safe motor vehicle operation
- the types of motor vehicles covered by safe operations regulations
- the types of cargo covered by safe operations regulations
- the effects of alcohol and drug use on safe motor-vehicle operations

Your CDL preparation must include a good review of these safety regulations. If you have never looked at the safety regulations, this should be more than a good review. It should be a close examination.

If you haven't gotten a copy of the FMCSR yet, now would be a good time to do so. We won't recreate the FMCSR in this book. In this chapter, we'll touch briefly on each of the FMCSR sections you must know for the CDL. We won't cover them in detail, though. We'll just cover the highlights. These regulations are covered in depth in the FMCSR. You should follow along in your copy of the FMCSR, or online at *http://www.fmcsa.dot. gov/regulations*. You must be completely familiar with the FMCSR to drive a CMV.

We will provide extra information, charts, and pictures that will help to explain these safety regulations. In reading these rules, you may discover some common driver practices are actually illegal.

You may be tested on any of this by the CDL Knowledge Test. The test will be looking for the right answer according to the law. You may find that some of the methods differ from what other drivers usually do. As far as the test is concerned, it doesn't matter that "nobody does it that way." Laws state what has been judged to be the best for everyone. The CDL test will test your knowledge of these laws.

The following are the Parts of the FMCSR the CDL requires you to know:

- Part 391—Qualifications of Drivers
- Part 392—Driving of Motor Vehicles
- Part 393—Parts and Accessories Necessary for Safe Operation
- Part 395—Hours of Service of Drivers

- Part 396—Inspection, Repair, and Maintenance
- Part 397—Transportation of Hazardous Materials; Driving and Parking Rules

You must know the driver-related parts. As you read the FMCSR, you will find regulations that apply more to employers than to drivers. You may certainly read them for your own information. However, the CDL does not require you to know those regulations.

In this chapter, you will focus on Group A, B, and C trucks hauling a single trailer, if any. Doubles and triples will get more attention in Chapter 11. Tankers will be covered in Chapter 12.

FMCSR PART 391

FMCSR Part 391 covers Qualifications of Drivers. You learned about this part in Chapter 4. It came up in the section on FMCSR 383.71. This is the CDL regulation that deals with the qualification of drivers for commercial driver licensing. There it stated that unless drivers are not covered by Part 391, they must meet the qualifications described in Part 391.

Part 391 covers the following subjects:

- Qualification and Disqualification of Drivers
- Background and Character
- Tests
- Physical Qualifications and Examinations
- Files and Records
- Exemptions

The point of Part 391 is to set minimum qualifications for motor vehicle drivers. This is different from the minimum standards set by the CDL regulations. These minimum standards refer to knowledge and skills. Minimum qualifications have to do with age, background, physical limitations, and so forth.

General Exemptions

Let's get the exemptions out of the way first. Certain drivers are not covered by Part 391. These are drivers of vehicles involved in custom-harvesting operations. Drivers of vehicles involved in beekeeping, and certain other farm vehicles, are also exempt. FMCSR 391.2 discusses these exemptions in greater detail.

Some intracity zone drivers are exempt from some of the regulations in FMCSR 391. For example, these drivers can be younger than 21 years old. They must meet nearly all the other conditions of Part 391, though.

Qualifications

Most CMV drivers, as you can see, must meet the qualifications. Briefly, these are:

- must be at least 21 years old
- must read and speak English enough to talk with the general public, understand highway signs, respond to officials, and fill out reports and records
- must be able to operate the vehicle safely
- must be able to tell if cargo is distributed properly
- must meet the physical qualifications set out in Subpart E of this section
- must have a single valid commercial motor vehicle operator's license
- must have notified the motor carrier of any violations or certified that there are none to report
- must not have been disqualified to drive a CMV
- must have completed an approved road test and received a certificate

To "certify" means to confirm that something is correct and true. You usually certify in writing. A "certification" is that written statement.

Disqualifications

Simply put, the regulations state that drivers who are disqualified shall not drive a CMV. If your license has been revoked, suspended, withdrawn, or denied, you are disqualified. You are disqualified for the length of time you are without your license. These periods vary with the offense.

A driver may be disqualified for criminal or other offenses. This would be the case under the following conditions. First, the driver must be convicted of the crime. Being charged with a crime and forfeiting bond ("skipping bail") before being convicted is treated the same as a conviction. Second, the offense must have been committed while the driver was on-duty. The third condition is that the driver was employed by a motor carrier or involved in commercial transport. If all these three conditions are met, the driver may be disqualified for criminal offenses.

What are those offenses? One is driving while under the influence of alcohol. This is defined as driving a CMV with an AC of 0.04 percent or more. Your state may have a stricter definition. Your state may set the AC even lower than 0.04 percent.

If you refuse to take a test for alcohol, you may be disqualified just as if you were DUI. Remember "implied consent"? If you don't, review "implied consent" in Chapter 4.

Another criminal offense is driving a motor vehicle while under the influence of a controlled substance or drug listed in 21 CFR 1308.11 Schedule I. These drugs are sometimes referred to as "scheduled" drugs.

Figure 5-1. Just because your doctor prescribed some medicine doesn't mean it's safe to take while driving.

Your doctor may prescribe a controlled substance for you. Just because your doctor prescribed it doesn't mean that it's safe to take it while driving. These drugs may have side effects. These side effects may make safe driving impossible. Ask your doctor about the side effects of any medicine prescribed for you. Find out if it's safe to take it while driving. Never take drugs prescribed for someone else.

Transportation, possession, or unlawful use of Schedule I drugs while on-duty as a driver are disqualifying criminal offenses.

You can get a list of Schedule I drugs from the Office of Motor Carrier Standards or any Regional Office of Motor Carrier and Highway Safety of the Federal Highway Administration. You can find addresses for these agencies in Appendix D at the back of this book.

Leaving the scene of an accident while operating a CMV is a disqualifying offense. So is committing a felony while using a CMV and violating out-of-service orders.

Background and Character

When you apply for a job as a motor vehicle driver, you must fill out a driver application form. This asks for the same type of information ordinary job applications do, and then some. You must supply the following information:

- the name and address of the motor carrier employer
- your full name, mailing address, birth date, and Social Security number
- addresses you have had within the past three years
- the date the application was made
- the name of the state that issued your valid motor vehicle license or permit, plus the number and expiration date of the license or permit
- your experience in motor vehicle operation, including the type of equipment you have driven
- a list of all motor vehicle accidents you have had in the last three years. This must include the date and type of each accident. Any

deaths or injuries that resulted must also be listed.
- a list of all violations of motor vehicle laws or ordinances (other than parking) for which you were convicted in the last three years. (Skipping bail counts as a conviction.)
- the details of any denial, revocation, or suspension of any motor vehicle license or permit. If you have never had a license or permit denied, revoked, or suspended, you must certify that.
- the names and addresses of employers you worked for in the last three years. You must include the dates you were employed and why you left the job or jobs.

If you wish to drive a **commercial** motor vehicle in one of the CDL vehicle groups, you must list the names and addresses of employers for whom you worked as operator of a CMV in the seven years prior. In other words, you must give the past 10 years of your employment history. Again, you must include the dates you were employed and why you left.

Last, you date and sign the certification at the end of the form. The statement reads "This certifies that this application was completed by me, and that all entries on it and information in it are true and complete to the best of my knowledge."

If your motor carrier asks you for yet more information, you must provide it. The law permits your motor carrier to request such information from you. The motor carrier is required to tell you that the information you provide may be used. You are also to be told that your background may be checked. Former employers may be contacted.

Every 12 months, your motor carrier must review your driving record. The motor carrier must check to see that all its drivers meet the minimum requirements for safe driving. The motor carrier must check that none have been disqualified. The motor carrier will look closely not only at accidents, but also at speeding, reckless driving, and DUI. All of these show the driver is not sincere about safe driving.

As a licensed CMV driver, you have notification responsibilities with regard to your motor carrier. You must report convictions for traffic law violations to your motor carrier. The motor carrier will require you to complete a form like the one in Figure 5-2.

```
MOTOR VEHICLE DRIVER'S CERTIFICATION
I certify that the following is a true and complete list of traffic
violations (other than parking violations) for which I have been
convicted or forfeited bond or collateral during the past 12 months.
   Date                          Offense
   of conviction

. . . . . . . . . . . . . . . . . . . . . . . . . . . . . . . .

. . . . . . . . . . . . . . . . . . . . . . . . . . . . . . . .

. . . . . . . . . . . . . . . . . . . . . . . . . . . . . . . .

. . . . . . . . . . . . . . . . . . . . . . . . . . . . . . . .

   Location                      Type of vehicle operated

. . . . . . . . . . . . . . . . . . . . . . . . . . . . . . . .

If no violations are listed above, I certify that I have not been
convicted or forfeited bond or collateral on account of any violation
required to be listed during the past 12 months. . . . . . . . . .

. . . . . . . . . . . . . . . . . . . . . . . . . . . . . . . .
   (Date of certification)      (Driver's signature)

. . . . . . . . . . . . . . . . . . . . . . . . . . . . . . . .
              (Motor carrier's name)

. . . . . . . . . . . . . . . . . . . . . . . . . . . . . . . .
              (Motor carrier's address)

. . . . . . . . . . . . . . . . . . . . . . . . . . . . . . . .
      (Reviewed by: Signature)        (Title)
```

Figure 5-2. Motor Vehicle Driver's Certification.

Examinations and Tests

FMCSR Part 391 describes the road test that CMV drivers must take. This road test is different from the CDL skills test. This road test is given to you by your motor carrier employer. The CDL skills test is given to you by the state that gives you your CDL license. You must take both.

The road test required by FMCSR Part 391 must include at least the following:

- a pre-trip inspection (described in FMCSR 392.7)
- coupling and uncoupling any combination units
- putting the vehicle in operation
- use of the vehicle's controls and emergency equipment
- operating the vehicle in traffic and while passing other vehicles
- turning the vehicle
- braking, and slowing the vehicle without using the brakes
- backing and parking the vehicle

Your motor carrier employer will give you the road test. It can also be given by someone the employer appoints to do this. (If you yourself are the motor carrier, you cannot give yourself the road test. You must assign someone else to give you the test.) Your performance will be rated on a form. The person who gave you the test will sign the form.

When you complete the road test, you will get a copy of the Certification of Road Test. It will look like the one in Figure 5-3.

If you recently had a road test elsewhere and have a valid CMV license, the motor carrier may waive (excuse you from taking) another road test. The motor carrier may also waive the road test if you have a valid certificate of driver's road test that is no more than three years old. The only exception to the road test waiver is if you were going to work for a motor carrier driving tankers or double/triple trailers.

On the other hand, even if you have such a certificate, the motor carrier may still have you take a new road test.

Physical Qualifications and Examinations

CMV drivers must meet certain physical qualifications. These are described in FMCSR Part 391. We won't list them here. A doctor will check for these qualifications by giving you a physical. This isn't an ordinary physical. You've probably heard it called a "DOT physical." What makes this CMV driver physical special is described in this section of the FMCSR.

Among other things, your hearing, sight, and reflexes will be tested. You'll be asked about your use of drugs and alcohol. You may be tested for drug and alcohol use. It's here that the doctor will advise you about any medicines you may be taking. The doctor must assure you that such drugs are safe to use while driving. If they are not, taking them can get you disqualified as we mentioned before. This is true even if the doctor prescribed them.

This physical must be completed by a medical examiner listed on the National Registry of Certified Medical Examiners. Vision tests (visual acuity, field of vision, and the ability to recognize colors) may be performed by a licensed optometrist.

CERTIFICATE OF DRIVER'S ROAD TEST

Instructions: If the road test is successfully completed, the person who gave it shall complete a certificate of the driver's road test. The original or copy of the certificate shall be retained in the employing motor carrier's driver qualification file of the person examined and a copy given to the person who was examined. (49 CFR 391.31(e)(f)(g))

CERTIFICATION OF ROAD TEST

Driver's Name _____

Social Security Number _____

Operator's or Chauffeur's License Number _____

State _____

Type of Power Unit _____

Type of Trailer(s) _____

If passenger carrier, type of bus _____

This is to certify that the above-named driver was given a road test under my supervision on _____, 20___, consisting of approximately _____ miles of driving.

It is my considered opinion that this driver possesses sufficient driving skill to operate safely the type of commercial motor vehicle listed above.

_____
(Signature of Examiner)

_____
(Title)

_____
(Organization and Address of Examiner)

Figure 5-3. Certification of Road Test. You'll sign the original, and the motor carrier will keep it on file.

The physical must follow the guidelines in FMCSR Part 391. The doctor will sign and date the examination form and a medical examiner's certificate. The doctor will keep the exam form on file. Both you and your motor carrier will be given a copy of the medical examiner's certificate. It will look like Figure 5-4 on page 64.

You must have the physical if you have never had a so-called "DOT physical" before. If your last DOT physical was more than two years ago, you must be examined again and whenever a physician specifies. Exempt intracity zone drivers must get a DOT physical if their last one was more than one year ago. You must also get an exam if your ability to drive has been reduced by injury or illness.

FMCSR Part 391 tells you what to do if you disagree with the doctor's findings. Also, certain physical problems that might disqualify you may be waived. The process of getting a waiver is also described in this part.

Some states require that a DOT physical be submitted to the Department of Motor Vehicles before it will administer the CDL Knowledge Test.

Files and Records

FMCSR Part 391 describes the files and records motor carriers must keep on each of their drivers. They're of concern mainly to motor carriers. Read them enough to be familiar with them. Unless you yourself are a motor carrier, though, you don't need to be concerned about complying with them.

Limited Exemptions

FMCSR 391.61, 391.62, 391.63, 391.65, 391.67, and 391.69 are also of interest mainly to the motor carrier. These regulations list the few drivers who are exempt from FMCSR 391. There are few indeed. Some drivers who only work now and then are exempt from some of the regulations. So are drivers who usually work for one motor carrier but are doing a job for another motor carrier. Drivers of farm vehicles are exempt from some of the regulations. So are some drivers in Hawaii and some intrastate drivers hauling combustible liquids.

When you read these sections, you will see that most drivers are covered by FMCSR 391.

FMCSR PART 392

The next part of the FMCSR that all drivers must know is Part 392, Driving of Motor Vehicles. This part covers:

- illness, fatigue, drugs, intoxicating beverages, driver schedules, equipment, safe loading, and other general topics
- driving of vehicles
- stopped vehicles
- safety while fueling
- unauthorized passengers
- coasting and other prohibited practices
- texting and cell phone use

This part introduces the concept of "higher standard of care." The FMCSR says that drivers must obey local laws first. If the federal law is more strict, though, the federal law must be obeyed instead. This results in a "higher standard of care." Simply put, the stricter rule applies.

You've read about higher standard of care before, although not in so many words. It's what allows states to have a more strict definition of DUI than the federal law. Federal law considers a driver with 0.04 percent AC as DUI. A state could make that 0.03 percent, or less. That would result in a higher standard of care. If a state set the AC at 0.05 percent, the federal law would apply, since that would result in the higher standard of care.

The following are highlights of the other sections of FMCSR 392.

Ill or Fatigued Operator

Illness or fatigue is not to be taken lightly. A sick or tired driver is not a safe driver. It's not merely being lazy to get rest when you need it. You're not being a hero by continuing to drive. You are in fact breaking the law. FMCSR Part 392 says a driver must not drive if too tired or ill to be fully alert and in control. Only in an emergency should you continue to drive. Then, you should go on only as long as needed to handle the emergency.

Once in a while, you may become so ill that you cannot operate a motor vehicle safely. If this happens to you, you must not drive. However, in case of an emergency, you may drive to the nearest place where you can safely stop.

MEDICAL EXAMINER'S CERTIFICATE

I certify that I have examined _____ in accordance with the Federal Motor Carrier Safety Regulations (49 CFR 391.41-391.49) and with knowledge of the driving duties, I find this person is qualified, and, if applicable, only when:

☐ wearing corrective lenses
☐ wearing hearing aid
☐ accompanied by a _____ waiver/exemption

☐ driving within an exempt intracity zone (49 CFR 391.62)
☐ accompanied by a Skill Performance Evaluation Certificate (SPE)
☐ qualified by operation of 49 CFR 391.64

The information I have provided regarding this physical examination is true and complete. A complete examination form with any attachment embodies my findings completely and correctly, and is on file in my office.

SIGNATURE OF MEDICAL EXAMINER	TELEPHONE	DATE	
MEDICAL EXAMINER'S NAME (PRINT)	☐ MD ☐ DO ☐ Physician Assistant	☐ Chiropractor ☐ Advanced Practice Nurse ☐ Other Practitioner	
MEDICAL EXAMINER'S LICENSE OR CERTIFICATE NO./ISSUING STATE	NATIONAL REGISTRY NO.		
SIGNATURE OF DRIVER	INTRASTATE ONLY ☐ YES ☐ NO	CDL ☐ YES ☐ NO	DRIVER'S LICENSE NO.	STATE
ADDRESS OF DRIVER			
MEDICAL CERTIFICATION EXPIRATION DATE			

Figure 5-4. Medical Examiner's Certificate.

PASS BILLBOARD
Fatigue

Driving a vehicle for long hours is tiring. Even the best of drivers will become less alert. To stay alert and safe, get enough sleep. Leaving on a long trip when you're already tired is dangerous. If you have a long trip scheduled, make sure that you get a good sleep before you go. Most people require seven to eight hours of sleep every 24 hours.

Schedule trips safely. Your body gets used to sleeping during certain hours. If you are driving during those hours, you will be less alert. If possible, try to schedule trips for the hours that you are normally awake. Many heavy motor vehicle accidents occur between midnight and 6 A.M. You can easily fall asleep at these times, especially if you don't regularly drive at those hours. Trying to push on and finish a long trip at these times can be very dangerous. See Figure 5-5.

Avoid medication. Many medicines can make you sleepy. Those that do have a label warning against operating vehicles or machinery. The most common medicine of this type is an ordinary cold pill. If you have to drive with a cold, you are better off suffering from the cold than from the effects of the medicine. Keep cool. A hot, poorly ventilated cab can make you sleepy. Keep the window or vent cracked or use the air conditioner, if you have one.

Take breaks. Short breaks can keep you alert. However, the time to take them is before you feel really drowsy or tired. Stop often. Walk around and inspect your vehicle. It may help to do some physical exercises.

When you are sleepy, trying to push on is far more dangerous than you think. It is a major cause of fatal accidents. When your body needs sleep, sleep is the only thing that will work. If you have to make a stop anyway, make it whenever you feel the first signs of sleepiness, even if it is earlier than you planned. By getting up a little earlier the next day, you can keep on schedule and not have to drive while you're not alert.

Take a nap. If you can't stop for the night, at least pull off the road and take a nap. A nap as short as a half-hour will do more to overcome fatigue than a half-hour coffee stop.

Avoid drugs. There are no drugs that can overcome being tired. While they may keep you awake for a while, they won't make you alert. Eventually, you'll be even more tired than if you hadn't taken them at all. Sleep is the only thing that can overcome fatigue.

Figure 5-5. Most heavy motor vehicle accidents happen between midnight and 6 A.M.

Drugs and Other Substances

The misuse of drugs and alcohol while on-duty is forbidden by several regulations. Some are found in FMCSR Part 40 and FMCSR Part 382. FMCSR 392.4 states that no driver shall be on-duty and possess, be under the influence of, or use:

- a Schedule I substance
- amphetamines ("pep pills" or "bennies")
- narcotics or their derivatives
- any substance that keeps a driver from being able to drive safely

This part reminds you that it is OK to use drugs a doctor has prescribed for you. This is only if the doctor has assured you the drug will not affect your driving ability.

This part also clears up the definition of "possession." It is legal to "possess" and transport these drugs as cargo. They must be listed on the manifest, however.

Alcohol

All drivers are affected by drinking alcohol. Alcohol affects judgment, vision, coordination, and reaction time. It causes serious driving errors, such as:

- increased reaction time to hazards
- driving too fast or too slow
- driving in the wrong lane
- running over the curb
- weaving
- straddling lanes
- quick, jerky starts
- failure to signal or use lights
- failure to stop at stop signs and red lights
- improper passing

All these can mean increased chances of a crash and chances of losing your driver's license. Accident statistics show that the chance of a crash is much greater for drivers who have been drinking than for drivers who have not.

FMCSR Part 392 states that no driver shall drink an intoxicating beverage within four hours before going on-duty, driving, or otherwise being in control of a motor vehicle. This means any and all intoxicating beverages. It doesn't matter how much alcohol is in them. A little or a lot, they are prohibited. Remember, alcohol is alcohol. Beer and wine are no safer than liquor.

You must not even be under the influence of such a beverage within four hours of going on-duty. This may mean that if you are drinking, you must stop much sooner than four hours before going to work. It's possible to be "under the influence" of drinking for more than four hours. Here's why.

You have seen that drunkenness is measured as a percentage of alcohol in the blood. Three factors affect this percentage. These factors are time, the amount of alcohol, and your weight.

Alcohol, like any food or drink, is used by the body over time. After a certain amount of time, your body has used up the alcohol. There is no more in your system. It no longer affects you.

12 ounces of beer   5 ounces of wine   1½ ounces of liquor

Figure 5-6. It takes most people about one hour to burn off one 12-ounce bottle of beer, one 5-ounce glass of wine, or one 1 1/2-ounce shot of liquor.

How much time must pass before you are free of the effects of alcohol? It depends on how much you drank and your weight. It takes most people about one hour to burn off one 12-ounce bottle of beer, one 1 1/2-ounce shot of liquor, or one 5-ounce glass of wine. Smaller people feel the effects of alcohol more than larger people.

Put together the facts that you have just learned so far about the effects of alcohol and the requirements of the law. The law requires you to be free of the effects of alcohol for four hours before you go on-duty. Say you have one beer at noon. You are due to go on-duty at 4 P.M. Is it safe? No, it's not. You had the beer at noon. At 1 P.M., you were still affected by it. By 4 P.M., only three hours have passed. You still have an hour to go. Only then have you been free for four hours of the effects of that one beer you drank at noon.

Right away, you can see why you must stop drinking far in advance of when you plan to drive.

What if you had more than one drink? What if you are lighter or heavier than the average person? The charts in Figure 5-7 help answer that question.

Read down the left side to find the row with the figure closest to the number of drinks that concern you. Read along the top to find the column with the body weight closest to yours. The number where the column and row come together is your blood alcohol content at that weight at that number of drinks.

Take, for example, a 200-pound man who has four drinks. The chart shows his AC would be 0.08 percent. He is clearly under the influence. As far as the law is concerned, he will be for some time.

One drink would take one hour to burn off. Four drinks would take far longer to burn off. For each hour that passes since that last drink was consumed, subtract 0.15 percent. That's a little less than the alcohol in one drink. Our 200-pound drinker will need at least four hours after he finally has that last drink to get his AC below 0.04 percent. (He will need eight hours to completely burn off all four drinks.) Then he needs the additional four hours required by FMCSR 392.5. This man cannot legally drive for eight hours after he has his last drink! If you have many drinks spread throughout an afternoon or evening, you can easily still be affected by them the next day. You may feel fine and think you are sober. Your judgment and reflexes simply won't be up to par.

Don't think you can hurry the process of getting sober. Coffee and cold showers will make you feel more alert. You will still be under the influence. Only the passing of enough time will return you to a normal state.

Certainly, you must not consume an intoxicating beverage while on-duty, much less before.

You may transport intoxicating beverages, of course. As with controlled substances, such cargo must be on your manifest and transported as part of a shipment to be legal.

A driver who violates these regulations is immediately put out-of-service for 24 hours. You would have to report this to your employer within 24 hours. You would also have to report this to the state that gave you your license. You must make this report within 30 days, unless you request a review.

If you want a review of the out-of-service order, you must request it within 10 days. You must write to the Division Administrator or State Director for

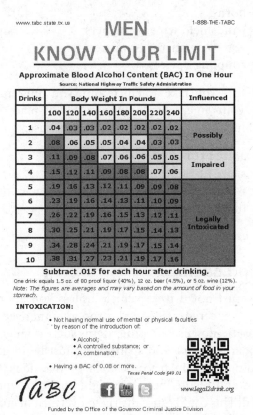

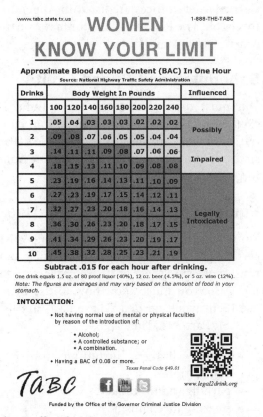

Provided to TABC by the National Highway Safety Traffic Administration

Figure 5-7. Blood alcohol content and weight. The numbers in the chart are AC percentages. Remember the legal limit? It's 0.04.

PASS BILLBOARD

There are many beliefs about how alcohol works in the body. So many of them are not based on fact. A common one is "I can handle it better than other people." It's simply not true.

The first thing that alcohol does is affect your judgment. It also gives you a false sense of competence. All it takes is one drink. After that, you simply can't trust yourself to decide if you are drunk or not. You must assume you are, even if you feel fine. Of course you'll feel great—you've been drinking! You won't be great, though. You'll be impaired. Most people are legally under the influence after having one drink.

Perhaps you wait until your reaction time slows before you decide you've had enough.

That's too late. By the time you notice your reactions have slowed, you are at about 0.05 percent AC. This is already over the legal limit. Maybe you decide to stop when you begin to stagger and your speech slurs. At this point you are at 0.2 percent AC. This is five times the 0.04 percent AC that the law considers the limit. You could go on drinking until you pass out. For most people this will happen by the time they reach 0.3 percent AC. If you drink enough to reach 0.4 percent AC, you will definitely not be driving. At 0.4 AC, you will fall into a coma, and likely die.

It's hard not to conclude that drinking and driving don't mix.

---

the geographical area in which the out-of-service order was issued. The addresses of the Motor Carrier Safety offices are listed in Appendix D of this book. The Regional Director may overturn the order or let it stand.

Schedules and Speed Limits

You must drive within the speed limit. It's no excuse to claim you can't complete your run without driving at an illegal speed. FMCSR Part 392 states your carrier must not schedule runs that force you to exceed the speed limit.

Equipment Inspection and Use

Your vehicle cannot be operated safely without certain parts. Before you take a vehicle out, you must make sure these parts are in place and working. Also, you must know how to use them, and use them when needed. These parts are listed in FMCSR Part 392. They are:

- service brakes, including trailer brake connections
- parking (hand) brake
- steering mechanism

- lighting devices and reflectors
- tires
- horn
- windshield wiper or wipers
- rear-view mirror or mirrors
- coupling devices

Before you take a vehicle out, you must make sure your emergency equipment is in place and working. Also, you must know how to use:

- a fire extinguisher
- spare fuses
- warning devices for stopped vehicles (like flares or reflectors)

and use them when needed.

Safe Loading

You must not drive a vehicle unless you are certain the cargo is loaded properly and secure. Tarps, tie-downs, and other items such as tailgates, tailboards, doors, and spare tires must be in good condition and in place. The cargo may look tied down and secure, but the weight could be unbalanced. This could cause the rig to overturn in curves and sharp turns. So you must know both proper cargo distribution and securement. FMCSR Part 392 also states the cargo must not block your view or your ability to move around. It must not block emergency equipment or the vehicle's controls.

All must be in good order before you start your trip. It must remain that way through the trip. You must inspect the cargo and how it's secured within the first 50 miles after you start your trip. Then you must check it again

- when you, the driver, change your duty status
- after the vehicle has been driven for three hours
- after the vehicle has been driven for 150 miles

whichever of these three comes first.

Slowing and Stopping

CMV drivers hauling certain cargo must take special action when crossing railroad tracks at grade. These regulations also apply to drivers hauling hazmat that requires placards. These steps must be taken at railroad crossings without signal lights, or a flagman, or some other way of telling you that it's safe to proceed.

FMCSR 392.10 states you must stop no farther than 50 feet and no closer than 15 feet of the track. Look both ways and listen for oncoming trains. When it's safe, you must proceed at a speed that will let you get all the way across without changing gears. You must not change gears while crossing the track.

You do not have to stop at a streetcar crossing or a track used only for industrial switching purposes within a business district. You also don't have to stop at a railroad grade crossing if a police officer or flagman directs you to proceed. You may cross tracks without stopping if they are controlled by a light that signals green for "go." A sign indicating that the rail line has been abandoned or is "exempt" gives you permission to cross without stopping.

If you are driving a vehicle other than the types just described, you must at least slow down when you approach a railroad crossing. Proceed at a speed slow enough that you could stop before reaching the nearest rail. You must not cross the track until you are sure it is safe to do so.

Hazardous Conditions

FMCSR Part 392 states you should slow down when hazardous conditions exist. This could be snow, ice, fog, mist, dust, smoke, rain, or slick roads—anything that reduces your ability to guide or control the vehicle. If conditions are bad enough, you are to stop altogether.

Seat Belts

You must not operate the CMV unless you have secured yourself with the seat belt.

Stopped Vehicles

Drivers must also take special action when their vehicles are stopped. They must use emergency signals—hazard flashers. Such signals and their use are described in FMCSR Part 392. You must use these signals when you stop on the traveled portion of the highway (other than traffic stops) or shoulder. Briefly, the rule is to turn on your four-way flashers when you first stop. Keep your flashers on until you put out your warning devices. These are the flares or reflective devices described in FMCSR Part 393. You'll learn about these devices later in this chapter.

You must put these warning devices out within 10 minutes of stopping. You must put one at the traffic side of your stopped vehicle. This placement is for traffic going both ways on an undivided highway. It must be within 10 feet of the vehicle in the direction of approaching traffic.

You must put another warning device 100 feet from your vehicle, in the center of the traffic lane in the direction of approaching traffic. Put a third device 100 feet from your vehicle in the opposite direction of the other two, away from approaching traffic.

This is almost easier to show than to describe. See Figure 5-8 on page 70.

You place your warning devices slightly differently when you are stopped on a hill, in a curve, or on a divided or one-way road. Figures 5-9 and 5-10 show the correct placement of warning devices under these conditions.

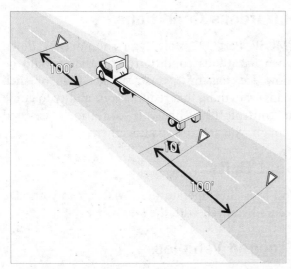

Figure 5-8. Placement of warning devices on the highway.

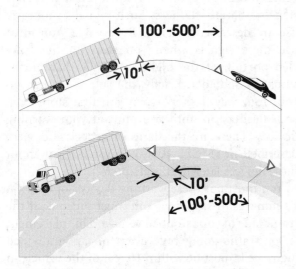

Figure 5-9. Placement of warning devices on a hill.

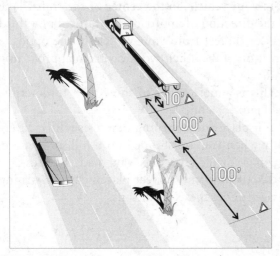

Figure 5-10. Placement of warning devices on a one-way road.

For older vehicles, flares and liquid-burning flares (called fusees or pot-torches) may be used instead of reflective triangles. However, when flammable liquids are leaking onto the road, the fusees may be used only at a distance where they won't create a fire or explosion. They also may not be used to protect vehicles hauling explosives. You must not use flame-producing flares to protect a vehicle used to haul flammable cargo, even if it's empty. Last, you may not use fusees to protect a vehicle fueled by compressed gas. Use reflective triangles, red lanterns, or red reflectors instead.

Fueling Procedures

The fuel your vehicle runs on can be dangerous if mishandled. The rules in FMCSR Part 392 describe how to handle fuel safely. Unless you need to run the engine to fuel your vehicle, you must not fuel your vehicle with the motor running. Do not smoke or have any open flame in the fueling area. Fueling lets off vapors you can't see. A cigarette or match could ignite these vapors.

Keep the nozzle of the fuel hose in contact with the fuel tank's intake pipe at all times. Do not allow anyone else to smoke or do anything else in the fueling area which would cause an explosion.

Prohibited Practices

The last part of FMCSR 392 describes actions that are forbidden. One is "no passengers." You must not transport anyone else in your motor vehicle. There are some exceptions to this rule. One is that you may transport other people who have been assigned to the vehicle, such as co-drivers. Another is an attendant charged with caring for livestock. You may take a passenger when your motor carrier has authorized it, in writing. This authorization must name the person, the beginning and end points of the transport, and the date on which the authorization expires.

The only time you may carry unauthorized people is to help them in an emergency. Then you may take passengers only as far as needed to get them to safety.

The last few rules in this section have to do with personal safety while inside the vehicle. First, motor vehicles must have an exit. The exit must be one that can easily be operated by the person inside.

Next, carbon monoxide can be dangerous in a closed area. If someone in the vehicle shows the effects of carbon monoxide poisoning, no one else may ride in or drive that vehicle. This is also true if carbon monoxide is found in the vehicle. No one may drive or ride in a vehicle if carbon monoxide poisoning could occur due to some mechanical defect.

Some drivers carry flame-producing heaters. They use them to melt ice from trailer doors or to warm the area when loading and unloading. This is permitted. It is, however, forbidden to use a flame-producing heater while the vehicle is moving.

Drivers may not use a radar detector or operate a CMV that has one.

Drivers are forbidden to text or use handheld mobile phones while driving except to contact police officers or emergency services.

That completes the overview of FMCSR 392.

FMCSR PART 393

FMCSR Part 393 describes parts and accessories needed to operate the vehicle safely. Briefly, these are:

- lighting devices, reflectors, and electrical equipment
- brakes
- windows
- fuel systems
- coupling devices and towing methods
- tires, windshield wipers, horn, and other assorted parts
- emergency equipment
- cargo tiedowns
- wheels, suspension systems, and other parts of the frame

Some of these were referred to by earlier FMCSR sections (see page 68). This part describes them in great detail.

Since you are driving a truck, not building one, you may think you don't need to know the technical specifications of lights and brakes. You must inspect these parts, though. You can't know if something is missing or broken if you don't know what's supposed to be there in the first place.

It would be helpful if you have access to a CMV while you go through FMCSR Part 393. Every time you get near that vehicle, see how many parts that you read about you can identify. When you next hit the books, review the section. Note any parts you missed in your identification. Go over those parts again and again until you can name every part of your CMV from memory.

We won't cover this part of the FMCSR in great detail now. That's not because the details aren't important. They are. You'll learn about them in future chapters on vehicle safety control systems and vehicle inspection.

The first two rules in this FMCSR Part 393 state you may have other parts and accessories on your vehicle unless they make the vehicle unsafe to drive. However, you must have at least the parts described in this section. They must meet the specifications stated in this section.

FMCSR 393.5 defines terms used throughout the rest of the section. For example, what you may call a "pole trailer" is termed a "pulpwood trailer" in the FMCSR. The amount you can turn your steering wheel before the front wheels also begin to move is called "steering wheel lash." Review these definitions to make sure you know the official names of the required parts.

Lighting Devices and Reflectors

FMCSR Part 393 describes the types of lights CMVs must have. It includes a table and several figures that show the color, location, and other specification of all these required lights. Study the table and the figures in your copy of the FMCSR. There are footnotes that give more details about items in the table on the page that follows the table. You can find the labels that identify the parts on the figures on the page that comes after all the figures.

We have recreated the figures here, with the labels right on the picture. That way you can use the scanning technique described in Chapter 3 to memorize the pictures.

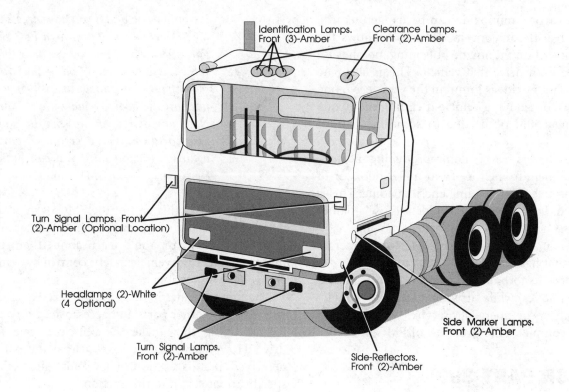

Identification Lamps.
Front (3)-Amber

Clearance Lamps.
Front (2)-Amber

Turn Signal Lamps. Front
(2)-Amber (Optional Location)

Headlamps (2)-White
(4 Optional)

Turn Signal Lamps.
Front (2)-Amber

Side-Reflectors.
Front (2)-Amber

Side Marker Lamps.
Front (2)-Amber

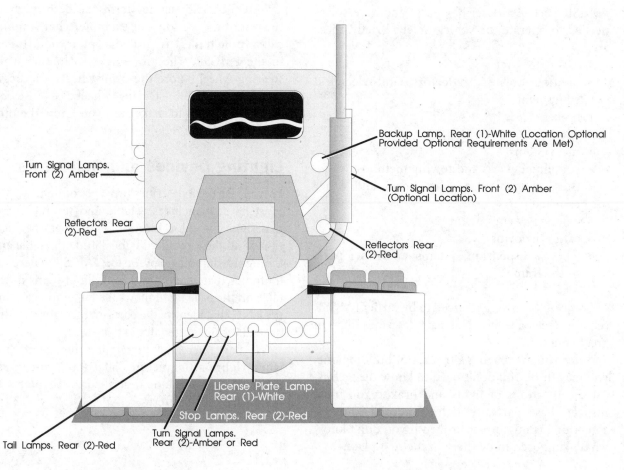

Backup Lamp. Rear (1)-White (Location Optional
Provided Optional Requirements Are Met)

Turn Signal Lamps.
Front (2) Amber

Turn Signal Lamps. Front (2) Amber
(Optional Location)

Reflectors Rear
(2)-Red

Reflectors Rear
(2)-Red

License Plate Lamp.
Rear (1)-White

Stop Lamps. Rear (2)-Red

Tail Lamps. Rear (2)-Red

Turn Signal Lamps.
Rear (2)-Amber or Red

Figure 5-11. Required lights and reflectors.

Front

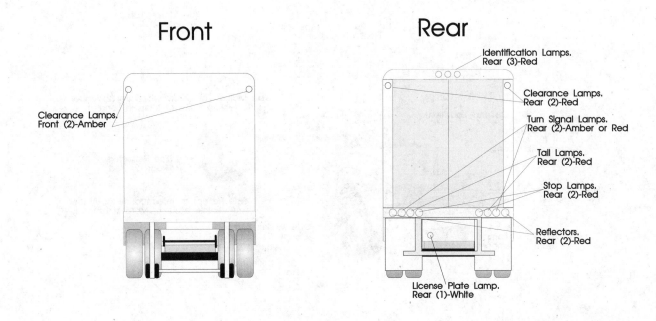

Clearance Lamps.
Front (2)-Amber

Rear

Identification Lamps.
Rear (3)-Red

Clearance Lamps.
Rear (2)-Red

Turn Signal Lamps.
Rear (2)-Amber or Red

Tail Lamps.
Rear (2)-Red

Stop Lamps.
Rear (2)-Red

Reflectors.
Rear (2)-Red

License Plate Lamp.
Rear (1)-White

Each Side

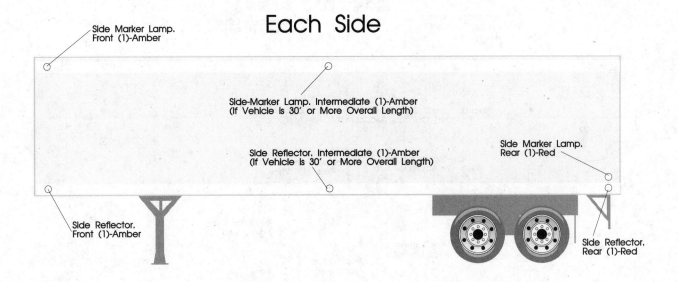

Side Marker Lamp.
Front (1)-Amber

Side-Marker Lamp. Intermediate (1)-Amber
(If Vehicle is 30' or More Overall Length)

Side Reflector. Intermediate (1)-Amber
(If Vehicle is 30' or More Overall Length)

Side Marker Lamp.
Rear (1)-Red

Side Reflector.
Front (1)-Amber

Side Reflector.
Rear (1)-Red

Figure 5-12. Required lights and reflectors.

Over 80 Inches

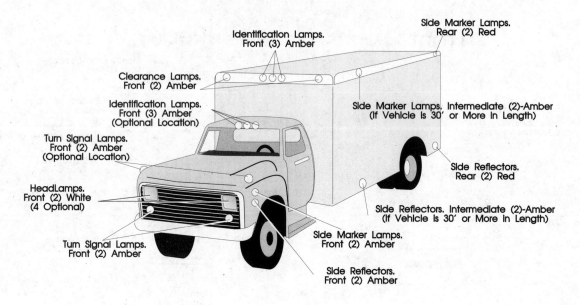

Side Marker Lamps.
Rear (2) Red

Identification Lamps.
Front (3) Amber

Clearance Lamps.
Front (2) Amber

Identification Lamps.
Front (3) Amber
(Optional Location)

Side Marker Lamps. Intermediate (2)-Amber
(If Vehicle Is 30' or More in Length)

Turn Signal Lamps.
Front (2) Amber
(Optional Location)

HeadLamps.
Front (2) White
(4 Optional)

Side Reflectors.
Rear (2) Red

Turn Signal Lamps.
Front (2) Amber

Side Reflectors. Intermediate (2)-Amber
(If Vehicle Is 30' or More in Length)

Side Marker Lamps.
Front (2) Amber

Side Reflectors.
Front (2) Amber

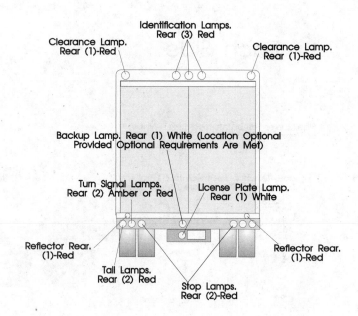

Identification Lamps.
Rear (3) Red

Clearance Lamp.
Rear (1)-Red

Clearance Lamp.
Rear (1)-Red

Backup Lamp. Rear (1) White (Location Optional
Provided Optional Requirements Are Met)

Turn Signal Lamps.
Rear (2) Amber or Red

License Plate Lamp.
Rear (1) White

Reflector Rear.
(1)-Red

Reflector Rear.
(1)-Red

Tail Lamps.
Rear (2) Red

Stop Lamps.
Rear (2)-Red

Figure 5-13. Required lights and reflectors.

Under 80 Inches

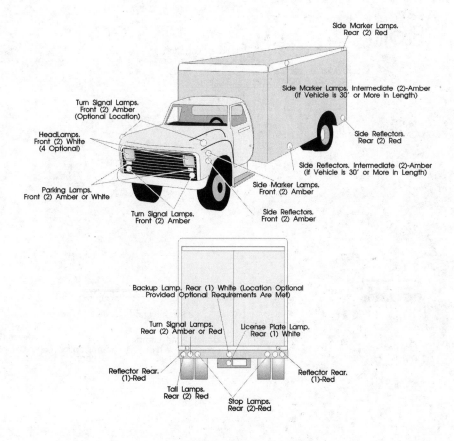

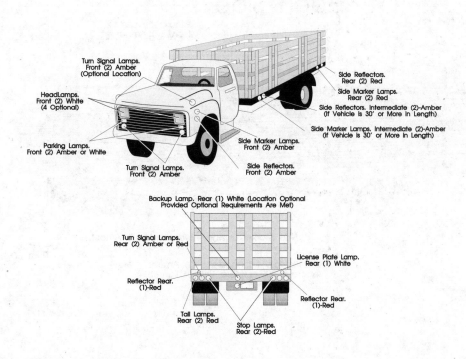

Figure 5-14. Required lights and reflectors.

Pole Trailers - All Vehicle Widths

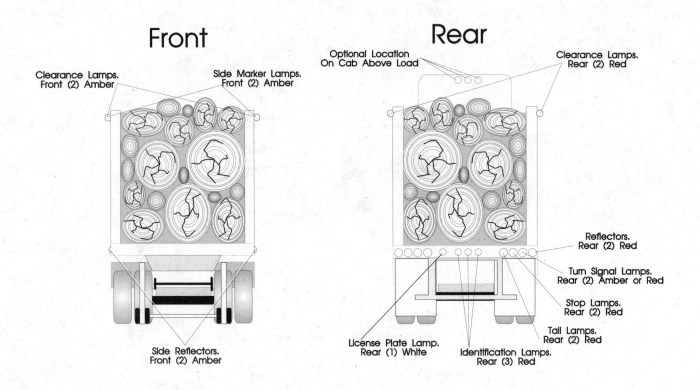

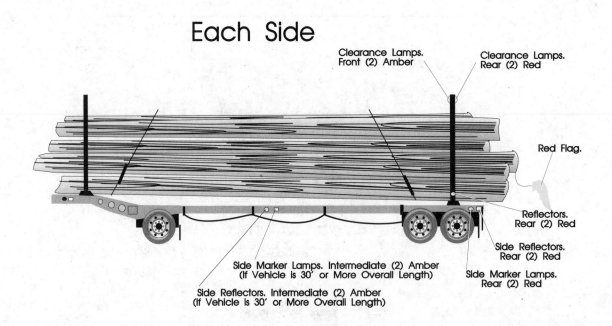

Figure 5-15. Required lights and reflectors.

Container Chassis

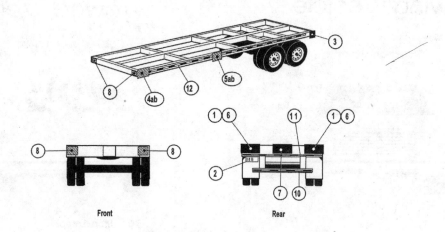

Front

Rear

Each Side

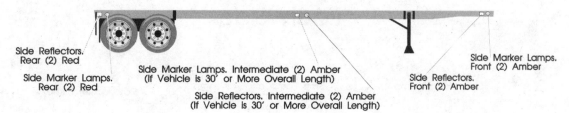

Side Reflectors.
Rear (2) Red

Side Marker Lamps.
Rear (2) Red

Side Marker Lamps. Intermediate (2) Amber
(If Vehicle is 30' or More Overall Length)

Side Reflectors. Intermediate (2) Amber
(If Vehicle is 30' or More Overall Length)

Side Reflectors.
Front (2) Amber

Side Marker Lamps.
Front (2) Amber

Converter Dolly

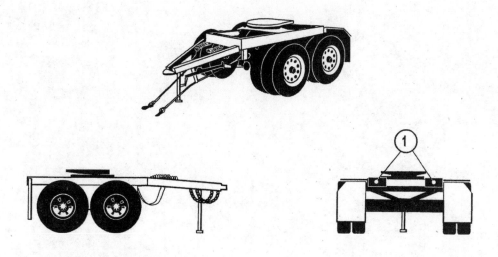

Side View of Dolly

Rear

Figure 5-16. Required lights and reflectors.

Front Of Towing Vehicle

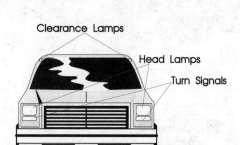

Rear Of Towed Vehicle

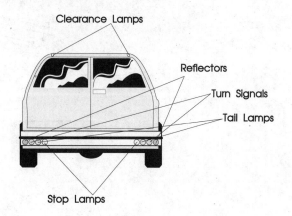

Each Side

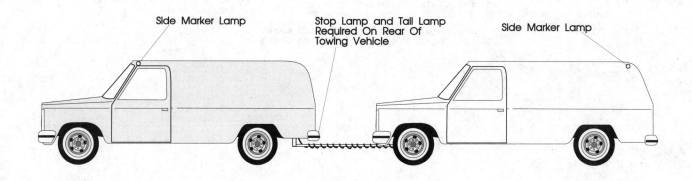

Figure 5-17. Required lights and reflectors.

Front Of Towing Vehicle

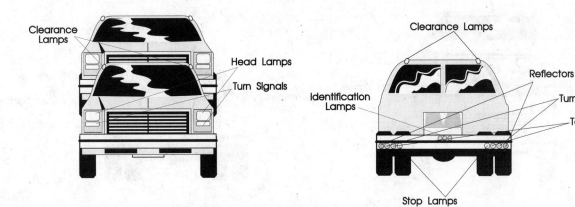

Clearance Lamps

Head Lamps

Turn Signals

Rear Of Rearmost Towed Vehicle

Clearance Lamps

Reflectors

Identification Lamps

Turn Signals

Tail Lamps

Stop Lamps

Each Side

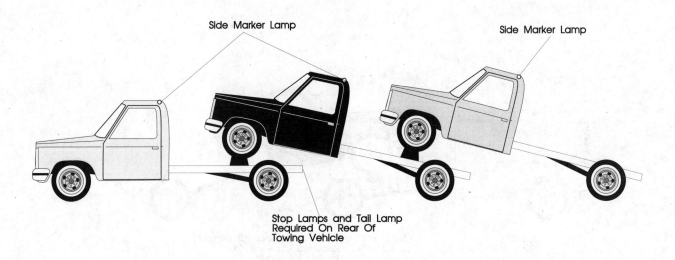

Side Marker Lamp

Side Marker Lamp

Stop Lamps and Tail Lamp Required On Rear Of Towing Vehicle

Figure 5-18. Required lights and reflectors.

Front Of Towing Vehicle

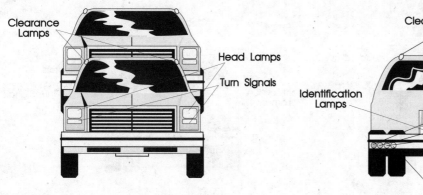

Clearance Lamps

Head Lamps

Turn Signals

Rear Of Towed Vehicle

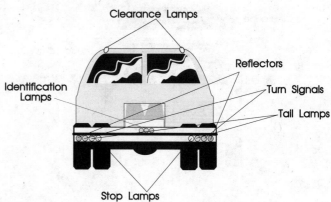

Clearance Lamps

Reflectors

Turn Signals

Tail Lamps

Identification Lamps

Stop Lamps

Each Side

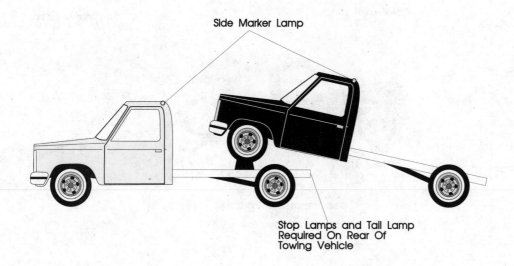

Side Marker Lamp

Stop Lamps and Tail Lamp Required On Rear Of Towing Vehicle

Figure 5-19. Required lights and reflectors.

FMCSR 393.17 describes the lights and reflectors to be used in driveaway-towaway operations (when a vehicle itself is the cargo). The figures in this section have the labels right along with the pictures. Use the scanning technique to get the information firmly in your mind.

Take the time now to learn what lights and reflectors are required. Pay special attention to those on the type of vehicle you plan to drive with your CDL. Learn the number, type, color, and location of all the required lights and reflectors. Then, you'll know what to look for when you inspect your vehicle.

Turn Signals

Many states still permit car drivers to use hand and arm signals when turning and stopping. Buses and trucks must have signaling systems as described in FMCSR Part 393. These lights must not only serve as turn signals. They must be able to work as hazard warning lights as well.

Clearance Lights

CMVs must have clearance lamps. These lights help outline the width and height of the vehicle. These are mounted at the highest and widest part of the sides, front, and back of the vehicle as shown in Figures 5-20 and 5-21.

Front

Figure 5-20. Required clearance lights.

Other Lights

You may have other lights and markers on your vehicle. However, they must not reduce the effectiveness of the required lights. For instance, an extra light that is brighter than the required light will make the required light hard to see. This extra light would be prohibited.

All the lights must be electric. Liquid burning lights may sometimes be used to mark the end of loads that stick out past the rear of the vehicle.

Headlights and Reflectors

FMCSR Part 393 describes the types of headlights and fog lights you must have on your vehicle and

Vehicle Without Permanent Top Or Sides

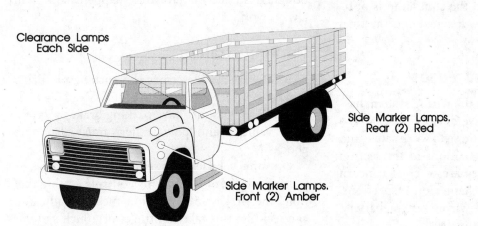

Rear

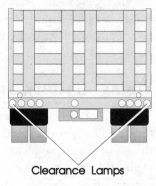

Figure 5-21. Required clearance lights.

lists the requirements for all other lights. Some states require that lights must be visible from a distance of 500 feet to 50 feet away in clear weather. This part describes how reflectors should be placed so as to give the greatest visibility.

Trailers may be equipped with conspicuity material. "Conspicuity" means having a quality of being very noticeable. Conspicuity material is usually a type of highly reflective tape. Trailers equipped with conspicuity material do not have to have the reflex reflectors required by Table 1 of FMCSR Part 393. However, this conspicuity material must meet the requirements stated in Part 393. In addition, it must be placed where reflex reflectors are required and meet the visibility requirements for the reflex reflectors.

Hazard warning signals on CMVs must work independently of the ignition switch. When the hazard warning signal is on, all the turn signals should flash at the same time.

All required lamps must be powered by the electrical system of the motor vehicle. An exception is battery-powered lamps used on projecting loads.

The CMV must have headlamps. These must offer the driver a choice of high- and low-beam settings. You cannot use auxiliary driving lamps and/or front fog lamps instead, although you can have them in addition to headlamps. All these lights have to be mounted and aimed according to specifications in the regulations. Other than turn signal lamps and warning lamps, all exterior lights must be steady-burning.

Stop lamps on each vehicle must light when the service brakes are applied. One exception is when the emergency feature of the trailer brakes is used. Another exception is when the stop lamp is optically combined with the turn signal and the turn signal is in use.

Wiring, Battery, and Fuses

FMCSR Part 393 discusses the wiring system, battery, and overload protective devices (fuses). As a driver, of course you want these to work safely and well. Still, many of the regulations in the section are not driver-related. They are of chief concern to manufacturers. In the chapter on inspections, you'll learn the parts of the wiring system, battery, and fuses that do concern the driver.

This part also discusses detachable connections.

Detachable connections are the electrical connection made between towing and towed vehicles, commonly called a "pigtail." This connection can't be made simply by twisting wires together. It must be with special, shielded cables.

Wires and cables must be attached to terminals with proper connectors.

Many of the details in the sections you just read may seem too small to be worth the bother. However, you are expected to be fully familiar with your equipment. When you inspect your vehicle as part of your CDL test, you may be asked questions about these parts. You must know the answers.

Brakes

The next group of sections in FMCSR 393 deals with the brakes. Many drivers fail the CDL tests because they don't know enough about their braking systems. A faulty braking system threatens your life and the lives of everyone else around you. You must know your braking system like the back of your hand.

FMCSR Part 393 states your CMV must have service, parking, and emergency brakes. The emergency brake controls must be placed so you can reach them when you are in the driver's seat with the seat belt on. The emergency brake control may be combined with either the service or parking brakes. However, all three controls may not be combined into one.

The brake systems on your CMV must be designed so no matter what happens, one will always work.

Parking Brakes

FMCSR Part 393 describes the parking or spring brakes. With few exceptions, all CMVs made after March 7, 1990, must have parking brakes. Certain farm vehicles and pole trailers must also have chocks.

On some vehicles, you set the parking brakes by pulling a lever or knob. This controls a cable that pulls the brakes into position mechanically. You can feel that this takes a little effort. Such systems must not require great strength to operate.

PASS BILLBOARD
Anti-Lock and Front Brakes

Some drivers believe that front brakes are dangerous. They think they can lock up in emergencies and cause the driver to lose control.

Experience and tests have shown this simply isn't true. Accident records have shown that front wheel brakes do not cause accidents or make accidents more serious. Working front brakes allow drivers to handle vehicles better and stop sooner.

Locked-up brakes are more the result of panic stopping than faulty equipment. When the front brakes do lock, the vehicle can get out of control, skid, and possibly jackknife, leaving a tractor and trailer at right angles to each other. To help control brake lock-up, most CMVs manufactured after March 1, 1999, have anti-lock brake systems and devices to warn the driver of system malfunctions.

The best anti-lock prevention is probably to avoid making panic stops in the first place. If you are a careful and alert driver and keep a good following distance, you won't have to stand on your brakes to avoid an accident.

On other vehicles, setting the parking brake requires more strength than most drivers have. Then the parking brake is set with the help of air pressure. When this is the case, the air supply for the parking brake must be separate from the service brake air supply. It must be reserved for the sole purpose of applying the parking brake.

Fluid or air pressure or electricity may be used to set the parking brakes. They must not be used to keep the parking brakes applied. Also, when you release the brakes, the system must be such that you can immediately reapply them. No other system may be used. Air pressure would not meet these requirements. Air pressure takes time to build. You might use up all your air pressure to keep the parking brakes on. Then, if you released them, you might not have enough air pressure to reapply them immediately.

The parking brake systems on most vehicles meet these requirements. (You'll see how in Chapter 6 and Chapter 10.) These regulations are mostly of concern to manufacturers. As a driver, your job will be to inspect and maintain the system so your brakes are always in working order.

All wheels on a CMV must have working brakes. This is required by FMCSR 393.42, which also lists the few exceptions. Among them are vehicles manufactured before 1980. If you have disconnected or removed the front brakes, you should know that since February 26, 1988, nearly all other CMVs have been required to have working front brakes.

Emergency Brakes

FMCSR Part 393 covers pulling a trailer that has brakes. It requires that the service brakes on the tractor will still work if the trailer breaks away.

Vehicles with air brakes must have two ways of starting the trailer emergency brakes. One must work automatically if the tractor air supply falls to between 20 and 45 pounds per square inch (psi). The other must be a manually controlled device. This must be within easy reach of the person in the driver's seat. The control must be clearly marked. It must be easy to see how to work the control.

Vehicles that pull trailers with vacuum brakes must have two controls. The first must be a single control that will operate all the brakes of the total combination. The second must be a trailer emergency brake control. This must be a manual control. It must be independent of brake air, hydraulic, and other pressure. It must be separate from all other controls. There's only one exception to this. That is if failure of the pressure on which the second control depends will cause the trailer brakes to come on automatically.

A trailer that is required to have brakes must have brakes that will apply automatically if the trailer breaks away from the tractor. These trailer brakes must stay on for at least 15 minutes.

The tractor braking air supply must be protected from air back-flow if the tractor air pressure falls. If a problem with the tractor air supply develops, the system must prevent the trailer air supply from flowing back to the tractor. If that happened, both the tractor and the trailer would be left without a supply of air. A relay or check valve takes care of this. You'll read about brake valves in more detail in the next few chapters.

These requirements don't apply to vehicles being towed as cargo.

Again, these regulations are mostly of concern to manufacturers. Your vehicle will probably have brakes designed to meet these requirements. Your job will be to inspect and maintain your brakes.

Brake Tubing and Hoses

FMCSR Part 393 sets the requirements for brake tubing and hoses. The brake lines on your vehicle will likely meet these manufacturing standards. Through use, however, they can become worn or damaged. As a CMV driver, you'll check to make sure your brake lines continue to meet these requirements.

Working Brakes

FMCSR Part 393 states that for the most part, all the brakes on the vehicle must work. There are very few exceptions, which are also described in this section. If you drive a vehicle manufactured before March 1, 1975, read this section of the FMCSR carefully. It lists specifications for the different parts of the brake system, such as the linings, pads, slack adjusters, and so forth. Brake systems that meet these specifications are brakes that work properly, ensuring safety. When you inspect your vehicle, check to see that your brakes meet these specifications.

Brake Reservoirs

FMCSR Part 393 describes the brake reservoirs (tanks). For air or vacuum brake systems, these tanks must be large enough to ensure a full service brake application with the engine stopped. This

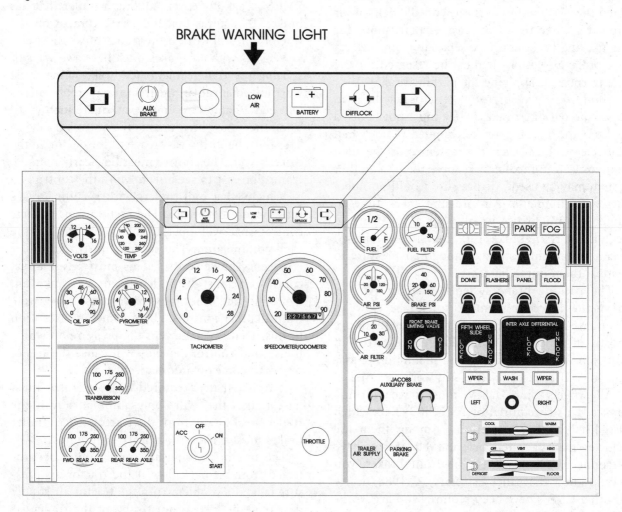

Figure 5-22. Brake systems have warning lights.

Type of motor vehicle	Service brake systems			Emergency brake systems
	Braking force as a percentage of gross vehicle or combination weight	Deceleration in feet per second per second	Application and braking distance in feet from initial speed of 20 mph	Application and braking distance in feet from initial speed of 20 mph
A. Passenger-carrying vehicles (1) Vehicles with a seating capacity of 10 persons or less, including driver, and built on a passenger car chassis	65.2	21.0	20.0	54.0
(2) Vehicles with a seating capacity of more than 10 persons, including driver, and built on a passenger car chassis, vehicles built on a truck or bus chassis and having a manufacturer's GVWR of 10,000 pounds or less	52.8	17.0	25.0	66.0
(3) All other passenger-carrying vehicles	43.5	14.0	35.0	85.0
B. Property-carrying vehicles (1) Single unit vehicles having a manufacturer's GVWR of 10,000 pounds or less	52.8	17.0	25.0	66.0
(2) Single unit vehicles having a manufacturer's GVWR of more than 10,000 pounds, except truck tractors. Combinations of a 2-axle towing vehicle and trailer having a GVWR of 3,000 pounds or less. All combinations of 2 or less vehicles in driveaway or towaway operation	43.5	14.0	35.0	85.0
All other property-carrying vehicles and combinations of property-carrying vehicles	43.5	14.0	40.0	90.0

Figure 5-23. Braking performance.

must not reduce the air pressure or vacuum below 70 percent of the pressure that was on the gauge just before the brakes were applied. A "full service brake application" is defined here as pushing the brake pedal to its limit.

This tank must be protected from leaks. If the connection to the air or vacuum supply is broken, some device must seal off the tank so its supply isn't lost. This device is usually called a "check valve." You'll test this valve to make sure it's working.

Some air brake systems have a wet tank and a dry tank. The check valve is found between these two tanks. In such systems, there's a manually operated drain cock on the wet tank. This can be used to inspect the check valve.

Test the check valve operation of vacuum systems with the engine off. The vacuum gauge should show the system still holds a vacuum. If the gauge shows the pressure rising, the check valve isn't working right.

Warning Devices and Gauges

Vehicles must have warning devices that will tell you when there's a problem with the service brake system. Hydraulic and vacuum brake system warning devices can be heard and seen. Air brake warning systems must have a low air pressure warning device.

Braking systems must have gauges, too. Air brake gauges tell how many psi of pressure are available for braking. Vacuum gauges give this information in inches of mercury.

Warning devices can't help you if they don't work. You will check to see the warning devices are working when you do your inspection. This section of the FMCSR states that warning devices and gauges must be kept in working order.

Some hydraulic braking systems are assisted by air or by vacuum. Some systems must have warning devices and gauges for the air or vacuum part as well as the hydraulic part.

Brake Performance

FMCSR 393 sets standards for brake performance. When you inspect your brakes, you will check to see that they perform to these standards.

This section of the FMCSR includes a vehicle brake performance table. It describes how service and emergency brakes should work for different vehicles. It gives braking force as a percentage of the GVWR or the GCWR. This is not something the driver is expected to check. Braking force is more a manufacturing specification.

The performance table also states how quickly the vehicle must slow and how far it may travel while slowing. As a driver, you could test the braking speed and distance. These are not the typical brake tests you perform in an inspection. You could perform tests that tell you if your vehicle's braking rate and distance meet the requirements. Here's how to read the rest of the table.

The speed and distance requirements assume you were going 20 miles per hour (mph) when you applied the brakes. The second measure of braking performance states the braking rate in feet per second per second. "Per second per second" isn't a misprint. Here's what it means. It means you should slow so that you travel at the rate of so many feet per second. Then, for every second it takes you until you are stopped, you should have traveled at that rate.

The third measure of braking performance is braking distance. This is a simple measure of how far the vehicle travels before it comes to a stop. The same Group C CMV must come to a stop from 20 mph in no more than 25 feet.

Required braking performance, then, is a combination of how quickly the brakes bring you to a stop and how far you travel before you stop. Performance is also measured by how much force the brakes apply.

These standards assume the vehicle is braking on a level, dry, smooth, hard surface that is free of gravel or other loose material. The vehicle starts in the center of a 12-foot wide lane and does not pull to either side when tested for this type of performance.

Figure 5-23 on page 85 is the brake performance table. Find the vehicle you plan to be licensed to drive. You may be interested to note the braking performance required for that vehicle. Know that the brake system tests you will make during your inspection will not have to yield such technical information, however.

The braking systems on most trucks, if properly installed and maintained, will provide all the functions required by FMCSR Part 393. In Chapter 6, you'll learn how brakes work. When you understand how brakes work, you will be better able to spot and correct problems. Then you'll have no trouble getting good performance from your brakes.

Anti-lock Brake System

As you've read, anti-lock brake systems and indicators that warn of malfunctions must be on most heavy vehicles, including:

- each truck manufactured on or after March 1, 1999, equipped with hydraulic brakes

- each truck tractor manufactured on or after March 1, 1997, except those used in driveaway-towaway operations
- each air-braked CMV other than a truck tractor, that was manufactured on or after March 1, 1998, except those used in drive-away-towaway operations
- each truck tractor and each single-unit vehicle manufactured on or after March 1, 2001, that is equipped to tow another air-braked vehicle
- each semitrailer, trailer converter dolly, full trailer, and trailer designed to tow another air-brake equipped trailer, manufactured on or after March 1, 2001, as well as each trailer

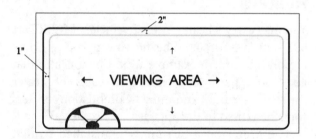

Figure 5-24. Required viewing area.

Windows

The next few regulations in this section of the FMCSR cover windows. There must be a certain amount of glass in your vehicle. This is so you have a good field of vision. The glass must be a certain material, for your safety and the safety of any passengers you may carry. Tinting to the immediate right and left of the driver is allowed, but it must not reduce the amount of light passing through to less than 70 percent of normal.

Your windshield must be clear, clean, and undamaged. FMCSR Part 393 sets the standards for windshield condition. You'll check for these items when you inspect your vehicle.

Devices mounted at the top of the windshield, like transponders, may not be mounted more than 6 inches below the upper edge of the windshield. They must be outside the area that the windshield wipers sweep. So as not to interfere with the driver's field of vision, they must be outside the sight lines to the road, highway signs, and signals.

You may not have any labels, stickers, decals, or other decorations on your windshield, or on your side windows. The only ones that are allowed are those stickers required by law. They must be placed at the bottom of the windshield. They must not extend higher than 4 inches from the bottom of the windshield.

Your truck must have a windshield, and a window on each side of the cab. Trucks with folding doors, or doors with clear openings in place of windows, don't have to be equipped with windows.

This has been a lengthy part of the FMCSR, and we're not done with it yet! Hang in there! As these chapters get longer, you're getting used to concentrating for longer spans of time. This practice will help you to stay sharp when you take your CDL tests.

Fuel Systems

The fuel that powers your vehicle can be a hazard if not stored and handled properly. Many of the regulations in this section exist for your safety. FMCSR Part 393 specifies how the fuel system should be installed on your vehicle. It states the specifications for fuel tanks. This section describes the leakage tests fuel tanks must past. These are not tests you can easily make by yourself. They are of greater concern to manufacturers.

FMCSR Part 393 covers liquid petroleum gas (LPG) systems and compressed natural gas (CNG). Some CMVs use CNG or LPG for fuel or to fuel auxiliary equipment, like the engine for the reefer unit.

Coupling Devices and Towing Methods

The next section in this part concerns coupling devices. It includes important definitions and descriptions of the fifth wheel. This is a part you will definitely check when you inspect combination vehicles.

You'll find these devices covered in more detail in Chapter 11. This section states limits on coupling and towing many tractors together and hauling them as cargo. When one tractor tows another as cargo, a special fifth wheel device is used. This is called a saddle mount. Only three saddle mounts can be used in any combination. Also, no more than one tow bar can be used.

In most cases of towing vehicles with saddle mounts or tow bars, the towed vehicles must have brakes. Brake lines must run all the way through the combination up to the tractor that's doing the towing. From your tractor, you must be able to apply the brakes on all vehicles in this combination. This is because bobtail tractors don't have much braking power by themselves. If all the vehicles in the combination have brakes, the stopping ability will be improved.

You may not use your tractor's bumper as a tow bar. Towed vehicles must face forward. In some cases, their front wheels must be controlled so they won't turn past the widest part of the tractor. The tow bar must give you some control over the towed vehicles' steering. The towed vehicles must track the towing vehicles.

FMCSR Part 393 has many detailed specifications for the construction of tow bars, saddle mounts, and U-bolts. These are mostly of interest to manufacturers. If you drive in a driveaway-towaway operation, you must know how to use the equipment. This part of the FMCSR tells you what is allowed, and how towed vehicles may be connected in a combination.

Tires

The next section of FMCSR Part 393 deals with tires. You may already know that you cannot drive on tires with tread so worn that the belts show through. Also prohibited are tires

- with tread or sidewall separation
- that are flat or that have leaks you can hear
- with a tread groove pattern of less than 2/32 of an inch (4/32 of an inch for front tires)

A regrooved tire with a load-carrying capacity equal to or greater than 4,920 pounds may not be used on the front wheels of any truck or truck tractor. With few exceptions, you can't use tires to support weights heavier than what the tire is rated for. (A number of different agencies rate tires. These agencies are listed in Federal Motor Vehicle Safety Standard [FMVSS] Number 571.119.)

Federal Motor Vehicle Safety Standards not included in the FMCSR can be found in the Code of Federal Regulations Title 49. You must also keep the tire inflated to the pressure specified for the load being carried.

Sleeper Berths

The size, shape, and structure of the sleeper berth is specified in FMCSR Part 393. Most of these specifications are geared to your safety and comfort. They're chiefly of concern to manufacturers. You should be aware of them, though, especially if you plan to customize your sleeper.

Heaters

You may carry a heater in your vehicle. FMCSR Part 393 states what type of heater you may have. Some are prohibited. An example would be a heater that gives off exhaust gases into your cab. The heater must be securely mounted. Combustion heaters must be vented to the outside.

The heater on a bus must be installed so that passengers can't get to the controls.

Windshield Wipers

Your CMV must have multiple-speed, power-driven windshield wipers and washers. In a driveaway-towaway operation, only the vehicle you drive has to have wipers and washers. The towed vehicles don't need working wipers and washers.

Defrosting Device

To prevent the buildup of snow, ice, or frost, your vehicle must have a windshield defroster.

Rear-view Mirrors

All trucks and buses made after 1980 must have rear-view mirrors, one on each side. Earlier models may have one outside mirror at the driver's side and one inside mirror if it gives a full view of the rear. Your mirrors should show you the highway behind you and the sides of the vehicle.

Horn

FMCSR Part 393 says trucks must have a horn.

Speedometer

Your vehicle must also have a speedometer. It must be accurate to within plus or minus 5 miles per hour when traveling at 50 miles per hour.

Exhaust Systems

Motor vehicles must have exhaust systems. The system must be installed so hot gases or surfaces don't damage wiring, the fuel supply, or any part of the vehicle that could burn.

This section of FMCSR Part 393 specifies how exhaust gases must be discharged. Exhaust systems that are properly installed and maintained will meet these requirements. Your job will be to ensure that the system continues to work correctly.

This section also notes that the exhaust system may not be repaired with wrap or patches.

Floor

FMCSR Part 393 even considers the floor necessary for safe vehicle operations. A floor with holes or openings can allow exhaust fumes into the cab. That could harm the driver or passengers. So, the floor must be sturdy and in good condition. It must also be free of oil or grease that would make it slippery.

Rear End Protection

FMCSR Part 393 requires certain vehicles to have rear bumpers. Vehicles higher than 30 inches from the ground (when empty) must have rear bumpers. This keeps shorter vehicles from running under the taller vehicle. The clearance between the bumper and the ground must be no more than 30 inches, again measured when the vehicle is empty. These bumpers must meet other size and mounting requirements described in this section. As with many of the regulations in this part, the driver's responsibility is to maintain the equipment.

Projecting Loads

Any part of the vehicle or load that extends beyond the rear or sides must be marked with a red flag. An extension more than four inches from the side or four feet from the rear would have to be marked. The flag must be 18 inches square and red.

Television Receivers

You are allowed to have a TV in your truck. You definitely must not be watching it while driving! Therefore FMCSR Part 393 states the TV must be set behind the driver's seat. It must be placed outside your line of vision when you are driving. The controls must be placed far enough away so you must leave your seat to get to them.

Seat Belts

Nearly every truck on the road today must have a seat belt for the driver. Some trucks must have seat belts for the passenger. Tractors with incomplete cabs don't have to have seat belts when they are being towed as cargo.

Noise Levels

Maximum levels for noise in your cab are set by FMCSR Part 393. Too much noise could be distracting and make you tired. As a driver, you'll be the first to notice unusually high noise levels. You're not expected to whip out a sound level meter and perform tests. You could report an unusually high level of noise to Maintenance, though. It could be a sign of some mechanical problem. Repairing that would give you relief from the noise, also.

Emergency Equipment

Most vehicles must have emergency equipment. One exception is vehicles towed as cargo. The required emergency equipment is as follows:

- fire extinguisher
- spare fuses
- warning devices for stopped vehicles

The fire extinguisher must be securely mounted and easy to access. You must ensure that it's always filled and ready to use.

You must have at least one spare fuse for each kind and size used. You don't need spare fuses if all your overload protective devices are circuit breakers that can be reset.

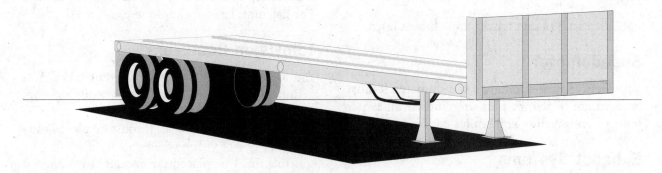

Figure 5-25. A bulkhead or header board on a flatbed trailer.

You must have warning devices to set out when you make emergency stops. Vehicles equipped with warning devices after 1974 must have three red reflective triangles. Older vehicles may have three electric lamps (or three red emergency reflectors) and two red flags instead. Another substitute can be three liquid-burning emergency flares or six fusees. This last choice is not an option for any vehicle hauling explosives or flammable liquids or gas. You may also not use these flares with a cargo tank, empty or full, that hauls compressed gas or uses it for fuel.

FMCSR Part 393 states the specifications for reflective triangles and other warning devices. Whatever you use must meet these specifications.

Cargo Securement

Trucks and trailers must be equipped to keep cargo from shifting, falling, blowing, or leaking. The cargo must be protected from damage. Additionally, cargo that shifts around can affect your vehicle's stability. Cargo securement rules are for both your safety as well as that of those on the road with you. Chapter 8 will cover cargo securement rules in detail. Here is simply a brief overview.

Some cargo can be completely and safely contained within the vehicle itself. Examples of such vehicles include tankers, vans, or trailers with sides or stakes and a rear endgate, endboard, or stakes. The regulations in this Part are not about cargo in those types of vehicles. The regulations concern cargo that has to be braced, blocked, or tied down. Tiedowns include such items as:

- chains
- straps
- cables
- binders
- webbing
- blocks and chocks
- load locks

In some cases, you will need to use combinations of securement devices.

The regulations set the specifications that cargo securement devices must meet. These are called working load limits. These limits ensure that you use cargo securement devices with adequate size and strength to do the job. The regulations also state what type of and how many devices you should use. They also guide you in how you should arrange these devices and attach them to the vehicle.

The regulations include specific cargo securement rules for particular types of cargo, such as:

- metal coils
- paper rolls
- concrete pipe
- intermodal containers, such as those that will be transported by ship or rail
- automobiles, light trucks, and vans
- heavy vehicles, equipment, and machinery
- flattened or crushed vehicles
- roll-on/roll-off or hook lift containers
- large boulders

The regulations state your trailer must have a "front-end structure." Most trailers have a header board. This keeps the cargo in place. Without such a structure, the cargo could shift and crush the tractor cab if you stop short.

If your trailer doesn't have a header board, your tractor must have a header board behind the cab. FMCSR Part 393 describes how. You'll return to this subject in the chapter on transporting cargo.

Body Components

This is the last section in FMCSR Part 393. It lists standards and specifications for body parts such as the frame, cab, wheels, steering, and suspension. You'll find those details covered in the chapters on vehicle control systems and inspections.

That brings us to the next part.

FMCSR PART 396

FMCSR Part 396 deals with the inspection, repair, and maintenance of motor vehicles. If you look at your copy of the FMCSR, you'll see this part is quite brief. That's because this FMCSR part doesn't give you the details of what or how to inspect. It mostly states when to inspect your vehicle. It describes the forms you should fill out. It also covers:

- lubrication
- unsafe operations
- inspection of motor vehicles in operation

So how do you know what to look for in an inspection? Well, as you learned earlier, certain vehicle parts must meet certain specifications. These are described in other parts of the FMCSR. You reviewed many of them quickly when you went through Part 393. Other parts of the FMCSR name the important vehicle components. Part 396 tells you to inspect those components.

For now, we'll just review what FMCSR Part 396 has to say about inspections. You'll learn how to do a proper inspection in Chapter 9.

Responsibility

FMCSR Part 396 states that vehicle inspection is the carrier's responsibility. Regulations call for carriers to have a regular system for inspecting all their vehicles. The parts to be inspected are as follows:

- parts and accessories mentioned in FMCSR Part 393
- frame and frame assemblies
- suspension system
- axles and attachments
- wheels and rims
- steering system
- other parts required for safe operation

Carriers must also keep a maintenance record on each of their vehicles. This account must include:

- vehicle identification
- type and date of inspection and maintenance
- lubrication record

If you have been driving for any time at all, you know that the driver, not the carrier, ends up doing the actual inspection and filling out the reports. FMCSR Part 396 states the carrier must inspect the vehicle or **cause it to be inspected**. That's how you as the driver end up with the job. So although these regulations are aimed at the carrier, we'll discuss them as if they were your responsibility. In effect, they are.

Lubrication

Motor vehicles must be lubricated. They must be free of oil and grease leaks.

No Unsafe Operations

The law prohibits you from abusing your vehicle. You may not drive it in a way that would cause it to break down or cause an accident.

In spite of your best efforts, your vehicle may develop a problem on the road. You must stop immediately if it won't cause a public hazard to do so. If stopping right where you are will create a hazard, you may continue to drive. Drive the vehicle only as long as it takes to get it to the nearest place where it can be repaired.

Official Highway Inspections

FMCSR Part 396 permits special agents of the Federal Motor Carrier Safety Administration (FMCSA) to stop and inspect your vehicle while you're out on the highway. It could be any official, such as a highway patrol officer or weigh master charged with enforcing transportation laws.

This agent will record the results of this inspection on a Driver Vehicle Examination Report.

If your vehicle doesn't pass inspection, it can be declared "out of service" by the special agent. Then you cannot drive the vehicle until it's been repaired and reinspected. If the vehicle cannot be repaired right where it is, it will have to be towed to a repair shop by a tow truck or wrecker.

If your vehicle does not pass the inspection, you must deliver the report to your carrier when you next arrive at the terminal. If you won't reach a terminal within 24 hours, you must mail the report. Once the motor carrier has received the inspection report, the carrier has 15 days to take care of the problems stated on the report.

Driver Vehicle Inspection Reports

As a driver, you will inspect your vehicle before and after your run. FMCSR Part 396 states you should inspect your vehicle at the end of your run and report any repairs that should be made. You probably call this a post-trip inspection.

Your report will show you inspected these items:

- service brakes, including trailer brake connections
- parking brake
- steering
- lights and reflectors
- tires
- horn
- windshield wipers
- rear-view mirrors
- coupling devices
- wheels and rims
- emergency equipment

You'll mark the form to report anything you feel would keep you from operating the vehicle safely. If nothing is wrong, you will state so on the form. In either case, you as the driver will sign the form. In the case of team operations, only one driver needs to sign the report. If you operate more than one vehicle in one day, you must fill out a report on each vehicle.

Any defects you noted must be repaired. The person who made the repairs signs the report. Then a copy of the report goes with the vehicle. A report doesn't have to be filed or maintained if no defects were found.

Before the vehicle goes out again, the driver performs another, pre-trip, inspection. In your pre-trip inspection, you verify repairs have in fact been made. If the post-trip inspection turned up some defects, and they are certified as repaired, you sign the report. This shows you noticed there were problems after the last run but you found they have been repaired.

This last signature isn't needed for defects that were part of a towed unit that is no longer in combination with the towing vehicle.

Driveaway-towaway Operations

Vehicles being towed as cargo don't have to be inspected from top to bottom. But you do have to inspect the saddle mount or tow bar. You must make sure the towed vehicles track as required by FMCSR Part 393.

In a driveaway operation, you are delivering the vehicle by driving it to its destination. Since you are actually driving the vehicle, you must make sure it's in a safe, road-worthy condition as required by law. It requires a complete inspection.

Periodic Inspections

Motor carriers must inspect their vehicles fully once every 12 months and keep reports. Inspectors have to meet certain qualifications. This inspection won't usually be made a driver responsibility. However, if you are a motor carrier as well as a driver, you are charged with inspecting your vehicles every year as described in FMCSR Part 396.

That's it. That's all FMCSR Part 396 has to say about vehicle inspections. Of course, actually doing the inspection takes a little longer to describe. That's a subject for another chapter.

FMCSR PART 397

You're almost done. Part 397 is the last part of the FMCSR that Part 383 requires you to know for your CDL. It covers the transportation of hazardous materials. This includes:

- attendance of motor vehicles
- parking

- fires
- smoking
- fueling
- tires
- instructions and documents

As you know, you must have placards on your vehicle if you are hauling materials that are hazardous. Some materials are not so hazardous by themselves. They become so when transported in certain amounts.

You also now know that you need a special endorsement to drive a placarded CMV. If you don't want or need the hazmat endorsement, you may wonder what's left for the basic CMV driver to know about hazmat.

FMCSR Part 383 states there are things you must know about hazardous materials to get your CDL. That is, you must know what hazardous materials are. You must know when you would need placards to haul them. You must know what you can haul without an endorsement, and when you would need an endorsement to haul hazardous materials. The law wants you to be aware that it takes special training to haul hazmat in placarded loads.

A knowledge of FMCSR Part 397 will fill the bill.

What Is Hazardous?

Certain materials are hazardous in any amounts. Others become hazardous when in large enough amounts, or when they are present with other materials. The Hazardous Materials Regulations, or Title 49 CFR Parts 171 through 185, define hazardous materials.

You might find the important sections of these parts summarized in the back of your FMCSR. This summary is called a Compendium of Hazardous Materials Regulations. If you don't find it there, visit the public library. Ask the reference librarian to help you find the right section of the CFR. You may also find these on the Internet, at http://www.phmsa.dot.gov/hazmat.

We'll highlight some of the main points of the hazmat regulations, then return to FMCSR Part 397.

Section 171.8 of the Federal Hazardous Materials Regulations (FHMR) defines hazardous materials as those the Secretary of Transportation has judged to pose "a risk to health, safety, and property when transported in commerce." This includes hazardous wastes, the dangerous by-products of many chemical processes. Hazardous materials include:

- marine pollutants
- poisonous gases or liquids
- elevated temperature materials
- poison inhalation hazards
- dangerous when wet
- organic peroxide
- flammable gases
- non-flammable gases
- flammable liquids
- oxidizers (cause other materials to react with oxygen, possibly with dangerous results)
- flammable solids
- corrosive materials (able to dissolve or wear away other materials)
- irritating materials
- spontaneous combustible liquids
- explosives
- etiologic agents (infectious substances, those able to cause disease)

These hazardous materials are grouped into classes. The classes are as follows:

Explosive (Class 1). Any substance or article, including a device, that is designed to function by explosion (for example, an extremely rapid release of gas and heat). Items that, by a chemical reaction, can function this way, even if not designed to do so, are included. Explosives in Class 1 are divided into six divisions:

- (1) Division 1.1 consists of explosives that have a mass explosion hazard. A mass explosion is one that affects almost the entire load very quickly.
- (2) Division 1.2 consists of explosives that have a projection hazard but not a mass explosion hazard.
- (3) Division 1.3 consists of explosives that have a fire hazard and either a minor blast hazard or a minor projection hazard or both, but not a mass explosion hazard.
- (4) Division 1.4 consists of explosives that present a minor explosion hazard. The explosive effects are largely confined to the pack-

age. No projection of fragments of significant size or range is expected. An external fire must not cause virtually almost the entire contents of the package to explode quickly.

- (5) Division 1.5 consists of very insensitive explosives. This division is comprised of substances that have a mass explosion hazard. They are so insensitive, however, that under normal conditions of transport, they are not likely to burn or explode.
- (6) Division 1.6 consists of extremely insensitive articles that do not have a mass explosive hazard. These are very insensitive detonating substances.

Flammable gas (Class 2, Division 2.1). These are materials that are a gas at certain temperatures and pressures and that can ignite under specific conditions.

Non-flammable compressed gas (Class 2, Division 2.2). Non-flammable, nonpoisonous compressed gas, including compressed gas, liquefied gas, pressurized cryogenic gas, compressed gas in solution, asphyxiant gas, and oxidizing gas.

Poisonous gas (Class 2, Divison 2.3). Gases that are poisonous when inhaled.

Flammable and combustible liquid (Class 3). Liquids that give off vapors that can ignite. Gasoline is a good example.

Flammable solid (Class 4, Division 4.1). Any of the following three types of materials:

- Desensitized explosives that when dry are explosives of Class 1 other than those of compatibility group A. These are wetted with enough water, alcohol, or plasticizer to keep them from exploding.
- Self-reactive materials that can heat up even without oxygen (air).
- Readily combustible solids that may cause a fire through friction, such as matches.

Spontaneously combustible material (Class 4, Division 4.2). A liquid or solid that, even in small quantities and without an external ignition source, can ignite within five minutes after coming in contact with air.

Dangerous when wet material (Class 4, Division 4.3). A material that, by contact with water, is liable to become spontaneously flammable or to give off flammable or toxic gas at a specific rate.

Oxidizer (Class 5, Division 5.1). A material that may, generally by yielding oxygen, cause or enhance the combustion of other materials. It can be solid or liquid.

Organic peroxide (Class 5, Division 5.2). Any organic compound containing a specific chemical structure. It presents an explosion danger.

Poisonous material (Class 6, Division 6.1). A material, other than a gas, that is known to be so toxic to humans that it could be a health hazard during transportation.

Infectious substance (Class 6, Division 6.2). A material known to contain or suspected of containing a pathogen. A pathogen is a virus, microorganism, or an infectious particle that has the potential to cause disease in humans or animals.

Radioactive material (Class 7). Any material with a certain specific activity that gives off radiation.

Corrosive material (Class 8). A liquid or solid that can destroy human skin at the site of contact within a specified period of time. Also included is a substance that has a severe corrosion rate on steel or aluminum.

Miscellaneous hazardous material (Class 9). A material that presents a hazard during transportation but that does not meet the definition of any other hazard class.

ORM-D material. A material such as a consumer commodity that, although otherwise subject to regulations, presents a limited hazard during transportation due to its form, quantity, and packaging.

Being able to recognize a hazardous-materials shipment is the first step in handling it correctly and safely. All licensed CMV drivers must be able to recognize a hazardous materials shipment. This is true even if they don't plan to haul it.

Unless otherwise provided in the regulations, certain materials may not be offered for transport. Examples of such materials are electrical devices that are likely to create sparks or generate a dangerous quantity of heat unless packaged in preventive packaging. Another example is material that could cause dangerous heat, flammable or poisonous gases, or vapors, or become corrosive when combined with something else.

As you can see, there are almost too many hazardous materials for you to remember. That's why you must take the time to review your cargo information. Two ways to do this are by the package labels and markings. Also, you must be able to recognize a hazardous shipment on the road. Vehicle placards are your clue that the vehicle is hauling hazmat.

Labels

The packages holding hazardous materials must be labeled. Shippers label hazmat packages, but drivers must recognize the labels and know what they mean. An example of a hazmat label is pictured in Figure 5-26.

Figure 5-26. Hazardous materials label.

Hazmat package labels are diamond-shaped and 3.9 inches on all sides. The labels show the class of hazardous materials to which the package belongs. Any numbers that might be on the labels are part of a worldwide system used to identify hazardous materials. Each number stands for a different hazmat.

Labels are not used on compressed gas cylinders. They don't stick well to slick metal surfaces. Instead, the label is placed on a hang tag hung around the neck of the cylinder. Or, a decal may be used. The cylinders themselves may be stamped as holding compressed gas. The Hazardous Materials Table, which is part of 49 CFR, lists materials that don't need labels.

Placards

Placards are required by regulations. A placard is similar to a label. However, a placard is attached to the vehicle's outside to show clearly that it contains a load of hazardous materials. The placard for a certain material will be of the same color and wording as its label. Look at Figure 5-27.

Figure 5-27. Hazardous materials placard.

Now, back to FMCSR Part 397. This part repeats the idea of "higher standard of care."

Recall "higher standard of care" from reading about FMCSR Part 392? Remember, when there are two or more sets of regulations, you should follow the strictest set. Local hazmat regulations may be more strict than the federal regulations. If so, you would follow the local regulations.

Routing

Vehicles hauling hazardous cargo don't always follow the easiest and fastest highways. Convenience does not determine the route. Instead, hazmat routing must avoid roads that go through heavily populated areas. The only exception to this is when there is no other possible way to get to the destination. The carrier will determine a route that complies with the requirements. Don't vary from this route except when a law enforcement official tells you to or directs you to an authorized detour.

Attendance

With few exceptions, a vehicle hauling explosives must never be abandoned. When a CMV is carrying Division 1.1, 1.2, or 1.3 materials on a public street or highway, there must always be a qualified person with the vehicle. FMCSR Part 397 refers to this as "attendance." To be "in attendance" means you must stay with it at all times. You must be in the vehicle, awake, and not in the sleeper. Or, you must be within 100 feet of the vehicle and have a clear view of it. The only exception is when you are performing duties necessary to operating the vehicle.

Parking

There are rules for where vehicles with Division 1.1, 1.2, or 1.3 materials may park or stop. These must not be on or within five feet of the traveled portion of a public street or highway. They must not be on private property unless someone in charge of the property knows what kind of hazmat the vehicle contains. Last, these CMVs must not park within 300 feet of a bridge, tunnel, dwelling, or place where people gather. You may only park there when absolutely necessary and when there is nowhere else to park. Vehicles hauling other types of hazmat may park under any of these conditions if there is nowhere else to park.

You may at times see a vehicle placarded for explosives that is unattended. This vehicle is probably parked in a "safe haven." Safe haven is a term used for a special parking area set aside just for this purpose.

Fire

Vehicles with hazardous cargo must be kept away from flames. Do not operate the vehicle near an open fire unless you know that you can pass the fire without stopping. Do not park within 300 feet of an open fire.

"Fire" includes flames used to light cigarettes. So, smoking is not allowed within 25 feet of a vehicle hauling Class 1, Class 5, or flammable materials classified as Division 2.1, Class 3, and Divisions 4.1 and 4.2. Don't smoke around an empty tank vehicle that carried Class 3, flammable materials, or Division 2.1 flammable gases.

Fueling

Special care must be taken when fueling a vehicle loaded with hazardous cargo. Don't operate the engine. Someone must be in control of the fueling process at the point where the fuel tank is filled.

Tires

A tire fire could be especially dangerous when you're hauling hazmat. Therefore, frequent tire inspections are required en route. Examine the tires at the beginning of each trip and each time the vehicle is parked. Repair a faulty tire before driving again (or drive only as far as you need to in order to repair the tire). Remove any overheated tires and put them at a safe distance.

Instructions and Documents

If you're hauling explosives, the motor carrier must give you, the driver, a copy of the rules in this part of the regulations and a document on what to do in case of an accident or delay. This document must include:

- contact names and telephone numbers
- the nature of the explosives being transported
- what to do in an emergency

You also must have shipping papers. Shippers certify on the shipping paper that the hazmat has been handled and packaged according to DOT regulations. Carry these shipping papers so they are easy to get to. They must be marked differently from shipping papers for regular cargo, or the hazmat shipping papers can be placed on the top of a stack of cargo documents. They must be within easy reach of a driver wearing a seat belt. They can also be kept in the holder mounted on the inside of the driver's door. They must be where they can be easily seen by someone looking into the cab.

Shipping papers are to be left in the cab, even when the driver leaves the tractor. They should be left in the holder or on the driver's seat in plain view. That is important in case of accidents. The driver may be hurt and not be able to speak. Emergency crews will know where to find the shipping papers that tell them what kind of hazmat is being hauled.

This has been a very general review of the hazmat regulations. Remember, even if you don't need or want a hazmat endorsement, you must know something about the subject. You must know what makes up a hazardous shipment so you don't haul it illegally by mistake. You should also know that not all dangerous cargo is hazardous in the legal sense.

Even without a hazmat endorsement, you may haul certain "hazardous materials." Those are, of course, dangerous materials that don't require the vehicle to have hazardous materials placards. There are also a few special times when a hazmat vehicle may be moved without placards. These are:

- when the vehicle is escorted by a state or local government representative
- when the carrier has permission from the Department of Transportation, or
- when the vehicle must be moved to protect property or people

You have now gone all the way through the FMCSR parts you must know to pass your CDL tests. This was a long chapter. But it was a very brief survey of these regulations. You'll return to some of these parts in the next few chapters on vehicle control systems, inspection, and driving. That will give you another chance to become well acquainted with them.

You should also spend some time reading through the FMCSR on your own. You must know the laws that govern your job. You may have heard the phrase "ignorance of the law is no defense." This means you cannot break a law then claim you were innocent because you didn't know the law. It's your job to know.

PASS POST-TRIP

Instructions: For each true/false test item, read the statement. Decide whether the statement is true or false. If it is true, circle the letter A. If it is false, circle the letter B. For each multiple-choice test item, choose the answer choice—A, B, C, or D— that correctly completes the statement or answers the question. There is only one correct answer. Use the answer sheet provided on page 99.

1. A state considers a CMV driver with an AC of 0.06 to be DUI. A driver found to be driving with an AC of 0.05 would be _____.
   A. asked for a urine sample
   B. disqualified
   C. allowed to continue driving
   D. put "out of service"

2. When you apply for a motor vehicle driving job, you must give your employer information on all your driving jobs for the last _____.
   A. seven years
   B. five years
   C. 10 years
   D. 20 years

3. If a drug or medicine you are taking was prescribed for you by your doctor, it is both safe and legal to take it while on-duty.
   A. True
   B. False

4. Drivers may be disqualified for criminal or other offenses even if they were off-duty when the crime was committed.
   A. True
   B. False

5. "Higher standard of care" means that drivers _____.
   A. must obey federal laws if they are more lenient than local laws
   B. must render first aid regardless of circumstances
   C. must obey local laws if they are stricter than federal laws
   D. must achieve higher than average scores on driving tests

6. The main reason for the post-trip inspection is to let the vehicle owner know about vehicle problems that may need repair.
   A. True
   B. False

7. No driver may drive a vehicle hauling any amount of hazardous material without a hazardous materials endorsement.
   A. True
   B. False

Instructions: Match the letter for the description in Column B with the number of the part in Column A that fits the description.

Column A Vehicle Parts	Column B Description
8. set of three reflective triangles	A. a part necessary for safe operation
9. clearance light	B. optional equipment
10. television receiver	C. emergency equipment necessary for safe operation

CHAPTER 5 ANSWER SHEET

Test Item	Answer Choices				Self - Scoring Correct	Incorrect
1.	A	B	C	D	☐	☐
2.	A	B	C	D	☐	☐
3.	A	B	C	D	☐	☐
4.	A	B	C	D	☐	☐
5.	A	B	C	D	☐	☐
6.	A	B	C	D	☐	☐
7.	A	B	C	D	☐	☐
8.	A	B	C	D	☐	☐
9.	A	B	C	D	☐	☐
10.	A	B	C	D	☐	☐

Cut Here

Chapter 6

Safety Control Systems

In this chapter, you will learn about:

- CMV safety control systems:
  —lights
  —horns
  —side and rear-view mirrors and their proper adjustment
  —fire extinguishers
  —instruments and gauges
- troubleshooting

To complete this chapter you will need:

- a dictionary
- a pencil or pen
- blank paper, a notebook, or an electronic notepad
- colored pencils, pens, markers, or highlighters
- a CDL preparation manual from your state Department of Motor Vehicles, if one is offered
- the Federal Motor Carrier Safety Regulations pocketbook (or access to U.S. Department of Transportation regulations, Parts 383 and 393 of Subchapter B, Chapter 3, Title 49, Code of Federal Regulations)
- the owner's manual for your vehicle, if available

PASS PRE-TRIP

Instructions: Read the statements. Decide whether each statement is true or false. If it is true, circle the letter A. If it is false, circle the letter B.

1. The proper way to use a fire extinguisher is to douse the flames with the dry chemical.
   A. True
   B. False

2. A vehicle will have warning lights or warning buzzers, but not both.
   A. True
   B. False

3. Vehicles with diesel engines don't have vacuum brakes because diesel engines don't create a vacuum on the intake stroke.
   A. True
   B. False

You just finished a long "tour" of FMCSR Parts 391, 392, 393, 396, and 397. It was quite an effort, but very important. Thorough knowledge of the FMCSR is vital to your success on the CDL tests.

Return now to Part 383. You'll recall this part contains the CDL laws themselves. If your copy of the FMCSR contains Subpart G, Required Knowledge and Skills, turn to it now. You'll see that item (a) states you must have knowledge of the safe operations regulations. Now that you've covered that requirement, let's tackle item (b), CMV safety control systems.

SAFETY CONTROL SYSTEMS

Know what equipment makes up your vehicle's safety control systems. Know where they are and how and when to use them.

Lights

Lights help you to communicate with other drivers. Lights help others to see you and help you to see others. You use lights to signal your plans to change lanes, slow, or stop. Many states require that you must have your lights on one-half hour after sunset until one-half hour before sunrise. Also turn on your lights any time you need more light to see clearly.

As you recall, FMCSR Part 393 describes the types of lights and reflectors CMVs must have. This part also describes the lights and reflectors needed in towaway operations. You studied tables and figures in Chapter 5 that illustrated these required devices. How much do you recall? Stop now and select the figure that fits the vehicle you plan to test in. Use the scanning recall technique to identify the lights and reflectors. Compare your recall with the completed figures that follow.

To check your drawings, refer to the illustrations on pages 72–81.

To review, you need the following:

Headlights. Two white headlights, one at the left and one at the right, on the front of the tractor. These are required on buses, trucks, and truck tractors. They must offer a high and low beam setting.

You may have fog lamps and other lights for bad weather or dark roads. These are in addition to, not instead of, headlights.

Front side-marker lamps. Two amber lights, one on each side, at the front of the side of buses and trucks, tractors, semitrailers, and full trailers.

Side-marker lamps. Two amber lights, one on each side, at or near the center between front and rear side-marker lamps of buses and trucks, semitrailers, full trailers, and pole trailers.

Front side reflex reflectors. Two amber reflectors, one on each side, toward the front of buses and trucks, tractors, semitrailers, full trailers, and pole trailers.

Side reflex reflectors. Two amber reflectors, one on each side, at or near midpoint between front and rear side reflectors of buses and trucks, large semitrailers, large full trailers, and pole trailers.

Front turn signals. Two amber signals, one at the left and one at the right side, on the front of the tractor, above or below the headlights. These are required on buses, trucks, and truck tractors.

Front identification lamps. Three amber lights, at the center of the vehicle or cab. These are required on large buses and trucks and truck tractors.

Front clearance lamps. Two amber lamps, one on each side of the front of larger buses and trucks, truck tractors, large semitrailers and full trailers, pole trailers, and projecting loads.

Rear side-marker lamps. One each at the lower left and lower right rear of the side of buses and trucks, semitrailers and full trailers, and pole trailers. These lights are red.

Rear side reflex reflectors. Found just below the rear side-marker lamps. These reflectors are red. These are required on buses and trucks, semitrailers, full trailers, and pole trailers.

Rear identification lamps. Three red lights centered at the top rear of large buses and trucks, large semitrailers and full trailers, pole trailers, and projecting loads. They're not required on smaller trucks and buses.

Rear clearance lamps. Two red lights, one at the top right and one at the top left of the rear of large trucks and buses, tractors, large semitrailers and full trailers, and projecting loads. These describe the overall width. They're not required on smaller trucks and buses.

Rear reflex reflectors. Two red reflectors, one at the lower right and one at the lower left of the rear of small and large buses and trucks, semitrailers, full trailers, and pole trailers.

Stop lamps. Two red lights, one at the lower right and one at the lower left of the rear. All vehicles must have these. Stop lamps are not required on projecting loads.

License plate lamp. One white light at the center rear, low, on trucks, tractors, semitrailers and full trailers, and pole trailers.

Backup lamp. One white light at the rear of buses and trucks and truck tractors.

Rear turn signal lamps. Two amber or red lights, one at the lower right and one at the lower left of the rear of buses and trucks, tractors, semitrailers, full trailers, pole trailers, and converter dollies.

Rear taillamps. Two red lights, one at the lower right and one at the lower left of all vehicles, even on projecting loads and converter dollies.

Parking lamps. Two amber or white lights, just below the headlights on small buses and trucks.

Four-way flashers. Two amber lights at the front and two amber or red lights at the rear of the vehicle. These are usually the front and rear turn signal lamps. They're equipped to do double-duty as warning lights. You can set them to flash simultaneously. They're required on buses and trucks, truck tractors, semitrailers and full trailers, pole trailers, and converter dollies.

Trailers and semitrailers with a width of 80 inches or more and a GVWR of 10,001 pounds or more must also have retroreflective sheeting and reflectors. Newer vehicles come equipped with these features. Older vehicles have been retrofitted to include them. They're usually red and white. Sheeting should run horizontally along the sides and across the full width of the lower rear area. On the upper rear area, it should run horizontally and vertically on the right and left corners. Reflectors

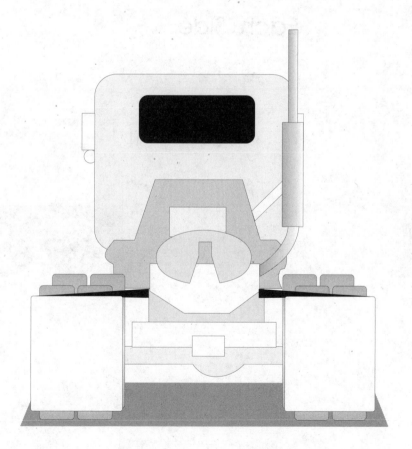

Figure 6-1. Select the drawing that looks most like your vehicle. Draw in the required lights and reflectors.

Front

Rear

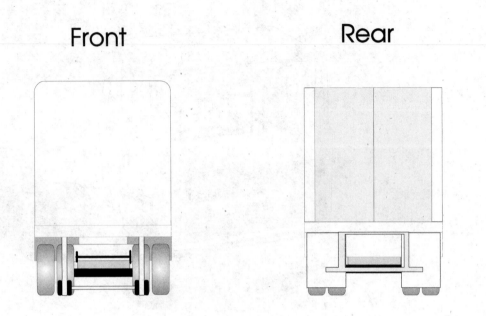

Each Side

Figure 6-2. Select the drawing that looks most like your vehicle. Draw in the required lights and reflectors.

Over 80 Inches

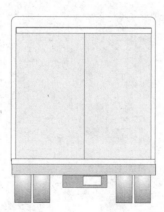

Figure 6-3. Select the drawing that looks most like your vehicle. Draw in the required lights and reflectors.

Under 80 Inches

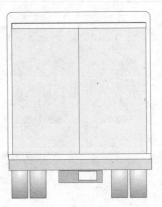

Figure 6-4. Select the drawing that looks most like your vehicle. Draw in the required lights and reflectors.

Under 80 Inches

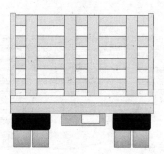

Figure 6-5. Select the drawing that looks most like your vehicle. Draw in the required lights and reflectors.

Pole Trailers - All Vehicle Widths

Front

Rear

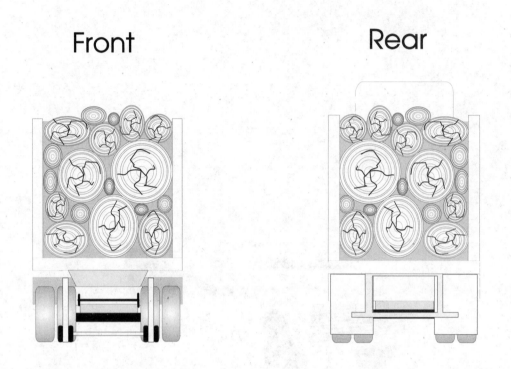

Each Side

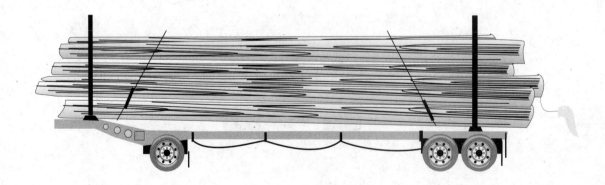

Figure 6-6. Select the drawing that looks most like your vehicle. Draw in the required lights and reflectors.

Container Chassis

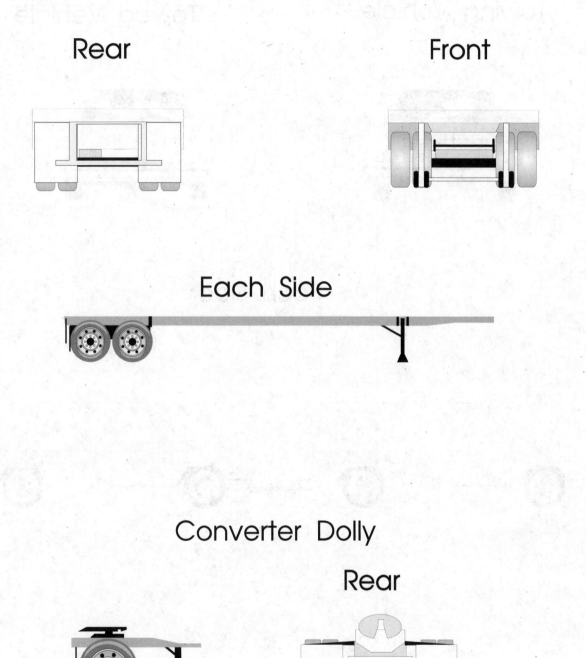

Rear

Front

Each Side

Converter Dolly

Rear

Figure 6-7. Select the drawing that looks most like your vehicle. Draw in the required lights and reflectors.

Front Of
Towing Vehicle

Rear Of
Towed Vehicle

Each Side

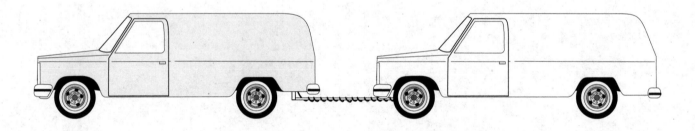

Figure 6-8. Select the drawing that looks most like your vehicle. Draw in the required lights and reflectors.

Front Of
Towing Vehicle

Rear Of Rearmost
Towed Vehicle

Each Side

Figure 6-9. Select the drawing that looks most like your vehicle. Draw in the required lights and reflectors.

Front Of
Towing Vehicle

Rear Of
Towed Vehicle

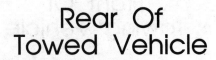

Each Side

Figure 6-10. Select the drawing that looks most like your vehicle. Draw in the required lights and reflectors.

Front

Vehicle Without
Permanent Top Or Sides

Rear

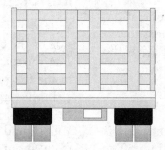

Figure 6-11. Select the drawing that looks most like your vehicle. Draw in the required lights and reflectors.

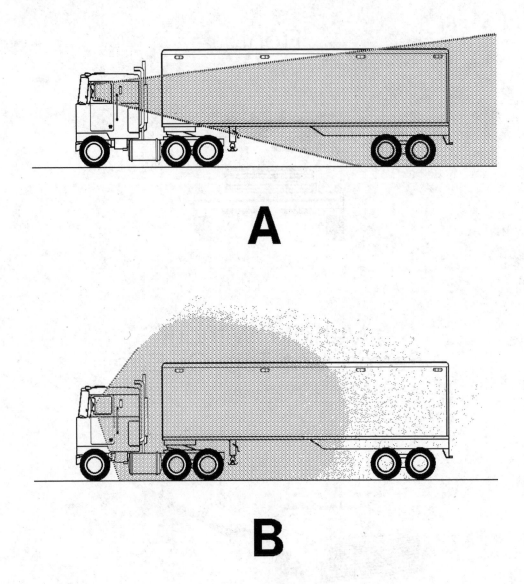

A

B

Figure 6-12. What your large flat mirrors let you see (A) and what your smaller convex mirrors let you see (B).

run horizontally on the trailer's sides from front to rear. On the lower rear area, they run horizontally across the full width. On the upper rear area, they are placed horizontally and vertically on the right and left corners.

Note that in a driveaway-towaway operation, both the towing vehicle and the rearmost towed vehicle must have lights and reflectors. Review the illustrations for descriptions and placement.

Horn

Your vehicle must have a horn. Like lights, this is a communication tool. Use it with restraint. The horn can often alarm other drivers needlessly.

As part of your inspection, make sure your horn works. If it doesn't, it could be a blown fuse or faulty wiring.

Mirrors

Your vehicle must have a rear-view mirror at each side of the cab. (Some vehicles may have one outside mirror on the driver's side. Another mirror inside the vehicle gives a view to the rear.)

In the large, flat mirror, you should see the traffic and the sides of your trailer, if you have one. You should be able to see the road behind you from about the middle of the trailer. Adjust the mirrors so you see the ground starting in front of the trailer wheels, all of both lanes next to your vehicle, and behind the trailer. In the small convex (spot) mirror, you should see the traffic and your drive wheels. The convex mirrors help you to see "blind spots" along the side of your vehicle.

Fire Extinguisher

As a professional driver, you have responsibilities in a vehicle fire. First, protect your life and the lives of others. Then, try to save your vehicle and its cargo. To do this, you need two things. You need to know something about fires. You also need a working fire extinguisher.

Causes of Fires

Fires can start after accidents as a result of spilled fuel or the improper use of flares. You could have a tire fire. Under-inflated tires and duals that touch create enough friction to burst into flames. Electrical fires can result from short circuits caused by damaged insulation or loose connections.

Carelessness around fuel can cause fires. Smoking around the fuel island, improper fueling, and loose fuel connections can lead to fires. Flammable cargo and cargo that isn't properly sealed or loaded can cause fires. So can poor ventilation in the cargo compartment. This is why it is so important to make a complete pre-trip inspection of the electrical, fuel, and exhaust systems, tires, and cargo.

Also, check the tires, wheels, and truck body for signs of heat whenever you stop during a trip. Fuel the vehicle safely as outlined later in this chapter.

Be careful with any part of the vehicle that creates heat or flames. Check your instruments and gauges often for signs of overheating. Use the mirrors to look for signs of smoke from tires or the vehicle. Fix whatever is causing the overheating, smoke, or flame before you have a big problem on your hands.

Use normal caution in handling anything flammable. Knowing how to fight fires is important. Fires have been made worse by drivers who didn't know what to do. Here's what to do if a fire does start:

- The first step is to get the vehicle off the road and stop. Park in an open area. Park away from buildings, trees, brush, other vehicles, or anything that might catch fire. Don't pull into a service station. There's too much flammable material there.
- Use your CB or cell phone if you have one to notify the highway patrol or police of your problem and your location.
- Keep the fire from spreading. Before trying to put out the fire, make sure that it doesn't spread any farther.
- If you have an engine fire, turn off the engine as soon as you can. Don't open the hood if you can avoid it. Aim your fire extinguisher through louvers, the radiator, or from the underside of the vehicle.
- If you have a cargo fire in a van or box trailer, keep the doors shut. This is especially important if your cargo contains hazardous materials. Opening the van doors will supply

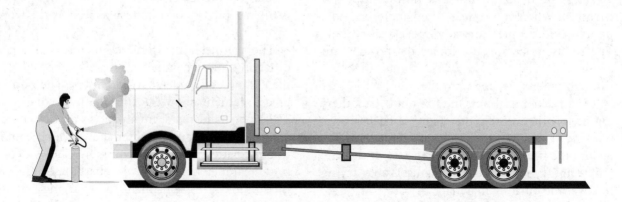

Figure 6-13. Aim your fire extinguisher through louvers, the radiator, or from the underside of the vehicle.

the fire with oxygen. That can cause it to burn very fast.

- Use water on burning wood, paper, or cloth. Don't use water on an electrical fire. You could get shocked! Don't use water on a gasoline fire! It will just spread the flames.
- A burning tire must be cooled. You may need a lot of water. If you don't have water, try throwing sand or dirt on the tire.
- If you're not sure what to use, especially on a hazardous materials fire, wait for qualified firefighters.
- Use the right kind of fire extinguisher for the fire.

Fire Extinguishers

FMCSR Part 393 states your truck must carry a fire extinguisher. This regulation also lists which fire extinguishers are acceptable. Make sure the extinguisher in your truck conforms to the regulations. It should be mounted securely on the vehicle where you can get to it easily. It should be designed so that you can easily see that it is fully charged.

Fires have been grouped according to class. When the fuel for the fire is wood, paper, cloth, trash, and other ordinary material, the fire is a Class A fire. When the fuel is gasoline, grease, oil, paint, and other flammable liquids, the fire is a Class B fire. Electrical fires are Class C fires.

Most trucks must carry a 5 B:C-type fire extinguisher or two 4 B:C-rated extinguishers. Vehicles hauling hazmat that is placarded must have a 10 B:C fire extinguisher. (The numbers 5 and 10 refer to the amount of fire the extinguisher will put out in square feet of flammable liquid.)

This type of extinguisher is filled with a dry chemical. When you squeeze the handle, a needle punctures an air pressure cartridge inside the tank. The released air pressure forces the powder out of the tank. The powder travels through the hose, through the nozzle, and onto the fire.

The dry chemical puts the fire out by smothering it. In other words, the chemical coats what's burning. That prevents air from fueling the fire. So aim the fire extinguisher at the base of the fire. You're not trying to douse the flames. You're trying to cover the burning material.

Know how the fire extinguisher works. Study the instructions printed on the extinguisher before you need it.

When using the extinguisher, stay as far away from the fire as possible. Position yourself upwind. Let the wind carry the extinguisher to the fire rather than carrying the flames to you.

Continue until whatever was burning has cooled. Just because there is no smoke or flame doesn't mean the fire is completely out or cannot restart.

Instruments and Gauges

The dashboard is also called a "dash" or "instrument panel." Figure 6-14 shows you an example of the gauges, switches, lights, and controls you'll find on a modern CMV dashboard. Not all dashboards are the same. However, this is typical of what you'll find on most vehicles.

Some of these monitor the operating condition of the engine. Some monitor other systems of the vehicle.

Besides gauges and warning lights, there are also switches and controls on the dash. They are used to operate the vehicle or its systems. An example is the switch for the air conditioner. When you use this switch, you turn on the air conditioner.

Start at the far left of our illustration. The first thing you'll see is an air conditioning vent. Next to the vent you'll notice four gauges:

- the voltmeter
- the engine temperature gauge
- the oil pressure gauge
- the pyrometer

The voltmeter shows the charge condition of the battery. The voltmeter is marked by the word "Volts" on the lower portion of the gauge. There is often a picture of a battery as well as the word "Volts."

The voltmeter has three segments, each for a different battery condition. The left-hand red segment shows an undercharged battery. The middle green segment shows normal battery condition. The right-hand red segment shows an overcharge condition. The gauge pointer shows which condition the battery is in. If the voltmeter shows a continuous undercharging or overcharging condition, there is likely a problem in the charging system. The voltmeter in Figure 6-14 is in the normal range.

You may have an ammeter on your dashboard. It tells you the amount of charge or discharge the battery is getting from the generator. It should read "zero" with the engine and electrical system turned off. When the engine starts, the ammeter needle should jump to the charge side and flutter. Once the engine warms up, the needle should drop back to zero. It is also normal for it to read slightly on the charging side.

The engine temperature gauge is usually marked "Water Temp" or "Temp." It shows the engine cooling system temperature. The gauge displays measurements in degrees. A typical gauge will have a range of about 100 to 250 degrees Fahrenheit. A normal reading is between 165 and 185 degrees Fahrenheit. The gauge will read higher if you are running in hot weather.

The engine oil pressure gauge shows how well the engine is being lubricated. Measurements are displayed in pounds per square inch (psi). This gauge will show a reading when the engine is running. When the oil is cold, the gauge will show a high reading. After the engine has warmed up, the reading should return to normal. When the engine is running at normal temperatures and the oil is hot, the normal idle pressure runs from 5 to 15

psi. The normal operating pressure runs from 30 to 75 psi. The oil pressure gauge in Figure 6-14 is at 55 psi. That means the vehicle is running in the normal range.

Always check this instrument immediately after starting the engine. If no pressure is shown, stop the engine at once. You can seriously damage the engine by running it with no oil pressure. A low reading could mean you're simply low on oil. This could be because of a leak. You could also have a clogged filter.

If you have an oil temperature gauge, make sure the temperature is within the normal range while you are driving.

The pyrometer tells you the engine exhaust temperature. The safe temperature range will be shown on the dashboard next to the gauge or on the gauge. High exhaust temperatures mean trouble. You could have a leak or a clog in the air intake or the exhaust system. There could be problems with the fuel ignition. Or you could simply be in the wrong gear ratio for the load, grade, or altitude.

Below this cluster of instruments are gauges that tell you the temperature of various vehicle parts:

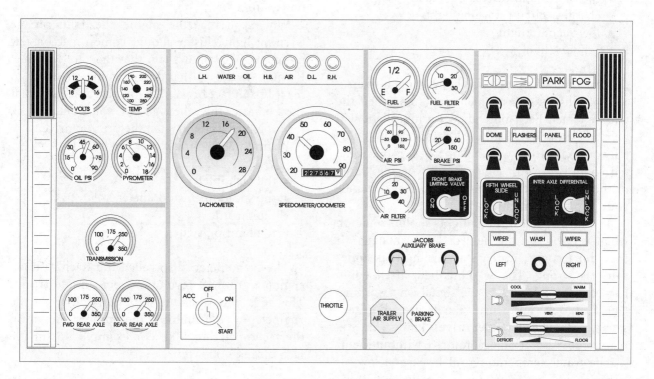

Figure 6-14. On a modern dashboard, you'll find switches and gauges grouped by function.

- the transmission temperature gauge
- rear rear axle and forward rear axle temperature gauges

Like the engine temperature gauge, these gauges display measurement in degrees. A high temperature reading will alert you to problems in that particular part. If the temperature is high, you should stop the vehicle before more damage occurs. Transmission oil temperatures range from 180 to 250 degrees Fahrenheit. The normal range for axles is from 160 to 220 degrees Fahrenheit. These are only guidelines. The range of a particular transmission, like a Fuller transmission, may be 180 to 225 degrees Fahrenheit. If the temperature gauge on this transmission reads 250 degrees Fahrenheit, you are almost to the critical range. Check your owner's manual to find out what's normal for your vehicle.

Next you'll find the warning lights:

- the left-hand turn signal (L.H.)
- the water level warning light (WATER)
- the oil level warning light (OIL)
- the high beam light (H.B.)
- the low air light (AIR)
- differential lock warning light (D.L.)
- the right-hand turn signal (R.H.)
- charging circuit warning light
- low vacuum light
- cab lock light

Sometimes there will also be warning buzzers.

This cluster of instruments is often called the telltale panel. Each individual warning light will be clearly marked. Some of these warning lights tell you a control is working or that it is on. Some tell you a control or gauge is not working. Some of them tell you there is a problem that demands your immediate attention.

The turn signal warning lights should come on whenever you use the right or left turn signal. If they don't come on, that's a warning that something is wrong. The warning light itself could be defective. It could be the signal isn't working. Check it out. Your turn signals are important safety tools. They let other drivers know what you plan to do. Also, it's illegal to drive without them.

The water temperature, oil level, and low air lights should not come on. If they do, something is wrong. The water temperature light will come on if the water temperature in the coolant system gets too high. The oil pressure light will come on when the oil pressure is too low. The low air light (on vehicles with air brakes) will come on if the air pressure in the braking system drops below a certain pressure, usually 60 psi. These three warning lights often have buzzers that sound, too.

The high beam and differential lock lights are reminders. The high beam light reminds you the high beam is on. The differential lock light reminds you the differential lock is in the locked position. You'll learn about the differential lock control later in this section.

Your vehicle may also have a charging circuit warning light. This light would come on if your battery wasn't charging. The light normally lights up when the starter switch is turned on. This tells you the light is working. It goes out when the engine starts, unless you have a problem.

Your vehicle may also have a low vacuum warning light. When this light comes on, it means the vacuum in the brake booster is below the safety limit. You could be dangerously low on braking power. Stop as soon as it is safe to do so. Don't drive again until the brake problem is fixed.

On a cabover, there may be a cab lock warning light. This light would tell you the cab tilt lock is not secure.

When you start the engine, the lights on the warning light cluster should come on for a few moments. That shows you that they are working. If you have a light that doesn't come on, check to see if it is in fact broken. If you have one light that stays on after the others go out, you should check for problems in that system before you drive.

Behind the steering wheel you'll find:

- the tachometer
- the speedometer/odometer
- the throttle
- the ignition switch

The tachometer (also called the "tach") shows engine crankshaft revolutions per minute (rpm). This tells you when to shift gears. To read the engine rpm, you multiply the number shown on the tachometer by 100. For example, 15 on the tach means 1,500 rpm.

The number of revolutions per minute that an engine can make differs from engine to engine. The average high horsepower diesel only goes to a maximum of 2,100 rpm. The range of the engine may go from 500 rpm (idle speed) to 2,100 rpm. The typical operating range of the engine is even shorter. Stay within the operating range to achieve good engine performance.

Engine speeds in most engines are governed. This means that there's a limited number of rpm the engine will make in any gear. If you want to see how many rpm your engine is doing, just look at the tach. If you've reached the top governed speed for your gear, it's time to shift. This is "driving by the tach." The owner's manual for your vehicle will often tell you what the top governed speed is.

The speedometer shows the vehicle's road speed in miles per hour (mph). It may also show a metric measurement, kilometers per hour (kph). Inside each speedometer is an odometer. The odometer keeps track of the total miles the vehicle has traveled. The mileage is shown in miles and tenths of miles.

The throttle is a kind of accelerator pedal on your dash. You pull it out to set engine speed or rpm. You would use the throttle in very cold weather to keep the engine warm when idling. You might also use it to get engine speed up and deliver more power to operate a power take-off (PTO) device. On newer trucks, the throttle has been replaced by the cruise control, which does the same thing, and more.

The starter or ignition switch turns on the electricity. It turns the engine over so it can start. When the key is straight up and down, the switch is "off." When you turn the key to the left, it turns on the accessory circuits. Turn the key to the right to turn on both the accessory and ignition circuits. Turn the key to the far right to engage the starter.

Release the key as soon as the engine turns over. After a false start, let the starter cool for 30 seconds before trying it again.

Next, moving from left to right, are

- the fuel gauge
- the fuel filter gauge
- the air filter gauge

The fuel gauge shows the fuel level in the supply tanks. Some vehicles have more than one fuel supply

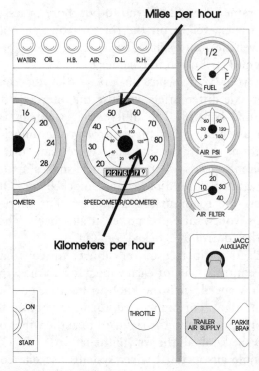

Figure 6-15. The speedometer.

tank. If your vehicle does, be sure to check the fuel level in all tanks before assuming you're out of fuel.

The fuel filter gauge shows the condition of the fuel filter. It has a colored band divided into two segments. The left segment is white. The middle and right segments are red. It also has numbered markings. You have a clogged fuel filter if the needle reads in the red range.

If your vehicle has air brakes, controls and gauges for this system come next. These are covered in detail in Chapter 10.

Next to the cluster of fuel instruments are the light switches.

On some vehicles, the light switches are on a stalk to the left of the steering wheel. You may find all your light switches here, including the dimmer switch, turn signals, and flashers. Below the light switches are other controls:

- the inter-axle differential control
- the windshield wipers
- the air conditioning controls
- cold start and warm-up switch
- exhaust brake switch

Tractors with sliding fifth wheels will have a slide control here. This allows you to move the fifth wheel into different positions on the tractor frame. That allows you to put more or less of the trailer weight on the tractor. You can make the vehicle more stable. You'll learn more about weight distribution in Chapter 8.

If your vehicle has dual rear axles that have inter-axle differentials, you use the inter-axle differential control. The differential is a rear axle gear assembly. With the control in the unlocked position, each axle shaft and wheel can turn at different speeds when the vehicle makes a turn. The control should be set at "Unlocked" or "Off" unless the road surface is slippery. For slippery roads, you want equal power at all wheels. Then, set the control to "Locked" or "On."

Electric wipers used on cabover tilt cabs have two controls, one for each wiper. To run the wiper on low speed, turn the knob to the first position to the right. For high speed, turn the knob to the second position to the right. To turn the wiper off, turn the knob all the way to the left.

Some air-operated wiper systems use one control knob for both wipers. It's located on the dash. Others have one control knob for each wiper. To turn the wipers on, turn the knobs to the right. To turn the wipers off, turn the knobs to the left. The position of the knobs controls the wiper speeds. Newer trucks sometimes have wiper controls that are mounted on stalks on the steering column instead of the dash and have variable speed control.

There is a separate knob to work the windshield washers.

Cab temperature is easily controlled in modern vehicles. Air speed controls include low, medium, and high. There are controls for cooling, heating, and defrosting.

Your vehicle may have a cold start switch and a warm-up switch. Use the cold start switch in cold weather when the engine is hard to start. Turn it "On" to preheat the engine. The switch will light up to tell you it's on. Always check the operator's manual for the engine and starting aid your vehicle has.

Turn the exhaust brake, if your vehicle has one, "On" with the exhaust brake switch. An exhaust brake light will go on to tell you the exhaust brake is "On." The exhaust brake provides extra slowing power. Use it on steep downgrades or when your vehicle is heavily loaded.

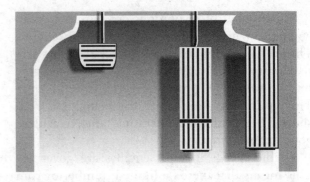

Figure 6-16. The controls on the floor under the steering wheel.

Next to these is another air conditioning vent. There are more controls on the floor of the cab:

- the accelerator pedal
- the brake pedal
- the clutch pedal
- the dimmer switch
- the transmission control lever

You'll find the accelerator on the floor of the cab under the steering wheel. You use your right foot to operate this pedal and control engine speed. When you depress the pedal, the speed of the vehicle increases (if the vehicle is in gear and the brakes are released). As you let your foot off the pedal, the speed decreases. If you take your foot off the pedal, the engine idles.

You'll find the brake pedal just to the left of the accelerator. You operate this pedal with your right foot. When you depress the brake pedal, the brakes are applied.

To the left of the brake pedal is the clutch pedal. You use your left foot to operate the clutch pedal. You disengage the clutch when you depress the clutch pedal. You engage the clutch when you release the clutch pedal.

Use low beams when driving in traffic. Use high beams on open roads when there is no oncoming traffic and no traffic closer than 500 feet in front of you.

The trailer brake hand-control valve is usually found on the steering column. It lets you apply the trailer service brakes without also applying the tractor service brakes.

Your vehicle may also have a power take-off lever. This control is really two knobs. Pull up on the first knob to connect the PTO to the transmission. Pull up on the second knob to use the PTO.

TROUBLESHOOTING

FMCSR Part 383 says you must be able to do a certain amount of troubleshooting. Your instruments and gauges tell you something's wrong. You must know how to read them and what you should do in response. You must be able to do some problem-solving when your vehicle fails to work as expected.

Take a look at the main systems that support your vehicle. When you understand what these systems do, you'll be better able to figure out what's wrong when they fail. (Plus, inspecting your vehicle will be easier and make more sense.)

Also, you'll be better prepared for the CDL tests. During the inspection, the examiner may ask you questions about vehicle parts and systems. You must know what makes them work. You must be able to show good knowledge about mechanical systems.

The systems and parts covered here are those required by FMCSR Part 393. You can learn a lot about these parts from the FMCSR. You can learn still more from the owner's manual for your vehicle. Ask your employer, safety supervisor, or maintenance personnel if there is one available. Ask at the reference desk of your local public library. Another place to try would be a vehicle dealer. Last, you could try writing to the vehicle's manufacturer or visit the manufacturer's web site online.

Maybe your knowledge of mechanical systems needs some serious shoring up. Check a nearby bookstore or library for books on diesel mechanics, electricity, hydraulics, motor vehicles, and so forth. You're sure to find a few good books on the mechanics of heavy vehicles. One we can recommend is *Bumper to Bumper®, The Complete Guide to Tractor-Trailer Operations.*

Wiring and Electrical Systems

The power that runs your CMV comes from the engine. The power behind that engine power is electrical power. You can't even start your vehicle without a starting circuit. You can't keep it going for long without a charging circuit. You can't run your lights without a lighting circuit and your dashboard instruments won't work without an instrument circuit.

As you can see, electricity serves many functions in vehicles. Unless you understand the electrical system, it just looks like a mess of wires. The first step in understanding the system is to understand some of the basics of electricity. An electron is a tiny particle that carries a negative charge of electricity. Electrical flow produces electrical current. Everything contains electrons, but some things conduct electricity better than others.

A good conductor of electricity has electrons that can easily be set in motion. Copper wire is a good conductor of electricity. Rubber is not a good conductor. That's why copper wire is used to move electric charges from one place to another. Rubber is used as an insulator around copper wire in places where you don't want to conduct electricity.

Wiring Systems

Insulated wires bring current to parts that need electricity to operate. Terminals are the connecting devices. They are found on the ends of the wires. They are on the electrical parts used to connect the wires to the components. There is also a main terminal from which the wires start and which contains all the system circuit breakers and fuses.

Measuring Electricity

Pressure gets the electrons flowing. "Voltage" is another name for this electrical pressure. Batteries and alternators or generators produce voltage. A voltmeter measures voltage. The term "amperage" or "amps" refers to the amount of electric current that is produced and carried by the wires. An ammeter measures amps. As you learned earlier, you may have an ammeter on your dashboard. It tells you the amount of charge or discharge the battery is getting from the alternator.

You also have a voltmeter on your dashboard. The voltmeter tells you about the charge condition of your battery. This in turn tells you how much power the battery has to get electrons flowing. Your battery could be undercharging. It's not getting enough electrons flowing to crank the engine. Undercharging causes the fluid in the battery to thicken. This shortens battery life.

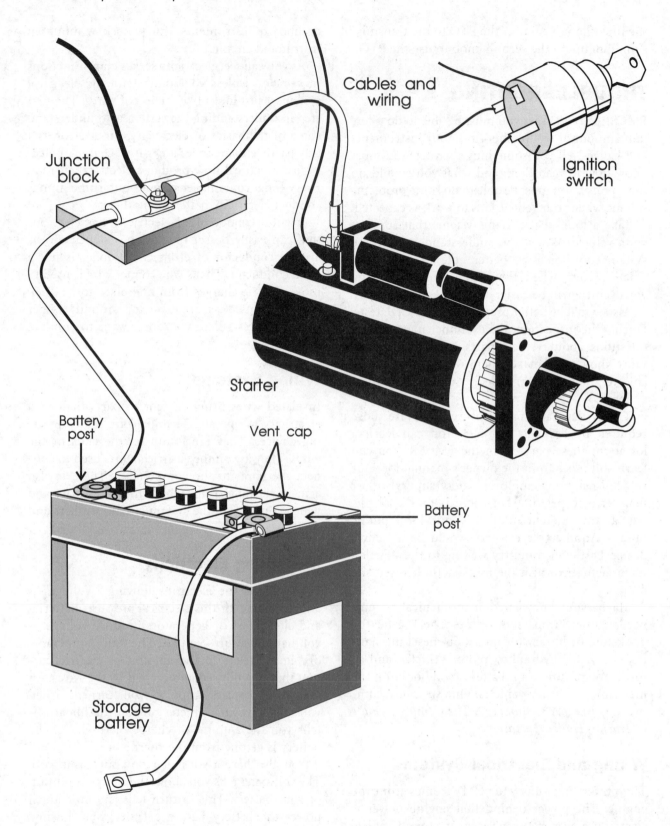

Cables and wiring

Ignition switch

Junction block

Starter

Battery post

Vent cap

Battery post

Storage battery

Figure 6-17. Battery and starter.

Your voltmeter could show the battery is overcharging. Too much pressure and too much electron activity will overheat the battery and cause the battery fluids to evaporate. This shortens battery life as well.

Your battery could be cycling. This means a rapid series of charges and discharges. This, too, shortens the life of the battery.

The desired voltmeter reading is a normal battery charge condition. Battery problems could be serious. If your voltmeter shows the battery is overcharging, undercharging, or cycling, have the battery checked.

When you inspect your vehicle, check the fluid level in the battery. It should be up to the filler rings in each cell. If you can't see the rings, just make sure the fluid covers the cell plates.

If you must add fluid, use distilled water. Avoid underfilling or overfilling. As you know, either of these can shorten battery life.

Gases form when the battery is charging. Vent caps over the cells let these gases escape. At the same time, they keep the fluid from splashing out of the battery.

Wiring

The electrons flow through electrical circuits. A circuit is a continuous path made up of a conductor (such as wire), and a source of power that drives the current around the circuit (the batteries and alternator or generator). The devices that use the electricity (your vehicle's starter and lights, for instance) are also part of the path. This type of circuit is called a "complete" or "closed" circuit. For current to flow, there must be a closed circuit. For a circuit to be closed, all the parts in that circuit must be grounded. That means there must be a wire or a conductor to bring the electrons back to where they started.

There are two other kinds of circuits: the open circuit and the short circuit. Electricity will not flow in either of these types of circuits. That usually means trouble, like your lights won't work.

An open circuit occurs when the normal flow of electrical current is stopped. A number of conditions can cause this. Corroded connections and broken wires account for the most open circuits. Open circuits account for most electrical problems. You can spot, and often fix, an open circuit caused by faulty connections and wires when

you inspect your vehicle. Reconnect loose wires. Broken wires must be replaced.

A short circuit occurs when the electrical current bypasses part of the normal circuit. This means that instead of flowing to a light bulb, for instance, the current stops short of its destination and flows back to the battery.

Shorts happen when the insulation has come off a section of a wire. The wire can touch something outside the normal circuit, like another wire or part of the frame. Then the current leaves the circuit. It takes the shortest route back to the source, along the other wire or frame. It never does make it to the light bulb.

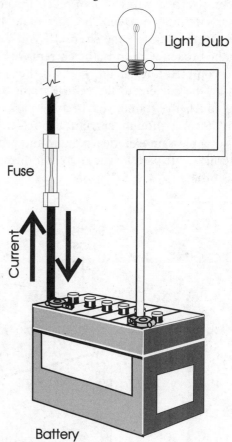

Figure 6-18. A short circuit will keep current from getting to the device it's meant to operate.

Any of these conditions can also cause a short circuit:

- the wires in an electrical coil (like the starter winding) lose their insulation and touch each other.
- a wire rubs against the frame or other metal part of the truck until the bare wire touches another piece of metal.

Regulations require that wiring be installed and insulated so shorts won't occur. During inspections, look for frayed, broken, loose, or hanging wires. These could result in a short circuit if not repaired. Electric power would then fail to reach the parts that need it, like your lights. Replace any wires that have worn insulation.

Grounding

A grounding circuit provides a short-cut path for the current, much the way a short circuit does. However, a grounding circuit works for your safety. Here's how.

When electrical wires burn or break, the normal path or circuit is broken. The current looks for some way to complete the circuit. If you were to grasp the broken wire ends, the current would use *you* to complete the circuit. Perhaps the short is directing the current to the vehicle frame. Again, if you touch the frame, you become part of the circuit. Strong-enough current can electrocute you. Not a pleasant experience. A ground provides an alternate, safe path for the current if the normal path is broken.

The connections on a grounding system should be easy to access.

Batteries

Batteries convert chemical energy into electrical energy. They then supply power to the rest of the electrical system. The major parts of a battery are a case, a number of individual cells, cell connectors, and two terminal posts.

The two posts on the top part of the battery are called "main battery terminals" or "battery posts." The positive (+) post is the larger one. The other is the negative (−) post. The battery cables are connected to these posts.

The vent caps are also on the top part of the battery. Gases build up when the battery charges. The vent caps let these gases escape. You will remove the vent caps to check the battery. In doing so, you may see the vents are clogged. They must be clean so the gases can escape.

Batteries are dry-charged, wet-charged, and maintenance-free. The dry-charged battery has no fluid in it when it leaves the factory. The dealer adds that to the battery when it's sold. The wet-

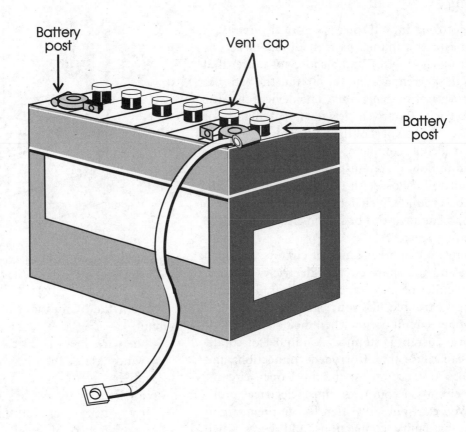

Figure 6-19. Battery.

type battery has fluid already in it when it leaves the factory. With these types of batteries, you must check the fluid level as part of your inspection. (The maintenance-free battery does not usually require this.)

Make sure each cell has a vent cap. Check to see that the vents are not clogged.

Check the battery mount. Check the hold-down bars to make sure the battery is snug. This keeps it from being damaged by vibration. The battery box cover should be in place. The battery box itself must not be cracked or leaking.

Look for battery cables that are frayed, worn, or cracked. They must be replaced. Check to make sure the battery connections are tight.

Electricity can be dangerous if you're not careful. Here's how to work around batteries safely. Disconnect the battery ground strap before you begin any electrical or engine work. Connect the ground strap last when you install a new battery. Do not lay metal tools or other objects on the battery.

Never hook up the battery backwards. Instead, be sure to connect the positive cable to the positive terminal post. Connect the negative cable to the negative terminal post. The positive cable clamp and terminal are usually larger than the negative.

Take care around battery acid. It's corrosive. Don't lean too closely to the battery when adding water. Battery fluid could splash up into your eyes.

Keep sparks and fires away from batteries. Gas from the electrolyte can catch fire. If you're a smoker, this is not a good time for a cigarette.

Overload Protective Devices

Circuit breakers and fuses protect the circuit from short circuits and from current overloads. A current overload happens when a circuit gets more current than it can handle. Wires are rated for how much current they can handle. Wiring can burn if it gets more current than it's designed to handle.

A short circuit is usually the cause of an overload. This is what happens. Turning on the lights, the starter motor or radio "uses" (actually, slows) the flow of the electricity. If there is a short circuit, the bulb, motor, or radio won't work. Since they're not working, they're not using, or slowing, the flow of current. This means there is more current in the wire than the wire can handle by itself. It

will overheat and burn. Fuses or circuit breakers are placed in each circuit to prevent this.

Here's how they work. Like wiring, fuses are rated by their ability to handle so much current. To protect a circuit, you use a fuse rated lower than the wiring. An overload that exceeds the fuse's rating will blow the fuse. This opens or breaks the circuit. The fuse will go before the current builds to a point that exceeds the rating of the wiring. The wiring is protected.

Once a fuse is blown, it must be replaced. You see why, if your vehicle uses fuses, you must have spares. That way, you can replace any fuses that might be blown and get back the use of your lights or starter. (Of course, you'll need to fix the cause of the overload before you replace the fuse. Otherwise, you'll just blow the new fuse.)

Circuit breakers are also rated by their ability to carry current. A circuit breaker used on a circuit will have the same current-carrying capacity as the circuit. If there is more current load than the circuit breaker can carry, the circuit breaker opens. This breaks the circuit. Once a circuit breaker has opened, it can be reset. This is an advantage circuit breakers have over fuses. Again, though, give some attention to whatever caused the circuit breaker to open. If you don't, it will just open again the next time the problem occurs.

Detachable Electrical Connections

There's one more important electrical part on your CMV. That's the detachable electrical connection between a tractor and trailer in a combination vehicle, often called a pigtail. It carries current from the tractor to the trailer, where it powers the trailer lights. Like any other circuit, this connection can short out or break. You'll make checking the electrical connection to the trailer part of your inspection.

If you're driving a straight truck and not pulling a trailer, you don't have detachable electrical connections.

The Electrical System

Wires, circuit breakers, fuses, terminals, and current-using parts make up the circuits in your vehicle. There is one main terminal block that contains all the circuit breakers and fuses. From this terminal block, wires run out in bunches to connectors.

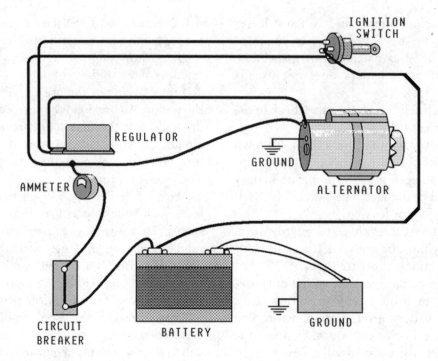

Figure 6-20. The electrical system.

At the connectors the wires split and go to other connectors or to the parts that use the electricity.

Most CMVs have a basic 12-volt electrical system. Many of the parts are the same as the ones in the humble family car. The battery is the power source for your truck's electrical system. It supplies the power that starts the truck's engine. Then the alternator supplies the power that keeps the battery charged and runs the truck's systems. Older trucks may have generators. Newer trucks will have alternators.

Figure 6-20 traces the path of current through a truck's electrical system. The path begins at the battery. When the ignition switch is turned to the start position, the current flows from the battery to the starter. The starter cranks the engine. After the engine has started running, current flows from the alternator back to the battery to replace the current used to start the engine. Current also flows to the other parts that need electrical power.

Brakes

What makes your vehicle stop? The brakes, yes. Dumb question.

Here's a better question. What makes the brakes work? Ah, that's a little tougher.

The answer is friction. In drum brakes, brake shoes move toward the brake drums. The shoes wear a lining or pad of coarse material. The shoes bring this lining into contact with the drum. This creates friction, which stops the truck. (In disc brakes, the friction pad moves into contact with a metal disc. The result is the same.) Varying the amount of pressure applied to the brakes changes the amount of force the brake shoe applies to the brake drum and the amount of friction that's created. The brake drum or disc is bolted to the wheel. If the drum (or disc) slows, so does the wheel. This is how you control the slowing and stopping of the truck.

In CMVs, the brake shoes and their linings or pads are brought into contact with the brake drums or discs and held there with pressure. The pressure is created in one of three ways:

- hydraulic pressure
- vacuum pressure
- air pressure

Sometimes, the pressure is used to apply the brakes when you need to stop. This is the case with hydraulic and air brakes. Vacuum brakes work the

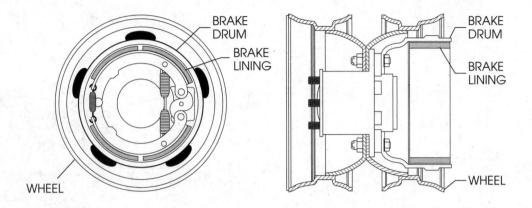

Figure 6-21. Friction between the brake lining and the brake drum slows the drum, and the wheel to which it's attached.

opposite way. There, pressure holds the brakes back so the wheels move freely. When you need the brakes, you take away the pressure. This releases the brakes. They apply and stop the vehicle.

In this chapter, you'll learn about hydraulic and vacuum pressure. We'll describe hydraulic and vacuum service brakes.

A knowledge of those systems will be enough to pass the CDL tests and drive vehicles without air brakes. However, you will have an air brake restriction on your CDL unless you also know about air brakes. Many CMV drivers will get along just fine with an air brake restriction. For those who wish to prepare for the air brake questions, air brake systems are covered separately in Chapter 10.

Hydraulic Brakes

Straight trucks often have hydraulic brakes. Hydraulic brakes use fluid pressure. Hydraulics use three facts about fluids:

- The nature of fluids is to flow. If fluids are not allowed to flow, they create pressure.
- Oil cannot be compressed (squeezed smaller).
- In a closed system, the pressure on fluid is equal at all points throughout the system.

Take the first point. The nature of fluid is to flow. Can you stop fluid from flowing? The only way to do this is to put it in a container. (You may argue that you can also stop a fluid from flowing by freezing it. It's not liquid anymore, then. It's considered a solid at that point.)

Take the second point. Oil cannot be compressed. You can't squeeze it into a smaller mass.

The third point is best explained by example. Look at Figure 6-22. This shows a simple closed hydraulic system. A pump or piston at the left and a cylinder supporting a weight at the right closes the system. Inside the pipeline is hydraulic fluid. Pressing the handle pushes the pump or piston down. This exerts pressure on the hydraulic fluid in the pipeline. The pressure flows through the system and pushes up on the cylinder at the right.

At no point in the system is there more or less pressure. There is just as much pressure at the right, pushing the cylinder up, as there is at the left, where the piston is pushing on the fluid.

This helps explain how pressure you exert at the brake pedal in the cab is transmitted in full force to the brake down at the wheels.

A hydraulic brake system looks a lot like Figure 6-23. There are lines holding hydraulic fluid between the brake pedal and the brakes. There are cylinders at both ends. When you press the brake pedal, you move a push rod and piston in a cylinder. This cylinder is called the master cylinder. This exerts pressure on the hydraulic fluid. The pressure moves through the fluid until it meets the cylinder at the brake end (the wheel cylinder). The pressure pushes on the cylinder. The cylinder moves the brake shoe and pad into contact with the drum or disc. Friction results. The wheels slow and stop.

If oil could be compressed, this wouldn't work. Pressing the master cylinder would just squeeze the oil. Oil can't be compressed, remember? Since the system is closed, the fluid has nowhere to go. Pressure results.

Also remember, this pressure will be equal at all points in the system. When the driver presses the

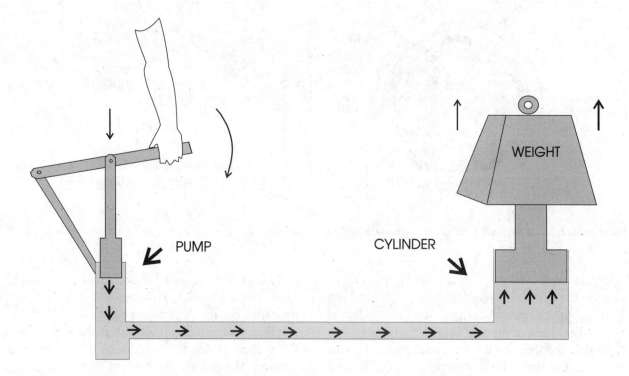

Figure 6-22. A simple hydraulic system.

brake pedal, equal pressure is applied at all wheels. By making the master and wheel cylinders different sizes, you can increase the pressure applied at the brake pedal when it reaches the wheels. That way, the driver doesn't have to press very hard at all to get a lot of braking force.

What if there are breaks or leaks in the system? Right—no brakes. Since fluid must flow, it will flow out of any leak it finds. If there is no fluid, there is nothing to transmit the pressure from the master cylinder to the wheel cylinders. (That's why inspecting for leaks in your braking system is so important.)

Dual brake systems help in such emergencies. They're designed with two fluid tanks and two pistons in the master cylinder. They're set up so that you will still have braking power at some wheels even if there's a leak in one of the brake lines.

Figure 6-22 shows a simple hydraulic system. A typical hydraulic brake system in a CMV would look like Figure 6-23.

Vacuum Brakes

A vacuum is a negative quantity. It's air pressure that is lower than the pressure of the atmosphere, the air around us.

You may have heard the phrase, "Nature hates a vacuum." It's true. Normal atmosphere will always move to fill a vacuum, an area with lower air pressure. It's as if the vacuum were sucking in the atmosphere, only in reverse. It's this pressure of the atmosphere forcing its way into the vacuum that does the work.

Vacuum brakes have a cylinder with a movable piston, much like hydraulic brakes. Vacuum brakes are often used to assist, or boost, hydraulic brakes. Atmospheric air is on one side of the piston. The vacuum is on the other side. The pressure of the atmosphere trying to fill the vacuum pushes the piston into the vacuum.

When the brakes are off, the piston is drawn back inside the cylinder. It's held in place by a spring. The piston has vacuum on both sides of it.

The vacuum cylinder has a control valve with four chambers. One chamber has atmospheric air. The second combines atmospheric air and vacuum to create different air pressure levels. The third chamber has a vacuum. The last chamber has hydraulic fluid.

When you press the brake pedal, hydraulic pressure from the master cylinder affects the control valve. It closes the vacuum valve and opens the

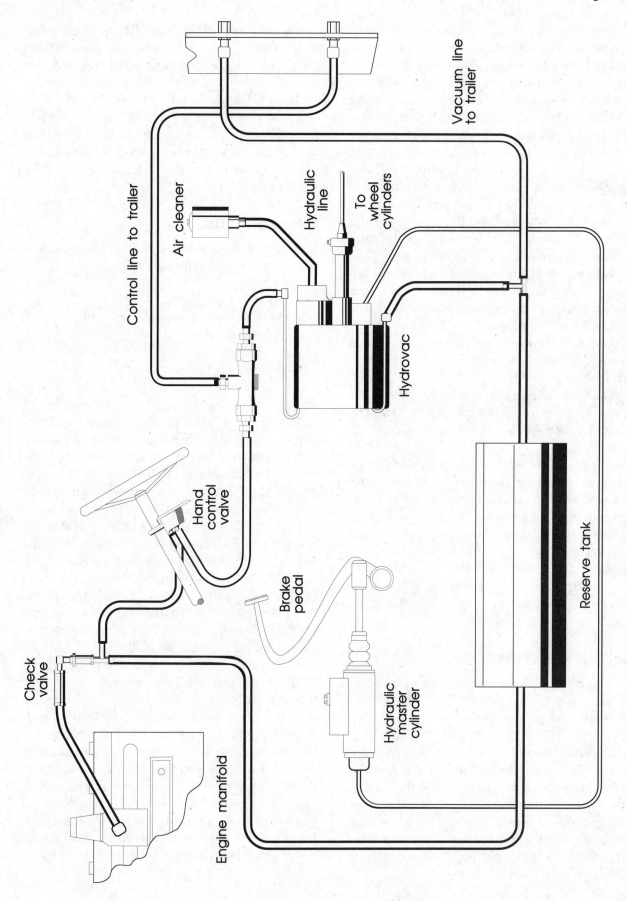

Vacuum line to trailer

Control line to trailer

Air cleaner

Hydraulic line

To wheel cylinders

Hydrovac

Hand control valve

Brake pedal

Hydraulic master cylinder

Reserve tank

Check valve

Engine manifold

Figure 6-23. A hydraulic brake system schematic.

atmosphere valve. This in turn allows atmospheric air to enter the second, mixing chamber. The result is lowered pressure. This closes the atmospheric valve, but you can reopen it by stepping on the brake pedal again. Letting up on the brake pedal will open the vacuum valve. You see that pressing and releasing the brake pedal changes the level of air pressure in the mixing chamber.

Pressure from the master hydraulic cylinder also affects the vacuum cylinder itself. It pushes the piston forward. The piston pushes hydraulic fluid beyond it into the brake lines. Pressure from the control valve adds to, or boosts, the pressure from the master cylinder. So pressure from both the main hydraulic brake system and the vacuum booster works to apply the brakes. A vacuum booster can reduce the effort required at the brake pedal by 30 to 70 percent.

Figure 6-24 shows the vacuum booster, control valve, and connections to the line from the master cylinder.

Should the vacuum system fail, the spring in the vacuum cylinder would hold the piston back. The hydraulic brakes would still work normally.

The vacuum is created by the intake manifold in the engine. Note that diesel engines don't create a vacuum on their intake strokes. A pump creates a vacuum when the engine doesn't or doesn't produce enough of a vacuum.

None of this will work if there's a break or crack in the vacuum lines or any part of the system. Can you state why? A break will allow the vacuum to leak out. Or more accurately, it will allow atmosphere in, filling the vacuum. No vacuum, no brakes.

Parking Brakes

The parking brake system comes into play when you park. You recall nearly all CMVs made after 1989 must have parking brakes.

You apply the parking brakes by pulling the parking brake control. All brake systems have this control. On some vehicles, pulling this lever or knob controls a cable. The cable pulls the parking brakes into position mechanically. To release the parking brakes, you push the control back in. This releases the cable and the brakes. Though it takes a little effort to "pull" the brakes on, it's nothing most drivers can't handle.

On other vehicles, setting the parking brake does require more strength than most drivers have. Then the parking mechanism is set with the help of air pressure. You'll learn about this system in detail in Chapter 10, Air Brakes.

Recall what FMCSR Part 393 requires of parking brakes? You must be able to set and release the manual parking brake as often as needed. The parking brake system just described meets these requirements.

Fuel Systems

The fuel system delivers the fuel to the engine. The parts of the fuel system are:

- the fuel tank
- the primary and secondary fuel filters
- the fuel pump
- the fuel lines
- the fuel injectors

and, of course, the fuel.

The fuel tank holds the fuel. Primary and secondary filters clean it before it reaches the fuel pump. The fuel pump delivers the fuel to the engine. Fuel lines carry the fuel from the pump to the cylinders. Fuel injectors spray the fuel into the combustion chambers.

Figure 6-25 traces the flow of the fuel through the fuel system. The fuel leaves the fuel tank and runs through the fuel lines to the check valve. From there, it flows through the primary and the secondary filters and into the fuel pump. The pump forces it through the fuel lines and into the fuel injectors. The injectors spray a measured amount of fuel into the cylinders.

Fuel is flammable. It can be dangerous if not handled with care. Fuel lines must not be allowed to touch hot surfaces. They must be supported so they don't drag on the ground. Keep in mind your vehicle could cross grade crossings that are higher than the regular road surface. Also, see that they're protected from objects that can bounce up from the road. See that fuel lines are not damaged or rubbing against other vehicle parts. There should be enough slack in the lines so they won't break as the vehicle moves. You can certainly check to see that the fuel tanks are mounted securely.

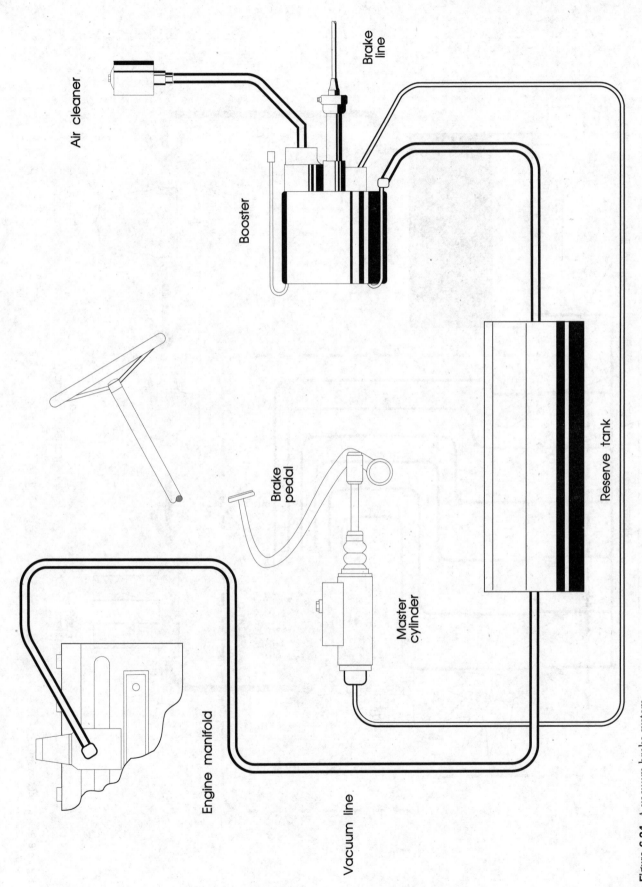

Air cleaner

Brake line

Booster

Brake pedal

Master cylinder

Reserve tank

Engine manifold

Vacuum line

Figure 6-24. A vacuum brake system.

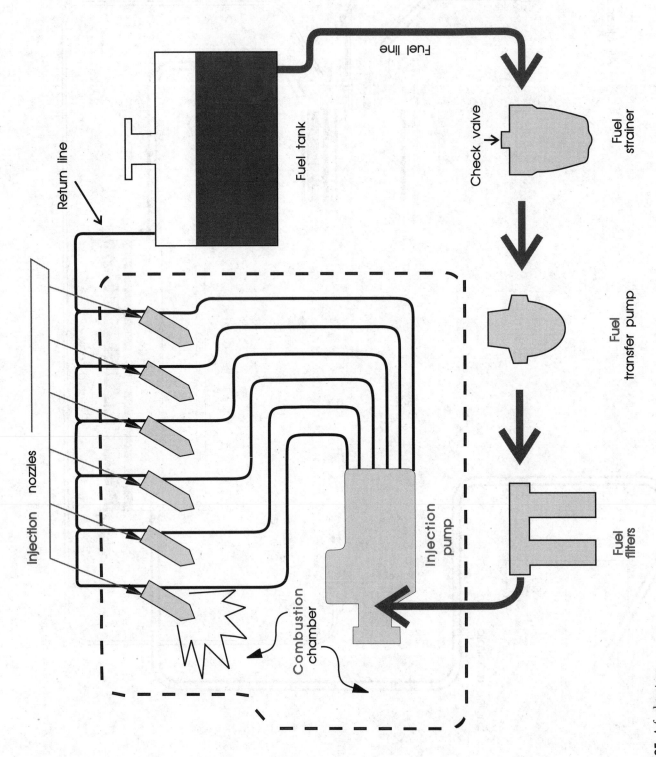

Figure 6-25. A fuel system.

Also, when you inspect your vehicle, check for leaks. If your fuel system is leaking, of course you'll lose precious fuel. But leaking fuel can also be a hazard. If it drips on a hot surface, it could ignite.

Coupling Devices and Towing Methods

Vehicles that tow other vehicles must do it in a way that meets the requirements of FMCSR Part 393. There are two main ways you will see vehicles towed. One is trailers being pulled by tractors. The other is tractors or other towing vehicles being towed as cargo. This is a driveaway-towaway operation. In this section we'll look at the parts used to pull trailers. Then we'll look at saddle mounts and drawbars.

Fifth Wheel

The fifth wheel allows the trailer to articulate (pivot). FMCSR Part 393 states that the towed vehicle must not trail more than three inches to either side of the tractor.

The fifth wheel also allows the trailer to oscillate (rise or drop) along with the tractor.

Of course, the fifth wheel allows you to separate the trailer from the tractor. The power unit isn't "married" to the cargo unit, as it is with a straight truck.

The fifth wheel is mounted on the rear of the tractor frame. The proper brackets and fasteners must be used to do this. The lower half of the fifth wheel (the mounting assembly) must not shift around on the frame. There must not be any bent parts or loose bolts in the fifth wheel lower half.

The placement of the fifth wheel is important. It controls how much weight is on each axle between the front and rear of the tractor. Poor weight distribution can reduce steering control. It also causes uneven wear on tires, brakes, and wheels.

You'll find fifth wheels on converter dollies as well as on tractor frames. A converter dolly converts a semitrailer to a full trailer.

Locking Device

The fifth wheel must have a locking device. This device keeps the towed and towing vehicles together until you release them on purpose. You might hear this locking device referred to as the "jaws." The jaws lock around the shaft of the trailer kingpin. This makes a secure connection. Any forward or backward movement of the tractor will be transmitted to the trailer. You have to get out of the cab and manually pull a lever to open the jaws.

Tow Bars

Full trailers must have tow bars. There must be a locking device on the towing vehicle. This keeps the two trailers coupled with a converter dolly from separating accidentally. You must use a pair of safety chains or cable with the tow bar connection to the towing vehicle.

Saddle Mounts

A saddle mount (see Figure 6-26) is a steel assembly used to couple a towed vehicle to the towing vehicle. The saddle mount has a kingpin that fits into the locking jaws and fifth wheel of the towing vehicle. Then it has a set of U-bolts or clamps that clamp securely to the front axle of the towed vehicle.

The kingpin of the saddle mount provides an articulation point the way the kingpin of a semi-trailer does.

You may need to prop up the fifth wheel so it's level. You may use hardwood blocks to do this. You may also have to raise the towed vehicle so it doesn't bump against the frame of the towing vehicle. You may use hardwood blocks for this as well.

Tires

Tires provide proper traction and reduce vibration. They absorb road shock. They transfer braking and driving force to the road. There are many different tire designs, but all tires are made about the same way.

Tires are made up of:

- plies
- bead coils and beads
- sidewalls
- tread

and the inner liner.

Plies consist of separate layers of rubber-cushioned cord. They make up the body of the tire. The plies are tied into bundles of wire called the bead coils.

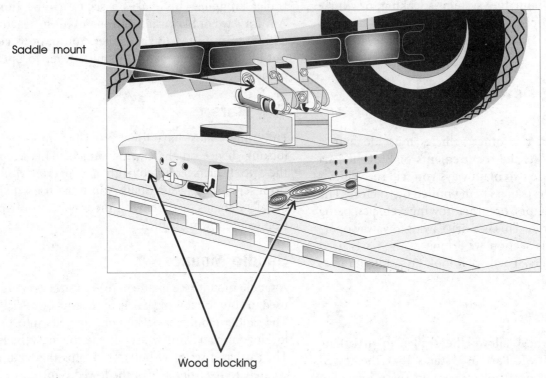

Saddle mount

Wood blocking

Figure 6-26. A saddle mount.

Plies can be bias, belted bias, or radial. Bias plies are placed at a crisscross angle. This makes the sidewall and tread rigid. In a belted bias ply tire, the plies cross at an angle. There's an added layered belt of fabric between the plies and the tread. The belts make the tread of this tire more rigid than the bias ply tire. The tread will last longer because the belts reduce tread motion when the tire is running.

In radial tires, the plies on this tire do not cross at an angle. The ply is laid from bead to bead, across the tire. Like the belted bias ply tire, the radial also has a number of belts. Radial construction supports the tread better than either the bias ply or the belted bias ply. The radial design means the sidewalls flex with less friction. That requires less horsepower and saves fuel. Radial tires also hold the road better, resist skidding better, and give a smoother ride than the bias types.

Bead coils form the bead. This is the part of the tire that fits into the rim. It secures the tire to the rim. Bead coils provide the hoop strength for the bead sections. This is so that the tire will hold its shape when it's being mounted on a wheel.

The sidewalls are layers of rubber covering. They connect the bead to the tread. Sidewalls protect the plies in the sidewall area.

The tread is the part of the tire that contacts the road. Treads are designed for specific driving jobs. Some jobs require that the tread provide extra traction. Others call for a tread designed for high-speed use. Tires on steering axles must roll well and provide good traction for cornering. Drive tires must provide good traction for both braking and acceleration. Trailer tires mostly need to roll well. Tire treads are also designed to face specific road conditions. For example, drive wheel position tires need maximum traction in rain, snow, sleet, and ice. The inner liner is the sealing material that keeps the air in the tire.

That's the basics of tire construction.

Tire size is shown by either a number or a series design designation. You find this information on the sidewall of your tires. An example of the number designation is a 10.00 × 22 tire. The first number is the tire's width. This means the inflated tire measured 10 inches from the farthest outside point on one sidewall to the farthest outside point

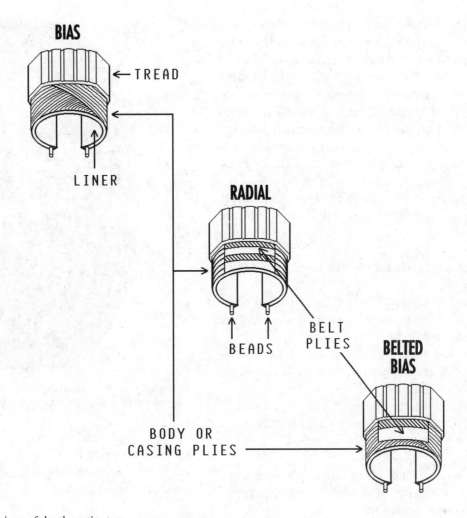

Figure 6-27. Cross sections of the three tire types.

on the other sidewall. The second number is the rim size. Our example tire will fit a 22-inch diameter rim.

The series designation system was developed because of the low profile tire. This tire is wide in relation to its height. Once again, let's take an example to look at this. Say a tire sidewall reads 295/75 R 22.5. This means the section width is 295 millimeters. (Low profile tires are measured in millimeters rather than inches.) The aspect ratio (the height compared with the width) is 75. The type is radial. The rim is 22.5 inches in diameter.

The government requires tire manufacturers to label all tires with several items of information. For example, tire labels must show the tire brand and manufacturer. Tire labels must also show the load rating and maximum load pressure.

Load rating refers to tire strength. This is rated from A to Z. "Z" is the strongest. The maximum load the tire can carry in terms of weight is shown in pounds. FMCSR Part 393 states you must not use a tire that is too weak to support the load. The load rating helps you select the right tire for the load.

The maximum pressure is shown in psi. It is given for cold tires that have been driven for less than one mile. This is why you check tire pressure before you drive. Don't try to check tire pressure by hitting your tires with a stick or a billy or kicking them. These methods will tell you a tire is not completely flat. They will not give you an accurate tire pressure reading. Instead, use a calibrated tire gauge. It wouldn't hurt to have one, and know how to use it, when you inspect your vehicle for the CDL test.

All this pertains mainly to new tires. Used tires can sometimes be made like new. These are regrooved tires and recaps or retreads.

A regrooved tire is one in which the grooves in the tread have been cut deeper into the surface of a tire that is nearly worn down to the legal minimum. A retreaded, or recapped, tire has had the old tread surface removed. Then new tread is bonded to the outside layers of the belts or body plies.

The legal minimum for tire tread depth is set in FMCSR Part 393. A motor vehicle must not use tires that:

- have any fabric exposed through the tread or sidewalls.
- have less than 4/32 of an inch of tread measured at any point in a major tread groove on the front axle.
- have less than 2/32 of an inch of tread measured at any point in a major tread groove on all other axles.
- have front tires that have been regrooved, if the tires have a load-carrying capacity equal to or greater than 4,920 pounds.

Most states do not allow any of the front tires to be regrooved. You must make yourself aware of any such local regulations.

You should measure tread depth when you inspect your vehicle. When you make groove measurements, you must not make them at a tie bar, hump, or fillet. A hump is a pattern of tire wear that looks a little like a cupped hand. The hump is the edge or higher part of the cup. Tie bars and fillets are not patterns of tire wear. They are design factors. So is a sipe. Sipes are cut across the tread to improve traction on wet road surfaces.

You must take the tread depth measurement on a major tread groove. Find a spot where the way is clear all the way to the body of the tire. See Figure 6-28.

You can use a special tread depth gauge for this. Or, you could use a common Lincoln-head penny. Insert the penny in the groove, with Lincoln's head upside down. The edge of the penny touches the body of the tire. The tread should come up to the top of Lincoln's head. That's about 2/32 of an inch. For 4/32 of an inch, the tread should come up to Lincoln's eyebrow.

Figure 6-28. Measure tread depth in a major tread groove, not on a hump or in a fillet or sipe.

Regrooved tires, recaps, and retreads, then, can be used on drive wheels and trailer wheels. They cannot be used on steering, or front, wheels on buses and most trucks.

Wheels

Wheels provide a mounting for the tires. The wheel supports and connects the tire to the truck axle. There are spoke wheels and disc wheels.

Spoke wheels use separate rims. They are clamped onto the wheels with wheel clamps. If the wheel clamps are not installed just right, the wheel may be out of round and wobble.

The rim and center portion of the disc wheel are one piece. The rim is part of the wheel and the wheel is bolted to the hub and brake drum assembly. No clamp is used. There is less chance of the wheel being out-of-round. A wheel that is out-of-round or wobbling is said to be not true running.

The rim's job is to support the tire bead and the lower sidewall.

Neither the wheel nor the rim should be cracked or broken. The stud or bolt holes on the wheels should be perfectly round. When you inspect your vehicle, look for holes that are egg-shaped (out-of-round). These are signs of defects. All the nuts and bolts should be in place, tightened down.

Cracked or broken wheels or rims could cause an accident. A damaged rim can cause the tire to lose pressure or come off. Missing clamps, spacers, studs, and lugs could cause problems. Mismatched, bent, or cracked lock rings are dangerous. If you see signs that the wheels or rims have been repaired by welding repairs, note that as a defect.

Rust around wheel nuts may mean the nuts are loose. Use a wrench to check tightness. You can't judge tightness well enough with just your hands. (If you have just had a tire changed, stop after you have driven on the new tire a while. Double-check to see that the nuts have not loosened.)

Make sure there is a good supply of oil in the hub. Check the oil level mark on the hub cap window. Check that there are no leaks.

Suspension System

The suspension supports the vehicle's weight. It keeps the frame from resting directly on the axles. It provides a smoother ride for the driver and the cargo. The suspension absorbs road shocks. Your vehicle may have one of the following suspension types:

- leaf spring
- coil spring
- torsion bar
- air bag

In a leaf spring suspension, layers of flexible metal are bolted together. The axle rests on the middle of the spring. The front and rear of the spring are attached to the frame.

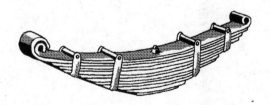

Figure 6-29. A multi-leaf front suspension.

A coil spring suspension puts a spiral of heavy-duty metal at each wheel. The top of the coil is attached to the frame, while the bottom is indirectly connected to the wheel.

Torsion bars absorb shock by twisting. They're made of special metal that returns to its original shape even after it's been twisted.

In these suspension types, the metal contracts and extends, or twists, in response to bumps or low spots in the road. This allows the wheel to move up and down with the road surface, while the frame remains level.

Your vehicle may have an air bag suspension instead of leaf springs or coils. The air bags are made of rubber fabric and are supplied with compressed air. The air supply allows the bag to expand or shrink, just as springs contract or extend.

Shock absorbers are attached to the springs. These are cylinders partly filled with hydraulic fluid. A piston moves up and down in the cylinder in response to the action of the spring. This helps to lessen the amount of springing motion that's transmitted to the driver.

Springs must always be in good condition. Suspension system defects can reduce your ability to control the vehicle.

Air bags are subject to leaks and valve problems. With air bag suspensions, the braking system should get charged with air before the air bag does. The air pressure valve on the air bag should not let air into the suspension until the braking system has at least 55 psi. The air bags should fill evenly all around. Otherwise, the vehicle will not be level. When the vehicle's air pressure gauge reads "normal," there should be no air leakage greater than 3 psi over five minutes.

Steering

The steering system enables the tractor to change direction and get around corners. A good steering system provides precise rolling, without slipping, when you turn a corner or take a curve.

The wheel in your hands is the steering wheel. It controls the steering wheels on the road that are connected to the steering axle. Between the steering wheel and the steering axle are the parts that make steering possible. Figure 6-30 shows you the basic steering components. You can see how the steering action flows through the system.

The steering wheel is connected to the steering column by a nut. The steering wheel transfers the driver's movements to the steering system.

As you turn the steering wheel, the column turns in the same direction. This turning motion continues through the U-joint to the steering gear shaft. From there the motion continues through another U-joint to the steering gear box. The steering gear box is also called the steering sector. It changes the rotating motion of the steer-ing column to the reciprocating (back and forth) motion, of the Pitman arm. The Pitman arm is a lever attached to the steering gear box. The drag link joins the Pitman arm and the steering lever.

The steering lever or steering arm is the first part of the steering axle. The steering, or front axle, does two jobs. First, it carries a load just like

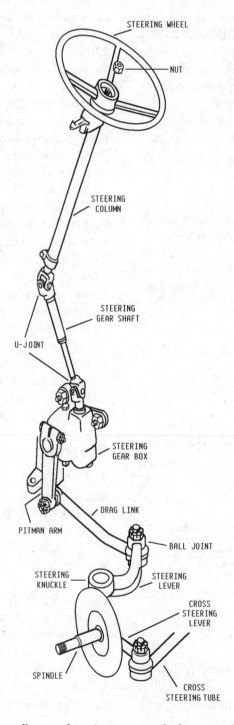

Figure 6-30. These components make up all types of steering systems, whether manual or power.

other axles. Second, it steers the vehicle. The steering lever or arm turns the front wheels left and right when the Pitman arm pulls it back and forth. The steering lever is connected to the steering knuckle. This is a moveable connection between the axle and the wheels. It allows the wheels to turn left or right.

There is a steering knuckle at the end of each axle. The steering knuckles contain the seals, bushings, and bearings that support the vehicle's weight. The steering knuckles transfer motion to the cross-steering lever, or tie rod arm, and the tie rod, or cross-steering tube.

The spindles are the parts of the steering axle knuckles that are inserted through the wheels.

The cross-steering tube, or tie rod, holds both wheels in the same position. As the left wheel turns, the right wheel moves in the same direction. A kingpin in the steering knuckle allows each wheel to have its own pivot point. The wheel rotates on a spindle. The spindle is also called the stub axle. It's attached to the kingpin.

Your vehicle may have manual or power steering. Power steering systems use hydraulic pressure or air pressure to assist the mechanical linkage in making the turn. This requires less of your strength to turn the steering wheel.

When hydraulic pressure is used to assist steering, a hydraulic unit replaces the steering gear box. A hydraulic pump is added to the engine to supply the pressure used to help turn the wheels. When you turn the steering wheel to the right, the hydraulic valve senses it. A valve opens. Fluid pressure helps turn the wheels to the right.

That's it—a quick tour through the major systems in most commercial motor vehicles. Now you know how wiring and electrical, braking, and fuel systems, coupling devices, tires, emergency equipment, body components, suspension, and steering systems support your vehicle.

All these systems and equipment are "necessary for safe operation" according to the FMCSR. The CDL examiner may, and probably will, ask you questions about some or all of these. You must be able to identify the parts on your vehicle. You must be able to give a basic description of how they work. You must be able to recognize defects and state how they affect safe operations.

EMERGENCIES

It's not enough to know how to operate your vehicle under normal conditions. FMCSR Part 393 states you must know how to use these systems and controls in emergencies. Two examples are mentioned. They are skids and loss of brakes. Of course there are others. The next chapter, on basic vehicle control, will include emergency maneuvers.

PASS POST-TRIP

Instructions: For each true/false test item, read the statement. Decide whether the statement is true or false. If it is true, circle the letter A. If it is false, circle the letter B. For each multiple-choice test item, choose the answer choice—A, B, C, or D—that correctly completes the statement or answers the question. There is only one correct answer. Use the answer sheet provided on page 141.

1. When you first start your engine, you should get a normal reading on the dashboard's _____ within a few seconds.
   A. oil pressure gauge
   B. fuel pressure gauge
   C. tire pressure temperature gauge
   D. all of the above

2. The proper way to use a fire extinguisher is to douse the flames with the dry chemical.
   A. True
   B. False

3. A high pyrometer reading could mean _____.
   A. a leak or a clog in the air intake
   B. problems with the fuel ignition
   C. you're in the wrong gear ratio for the grade
   D. any of the above

4. A vehicle will have warning lights or warning buzzers, but not both.
   A. True
   B. False

5. Warning lights tell you _____.
   A. a control is working
   B. a control or gauge is not working
   C. there is a serious problem with a part or system
   D. any of the above

6. If the low vacuum warning light comes on while you are driving, you should _____.
   A. note it and plan to fix it when you arrive at your destination
   B. feel confident the brake system is working properly
   C. stop as soon as it is safe to do so
   D. look for a blown fuse or loose wire

7. _____ at the top governed speed is the proper way to operate your vehicle.
   A. Driving
   B. Stopping
   C. Idling
   D. Shifting

8. Broken wires should be _____.
   A. reconnected
   B. replaced
   C. restripped
   D. regauged

9. You can skip checking the battery fluid level during your pre-trip inspection if _____.
   A. you are confident the level is at "FULL"
   B. you added fluid within the past 24 hours
   C. your battery is the "maintenance-free type"
   D. none of the above

10. Vacuum brakes in a vehicle with a diesel engine use the vacuum created on the intake stroke.
    A. True
    B. False

CHAPTER 6 ANSWER SHEET

Test Item	Answer Choices				Self - Scoring	
					Correct	Incorrect
1.	A	B	C	D	☐	☐
2.	A	B	C	D	☐	☐
3.	A	B	C	D	☐	☐
4.	A	B	C	D	☐	☐
5.	A	B	C	D	☐	☐
6.	A	B	C	D	☐	☐
7.	A	B	C	D	☐	☐
8.	A	B	C	D	☐	☐
9.	A	B	C	D	☐	☐
10.	A	B	C	D	☐	☐

Cut Here

Chapter 7

Basic Vehicle Control

In this chapter, you will learn about:

- basic vehicle control
  — shifting gears
  — backing
  — visual awareness while driving
  — communicating with other drivers
  — controlling speed
  — managing space
  — driving at night
  — driving in adverse conditions
  — road hazards
  — emergency maneuvers
  — controlling skids

To complete this chapter you will need:

- a dictionary
- a pencil or pen
- blank paper, a notebook, or an electronic notepad
- colored pencils, pens, markers, or highlighters
- a CDL preparation manual from your state Department of Motor Vehicles, if one is offered
- the Federal Motor Carrier Safety Regulations pocketbook (or access to U.S. Department of Transportation regulations, Parts 383, 393 and 396 of Subchapter B, Chapter 3, Title 49, Code of Federal Regulations)
- the owner's manual for your vehicle, if available

PASS PRE-TRIP

Instructions: Read the statements. Decide whether each statement is true or false. If it is true, circle the letter A. If it is false, circle the letter B.

1. Rough acceleration will not only damage the vehicle, it can cause you to lose control.
   A. True
   B. False

2. "Fanning the brakes" is a recommended downhill braking tactic.
   A. True
   B. False

3. You may hear a tire blow before you feel it.
   A. True
   B. False

As you know, Part Two of this book is about the knowledge and skills you must have to get your CDL. Chapters 4, 5, and 6 contained the facts you must have to have the "knowledge." This chapter, and the chapters to come, are more about "skills." As you read them, keep this in mind. You must practice to become skilled. We can describe a procedure or maneuver for you. To be skilled at that procedure or maneuver, you must practice it. The more you practice, the better you become. The goal is for the skill to become second-nature to you. Practice what you read about in Chapters 7, 8, and 9.

This chapter began with a long list of skills licensed CMV drivers must have. Who says you need to know this stuff to be a good driver? If you have a copy of the FMCSR, turn to Part 383 now. Does it include Subpart G? This is a list of the required knowledge and skills CMV drivers must have to earn a CDL. It tells you that you must know:

- the importance of visual search,
- techniques for controlling the space around the vehicle, and
- the causes and major types of skids, among other things.

Unless you're already an expert at everything in Subpart G, you'll find this chapter helpful. This chapter "fleshes out" the required knowledge and skill areas listed in Subpart G.

By reviewing the FMCSR in Chapter 5, you've become well acquainted with the safe operating regulations. In Chapter 6, you became familiar with CMV safety control systems. Let's continue with Subpart G and cover safe vehicle control.

SAFE VEHICLE CONTROL

To drive a vehicle safely, you must be able to control its speed and direction. Safe operation of a commercial vehicle requires skill in:

- accelerating
- steering
- shifting gears
- braking

Preparing to Drive

You must be in complete command of the vehicle, its systems, and controls. If the vehicle is new to you, prepare to drive by getting to know the vehicle first. Look it over before you start it. A good driver never starts the engine without first inspecting the vehicle. You'll learn about vehicle inspections in more detail in Chapter 9.

Before you climb in, look at the handhold and the steps to make sure they are free of dirt and grease. For large vehicles, use the three-point stance to enter the cab. Use either both hands and one foot or both feet and one hand to enter the cab.

Pull yourself in behind the wheel. Some driver's seats have weight and height controls and suspension systems that allow for a smooth ride even on bumpy roads. If you are too high or too low to see and reach for the controls, look for the seat adjuster.

Find the ignition switch. Put the key in the switch, but don't turn it on. Find the parking brake. Be sure it's set. Look at the transmission shift lever to see what type it is. Look for the shifting pattern. It will be posted in the cab. You'll also find a picture of the shift pattern in the owner's manual.

If you have an automatic transmission, put it in "PARK." If it's manual, grab the gear shift lever, depress the clutch, and place the lever in the neutral position. Let the clutch pedal out, sit back, and locate the cab controls and gauges described in Chapter 6.

Fasten your seat belt.

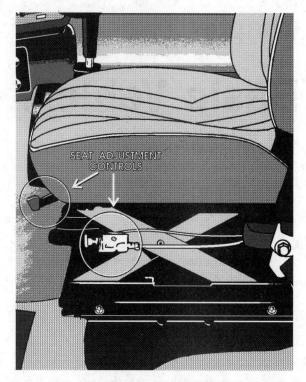

Figure 7-1. Adjust your seat so you have clear vision and your hands and feet can reach all the controls.

Starting

Now you're ready to start the engine. To start a gasoline engine, depress the gas pedal at least once to the floor. When the engine starts, press the accelerator to increase the rpm and keep the engine running. This pumps gas into the engine.

Do not depress the fuel pedal when you start a diesel engine. Do not depress the pedal when the engine starts either. You don't need to because fuel injectors meter diesel fuel into the cylinders in exact amounts. Instead, press the clutch pedal to the floor and hold it there. Turn the key. This causes electricity from the batteries to flow to the starter motor. The starter motor turns the flywheel and cranks the engine. The air and fuel ignite in each cylinder in turn, driving down the pistons and turning the crankshaft. As soon as the engine fires, release the key.

The engine is now running. Before you let the clutch out smoothly and slowly, make sure the gear shift lever is in neutral position ("P" or "PARK" for automatic transmissions). If the transmission is in gear when you let the clutch out, the vehicle could lurch forward or backward. You could cause an accident thinking the transmission is in neutral or park when it isn't.

Another reason you depress the clutch when you start a diesel engine is to let the starter turn as fast as it can. The transmission is filled with heavy gear oil that thickens in cold weather. If you don't depress the clutch pedal, the shaft will turn inside the transmission. This creates resistance, or drag, and slows the starter motor. If the starter motor turns the engine too slowly, the engine may not start.

The best rule is to start the engine with the clutch pedal down to the floor. Doing it every time will ensure safety and better starting.

Warm Up

Engines work best when they're warm. Give gasoline engines about five minutes for the engine to warm up before you put the vehicle into motion.

A diesel engine should warm up to at least 120 degrees Fahrenheit before you engage the clutch and start rolling. Recall the normal operating temperature? It's between 165 and 185 degrees Fahrenheit.

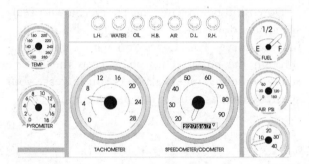

Figure 7-2. This engine is idling at 650 rpm. The air pressure and the operating temperature show this engine is about ready to go.

The diesel fuel and air mixture ignite best at high temperatures. The lubricating oil in the engine flows best. You can start the truck moving as soon as the air and oil pressures are up. Do not try to use full engine power until the engine has reached its operating temperature.

Your owner's manual will tell you the correct idling rpm. Automatic transmissions idle at higher rpm than manual. A "warm-up" switch will cause an even higher idling rpm. If you don't have a warm-up switch, adjust engine speed with the hand throttle rather than the accelerator. (Newer trucks have cruise control, which can do the same

thing.) Find the lowest speed at which the engine will run smoothly.

Don't roll back when you start. You may hit someone behind you. Partly engage the clutch before you take your right foot off the brake. Put on the parking brake if that helps to keep from rolling back. Release the parking brake only when you have applied enough engine power to keep from rolling back.

Accelerating

Increase your speed smoothly and gradually so the vehicle does not jerk. Rough acceleration can cause mechanical damage. If you're pulling a trailer, rough acceleration can damage the coupling. Accelerate very gradually when traction is poor, as in rain or snow. If you use too much power, the drive wheels may spin. You could lose control. If the drive wheels begin to spin, take your foot off the accelerator.

Steering

Hold the wheel correctly. Use both hands. Your hands should be on opposite sides of the wheel. Unless you have a firm hold, the wheel could pull away from your hands if you hit a curb or pothole.

Imagine the steering wheel is a clock. Put one hand at 9:00 and one hand at 3:00. Rest your thumbs along the top of the wheel, don't wrap them around it. Your hand could be injured if the wheel were to turn sharply or quickly.

When turning, don't cross your arms over the steering wheel. Pick up your hands one at a time and reposition them. Your left hand should not go past 12:00 when you turn right. Your right hand should not go past 12:00 when you turn left. You'll have good control of the wheel using this method.

Shifting

It's important to shift correctly. If you can't get your vehicle into the right gear while driving, you will have less control.

Use your tachometer as a guide. Your owner's manual will tell you what the operating rpm range is for your vehicle. Engine rpm is not the same thing as road speed (mph). Still, you may hear proper shifting technique called "matching" your engine rpm to your road speed.

Watch your tachometer, and shift up when your engine reaches the top of the range. (Some vehicles use "progressive" shifting. The rpm at which you shift becomes higher as you move up in the gears. Find out what's right for the vehicle you will operate.)

Or, you could use road speed (mph). Learn the range of speeds appropriate for each gear. Then, by using the speedometer, you'll know when to shift up.

With either method, you may learn to use engine sounds to know when to shift.

Most heavy vehicles with manual transmissions require double clutching to change gears. This is the basic upshifting method:

1. Release the accelerator, push in the clutch, and shift to neutral at the same time.
2. Release the clutch.
3. Let the engine and gears slow to the rpm required for the next gear.
4. Push in the clutch and shift to the higher gear at the same time.
5. Release the clutch and press the accelerator at the same time.

Double clutching well requires practice. If you remain too long in neutral, you may find it hard to put the vehicle into the next gear. If so, don't try to force it. Remain in neutral. Increase the engine speed to match the road speed, and try again.

To downshift:

1. Release the accelerator, push in the clutch, and shift to neutral at the same time.
2. Release the clutch.
3. Press the accelerator, increase the engine and gear speed to the rpm required in the lower gear.
4. Push in the clutch and shift to lower gear at the same time.
5. Release the clutch and press the accelerator at the same time.

Again watch the tachometer or the speedometer and downshift at the right rpm or road speed.

Downshift before starting down a hill. Slow down and shift down to a speed that you can control without using the brakes hard. Otherwise the brakes can overheat and lose their braking power. Usually you'll shift to a gear lower than the one you used to climb the same hill.

Downshift before entering a curve. Slow to a safe speed, and downshift to the right gear before entering the curve. This leaves you with some power through the curve. That helps keep the vehicle stable while turning. It also lets you increase your speed as soon as you are out of the curve.

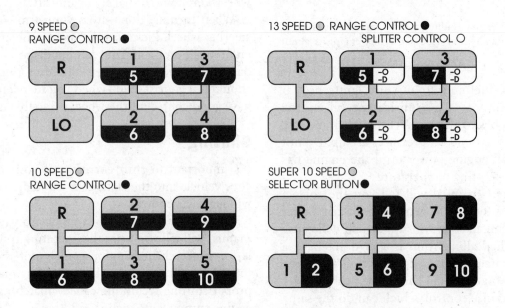

Figure 7-3. Shift patterns.

Many vehicles have multi-speed rear axles and auxiliary transmissions. These provide extra gears. You usually control them by a selector knob or switch on the gearshift lever of the main transmission. There are many different shift patterns. The owner's manual will picture your vehicle's shift pattern. Learn the right way to shift gears in the vehicle you will drive.

Some vehicles have automatic transmissions. You can choose a low range to get greater engine braking when going down grades. The lower ranges prevent the transmission from shifting up beyond the selected gear (unless the governor rpm is exceeded). It is very important to use this braking effect when going down grades.

Backing

Most backing accidents are those that occur on the right side, at the rear, and at the top of your vehicle. Yes, backing into something at the top of the vehicle is a common backing accident! Low hanging wires and eaves can damage the overhead area of your vehicle. Make sure the area above is clear of anything that might tear off an exhaust stack or otherwise damage the top of your cab.

Backing accidents often occur at the rear of the trailer. Check the area behind you before you back. Then check again.

Most backing accidents occur on the right side, because of the right side blind spot. The blind spot reaches from the right side door to the rear trailer axle and from about midway down the door to the ground. A spotter mirror helps you see into the blind spot. Even a spotter mirror might not show you everything you need to see.

Backing takes your complete attention. Turn off the radio and the CB, too. The less there is to distract you, the less likely you will be to lose concentration and get into an accident.

Always check the area you're backing into before you begin backing. Get out of the truck, walk behind it, and visually check the area. Even if you're just backing in a straight line, get out and take a good look at the area. Never assume you will catch everything with your mirrors. Don't forget to look at the top and under your vehicle. Always check just before you begin backing. You may have to get out and check the area after each few feet of backing.

Use your mirrors. Watch both sides of the vehicle. Don't open your door and lean out of it. That makes it impossible to use the right side mirror.

Never back up without putting on your emergency flashers. Your emergency flashers warn others to be alert to what you are about to do.

Part of the CDL Skills Test will test your backing ability. You'll be asked to perform several backing maneuvers, from straight-line backing to a backward serpentine. A backward serpentine requires you to turn to the left, then to the right, then to the left again, all while going backward. You'll find guidelines that will help you practice backing in Chapter 15.

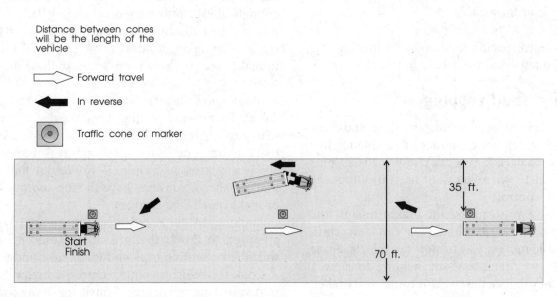

Figure 7-4. The CDL Skills Test will definitely test your backing ability.

Backing with a Trailer

If you plan to take your CDL test in a tractor-trailer combination, it's vital you master these backing techniques. Read Chapter 11 carefully.

Briefly, though, here's the idea behind backing with a trailer. When backing a car or straight vehicle, you turn the top of the steering wheel toward the direction you want to go. When backing a trailer, you turn the steering wheel opposite the direction in which you want the trailer to go. This is called "jacking the trailer." Once the trailer starts to turn, you must turn the wheel the other way to follow the trailer. This is, in fact, called "following the trailer" or "chasing the trailer."

Turning

Steering around a corner with a long vehicle, particularly a tractor-trailer, takes some planning. Because of the length, the rear wheels may not follow the same path as the front wheels, particularly when you are pulling a trailer. This is called "off-tracking." To make the perfect turn without scraping the corner, first:

- Position the vehicle for turning
- Signal well in advance
- Check traffic conditions and turn only when the way is clear
- Avoid swinging wide or cutting short
- Adjust the speed so you can stop if need be
- Check for cross traffic
- Yield the right-of-way if safety demands it
- Use your mirrors

You'll learn specific techniques for making safe right- and left-hand turns later in this chapter.

Slowing and Stopping

It's one thing to be able to start, shift, and steer your vehicle. It's quite another to be able to slow and stop it safely. If you can't stop a 20,000-plus pound vehicle when you want to, you cannot say you are in control.

Of course you can use the brakes to slow the vehicle. In some situations, you can downshift. There's still another way to slow the vehicle. Some vehicles have "retarders." By helping to slow a vehicle, they reduce the need for using your brakes.

They reduce brake wear and give you another way to reduce speed.

There are many types of retarders (exhaust, engine, hydraulic, electric). All retarders can be turned on or off by the driver. On some, the retarding power can be adjusted. When turned "on," retarders apply their braking power (to the drive wheels only) whenever you let up on the accelerator pedal all the way.

Caution. *When your drive wheels have poor traction, the retarder may cause them to skid. Therefore you should turn the retarder off whenever the road is wet, icy, or snow covered.*

Shutting Down

Allow the engine to idle for at least three minutes before you turn the engine off. This keeps lubrication flowing while hot engine parts cool off. Never stop the engine with the accelerator pedal pressed. You could damage the governor. Always stop the engine from idle.

Apply the parking brake before you leave your vehicle.

VISUAL ALERTNESS

To be a safe driver you need to know what's going on all around your vehicle. Not being visually alert is a major cause of accidents.

Looking Ahead

All drivers look ahead, but many don't look far enough ahead. Stopping or changing lanes can take a lot of distance. You must know what the traffic is doing on all sides of you. Look well ahead to make sure you have room to make these moves safely.

Most good drivers "look" 12 to 15 seconds ahead. That means looking ahead the distance you will travel in 12 to 15 seconds. At lower speeds, that's about one block. At highway speeds, it's about a quarter of a mile. If you're not looking that far ahead, you may have to stop too quickly or make quick lane changes.

Looking that far ahead doesn't mean not paying attention to things that are closer. Good drivers shift their attention back and forth, near and far.

Look for vehicles coming onto the highway or into your lane or turning. Watch for brake lights from slowing vehicles. By seeing these things in enough time to react, you can prepare to slow or change lanes if necessary to avoid a problem.

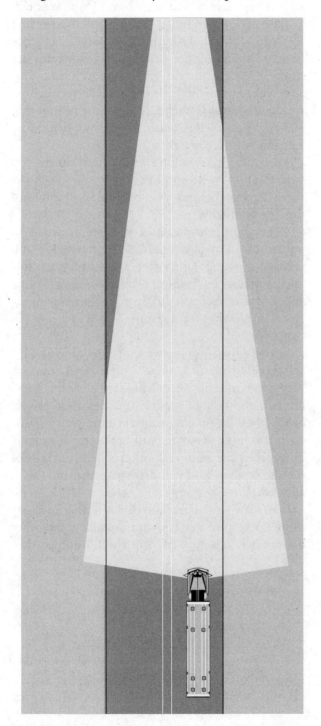

Figure 7-5. Be alert to the road ahead.

Observe the shape and condition of the road ahead. Look for hills and curves, or objects in the road. Any of these could mean you'll have to slow

or change lanes. Pay attention to traffic signals and signs. A light that has been green for a long time is often called a "stale" green light. It will probably change before you get there. Start slowing and be ready to stop. Traffic signs may alert you to road conditions where you may have to change speed.

Looking Behind and to the Sides

It's important to know what's going on behind and to the sides. Keep track of other vehicles when they enter the highway. Are they still behind you? Have they passed you or exited? In an emergency, you must know whether you can make a quick lane change. Use your mirrors to spot overtaking vehicles. There are "blind spots" that your mirrors cannot show you. Check your mirrors regularly to know where other vehicles are around you. Perhaps they have moved into your blind spots.

Use the mirrors to keep an eye on your tires. It's one way to spot a tire fire. If you're carrying open cargo, you can use the mirrors to make sure it's still secure. Look for loose straps, ropes, or chains. Watch for a flapping or ballooning tarp.

Check your mirrors even more often when making lane changes. Make sure no one is alongside you or about to pass you. Check your mirrors before you change lanes to make sure there is enough room. You have "enough" space when other drivers won't have to slow or brake to allow you to complete the lane change.

After you have signaled, check that no one has moved out of your blind spot. Use your mirrors right after you start the lane change to double-check that your path is clear. Use your mirrors again after you complete the lane change.

When turning, check your mirrors to make sure the rear of your vehicle will not hit anything. When merging, use your mirrors to make sure the gap in traffic is large enough for you to enter safely. Other drivers shouldn't have to slow or brake to let you in.

Proper Use of Mirrors

Use mirrors correctly by checking them quickly and understanding what you see.

When you use your mirrors while driving on the road, check quickly. Look back and forth between the mirrors and the road ahead. Don't focus on the mirrors for too long. Otherwise, you will travel quite a distance without knowing what's happening ahead.

Understand what you see. Many large vehicles have curved (convex, "fisheye," "spot," "bug-eye") mirrors that show a wider area than flat mirrors. This is often helpful. Everything appears smaller in a convex mirror than it would if you were looking at it directly. Things also seem farther away than they really are. Be aware of this and allow for it.

COMMUNICATING WITH OTHER DRIVERS

Other drivers can't know what you are going to do until you tell them. You must know how to use your safety control systems to communicate with other drivers.

Turn Signals

Use your turn signals. There are three good rules for using turn signals:

Signal early. Signal well before you turn. It is the best way to keep others from trying to pass you.

Signal continuously. You need both hands on the wheel to turn safely. Don't cancel the signal until you have completed the turn.

Cancel your signal. Don't forget to turn off your turn signal after you've turned. (Some vehicles do not have self-canceling signals.)

Use your turn signal to show that you plan to change lanes. Change lanes slowly and smoothly. That way you can cancel your plans and avoid hitting a driver you didn't see at first.

Brake Lights

Use the brake pedal to show you plan to slow or stop. A few light taps on the brake pedal (enough to flash the brake lights) warn other drivers that you are slowing. Use the four-way emergency flashers when you are driving very slowly or are stopped. Warn other drivers of trouble ahead. Your large vehicle may make it hard for drivers behind you to see hazards ahead. If you see a hazard that will require slowing, warn the drivers behind by flashing your brake lights.

Most car drivers don't know how slowly you have to go to make a tight turn in a large vehicle. Warn drivers behind you by braking early and slowing gradually.

CMV drivers sometimes stop in the road to unload cargo or passengers. Some must stop at a railroad crossing. Don't stop suddenly. Warn following drivers by flashing your brake lights. Some states require that you put on your four-way flashers when slowing for a required stop at a railroad crossing.

Drivers are often unaware of how quickly they are catching up to a slow vehicle until they are very close. You may be driving far below the speed limit or the flow of traffic. You can alert drivers behind you by turning on your emergency flashers if it is legal. (Laws regarding the use of flashers differ from one state to another. Check the laws of the states where you will drive to make sure this is allowed.)

Don't use your signals to direct traffic. You may think it's helpful to flash your lights and tell other drivers it is safe to pass. You should not do this.

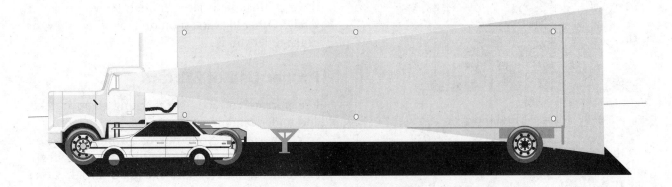

Figure 7-6. Your convex mirror will help you see into your blind spot.

If it turned out not to be safe, you could cause an accident. You would be blamed.

Communicate Your Presence

Other drivers may not notice your vehicle even when it's in plain sight. Prevent accidents by making sure they know you're around.

When you are about to pass a vehicle, pedestrian, or bicyclist, assume they don't see you. They could suddenly move in front of you. When it is legal, tap the horn lightly. At night, flash your lights from low to high beam and back. Then proceed carefully. Be prepared to avoid a crash because they may still not see or hear you.

At dawn or dusk or in rain or snow, you must make yourself easier to see. If you are having trouble seeing other vehicles, other drivers will have trouble seeing you. Turn on your lights. Use the headlights, not just the identification or clearance lights. Use the low beams. High beams can bother people in the daytime as well as at night.

Turn on your four-way flashers when you pull off the road and stop. This is important at night. Don't trust the taillights to give enough warning. Drivers have crashed into the rear of a parked vehicle because they thought it was moving normally.

Remember, if you stop on a road or the shoulder of a road, you must put out your reflective triangles within 10 minutes. Recall where to place your warning devices?

- on the traffic side of the vehicle, within 10 feet of the front or rear corners—to mark the location of the vehicle.
- about 100 feet behind and ahead of the vehicle, on the shoulder, or in the lane in which you are stopped.
- back beyond any hill, curve, or other obstruction that prevents other drivers from seeing the vehicle within 500 feet.
- or, if you must stop on or by a one-way or divided highway, place warning devices 10 feet, 100 feet, and 200 feet toward the approaching traffic.

As you are putting out the triangles, hold them between yourself and the oncoming traffic. That way other drivers can see you as you walk on the road.

You can use your horn to let others know you're there. It can help to avoid a crash. Use your horn when needed. However, it can really startle others. Use the horn only when there's no other way to get

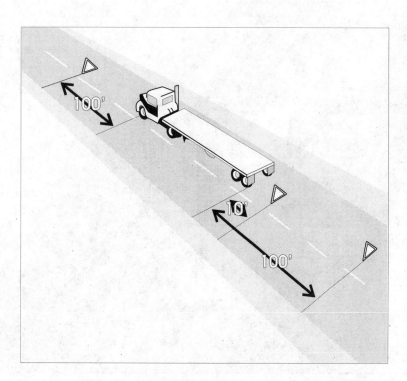

Figure 7-7. Placement of warning devices on the road.

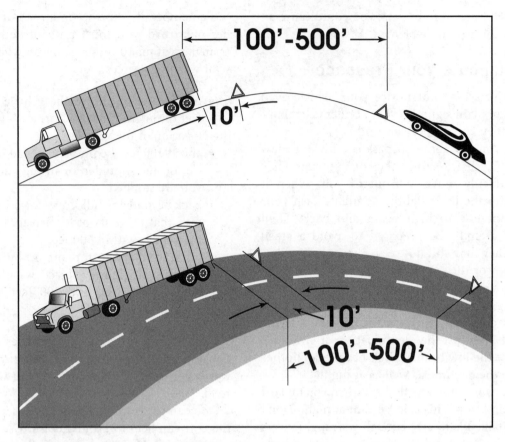

Figure 7-8. Placement of warning devices on a hill or curve.

Figure 7-9. Placement of warning devices on a one-way or divided highway.

their attention. It could be dangerous when used unnecessarily.

SPEED MANAGEMENT

Driving too fast is a major cause of fatal crashes. You must base your speed on the driving conditions. These include traction, curves, visibility, traffic, and hills.

Stopping Distance

Know how much distance you need to come to a complete stop. Stopping distance is:

> Perception Distance
> \+ Reaction Distance
> \+ Braking Distance
> = Total Stopping Distance

Perception distance is the distance your vehicle travels from the time your eyes see a hazard until your brain recognizes it. The perception time for an alert driver is about 3/4 second. At 55 mph, you travel 60 feet in 3/4 second.

Reaction distance is the distance traveled from the time your brain tells your foot to move from the accelerator until your foot actually pushes the brake pedal. The average driver has a reaction time of 3/4 second. This accounts for an added 60 feet traveled at 55 mph.

Braking distance is the distance it takes to stop once the brakes are put on. At 55 mph on dry pavement with good brakes it can take a heavy vehicle about 170 feet to stop. It takes about 4 1/2 seconds.

Total stopping distance is the sum of all three distances. At 55 mph it will take about six seconds to stop and your vehicle will travel about the distance of a football field. (60 + 60 + 170 = 290 feet.)

Increasing the speed greatly increases the distance you need to come to a stop. Whenever you double your speed, it takes about four times as much distance to stop. Not double—four times. Your vehicle will have four times the destructive power if it crashes. By slowing a little, you can gain a lot in reduced braking distance.

Weight also affects stopping distance. The heavier the vehicle, the more work the brakes must do to stop it. They also absorb more heat. The brakes, tires, springs, and shock absorbers on heavy vehicles are designed to work best when the vehicle is fully loaded. Empty trucks require greater stopping distances. That's because an empty vehicle has less traction. It can bounce and lock up its wheels. You could more easily go into a skid.

Speed and Road Conditions

You can't steer or brake a vehicle unless you have traction. Traction is friction between the tires and the road. There are some road conditions that reduce traction and call for lower speeds. Examples are slippery and curving roads.

Slippery Roads

When the road is slippery, it will take longer to stop and it will be harder to turn without skidding. You must drive slower to be able to stop in the same distance you could on a dry road. Wet roads can double stopping distance. Reduce speed by about one third on a wet road. For example, slow from 55 to about 35 mph. On packed snow, reduce speed by a half, or more. If the surface is icy, reduce speed to a crawl. In fact, stop driving as soon as you can safely do so. It's just too dangerous to go on.

Sometimes it's hard to know if the road is slippery. Keep these facts in mind. Shady parts of the road don't get as much sun as well-lit areas. So they will remain icy and slippery long after open areas have melted. When the temperature drops, bridges will freeze before the road will. Be especially careful when the temperature is close to 32 degrees Fahrenheit.

Slight melting will make ice wet. Wet ice is much more slippery than ice that is not wet.

Black ice is a thin, clear layer. It's clear enough that you can see the road underneath it. It makes the road look wet. You might not think it was icy. Any time the temperature is below freezing and the road looks wet, watch out for black ice.

An easy way to check for ice is to open the window and feel the front of the mirror, mirror support, or antenna. If there's ice on these, the road surface is probably starting to ice up.

Right after it starts to rain, the water mixes with oil left on the road by vehicles. This makes the road very slippery. If the rain continues, it will wash the oil away. As you read earlier, roads are more slippery at the beginning of a rainstorm than during it.

In some weather, water or slush collects on the road. When this happens, your vehicle can hydroplane. It's like water skiing. The tires are no longer riding on the road. They are riding on top of water. As you can imagine, they have little or no traction. You may not be able to steer or brake.

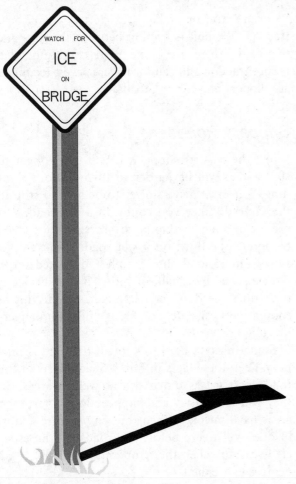

Figure 7-10. In cold weather, bridges can be icy even when roads are not.

You can regain control by releasing the accelerator and pushing in the clutch. This will slow your vehicle and let the wheels turn freely. If the vehicle is hydroplaning, do not use the brakes to reduce your speed. If the drive wheels start to skid, push in the clutch to let them turn freely.

It does not take a lot of water to cause hydroplaning. Hydroplaning can occur at speeds as low as 30 mph if there is a lot of water. Hydroplaning is more likely if tire pressure is low or the tread is worn. (The grooves in a tire carry away the water; if they aren't deep they don't work well.) Be especially careful driving through puddles. The water is often deep enough to cause hydroplaning.

Curving Roads

Adjust your speed for curves in the road. If you take a curve too fast, the wheels can lose their traction and continue straight ahead. You'll skid off the road. If the wheels keep their traction, the vehicle may roll over. Tests have shown that vehicles with a high center of gravity can roll over even if they are going at the posted speed limit for a curve.

Slow to a safe speed before you enter a curve. It's dangerous to apply the brakes in a curve. It's easy to lock the wheels and cause a skid. Reduce your speed as needed. Don't ever exceed the posted speed limit for the curve. You probably need to be going much slower. Be in a gear that will let you accelerate slightly in the curve. This will help you keep control.

Figure 7-11. Don't ever exceed the posted speed limit for a curve.

Speed and Distance Ahead

You should always be able to stop within the distance you can see ahead. In fog or rain, you can't see clearly that far ahead. Reduce your speed so you can stop within the distance you can see.

At night, you can't see as far with low beams as you can with high beams. When you must use low beams, reduce your speed.

In heavy traffic, the safest speed is the speed of other vehicles. Vehicles going the same direction at the same speed are not likely to run into one another. Drive at the speed of the traffic, if that is safe and legal. Keep a safe following distance.

The main reason drivers exceed speed limits is to save time. Driving faster than the speed of traffic doesn't save that much time, though.

Even if you do save time, the risks involved are not worth it. If you go faster than the speed of other traffic, you'll have to keep passing other vehicles. This increases the chance of a crash.

Trying to maintain the faster speed and making all those lane changes is tiring. Fatigue increases the chance of a crash.

Going with the flow of traffic is safer, and easier.

Speed and Downgrades

The most important thing you can do to go down long steep hills safely is to reduce your speed. If you do not go slowly enough, your brakes can become so hot they won't slow your vehicle. Pay attention to signs warning of long downhill grades. Check your brakes before starting down the hill. Shift your transmission to a low gear before starting down the grade.

Use your retarder if you have one. Shift to a gear that lets you go downhill at a steady, controlled speed without having to use the brakes. If you must use the brakes, use "snub braking." This is moderate and intermittent braking. Snub the brakes moderately using a brake application of 20 to 30 psi until the speed has been reduced five to six miles per hour. Then release. If your speed increases again, repeat the snub braking. Do not "fan" the brakes.

Going down steep hills safely is described in great detail in "Mountain Driving." Read that section carefully.

SPACE MANAGEMENT

To be a safe driver, you need space all around your vehicle. When things go wrong, space gives you time to think and to take action.

To have space available when something goes wrong, you need to manage space. While this is true for all drivers, it is very important for large vehicles. They take up more space. They require more space for stopping and turning.

The Space Ahead

The area ahead of the vehicle (the space into which you're driving) is most important. You need space ahead in case you must stop suddenly. According to accident reports, the vehicle that trucks and buses most often run into is the one in front of them. The most frequent cause is following too closely. Remember, if the vehicle ahead of you is smaller than yours, it can probably stop faster than you can. You may crash if you are following too closely.

Keep at least one second of space between you and the vehicle ahead of you for each 10 feet of vehicle length at speeds below 40 mph. What does that mean, "one second of space?" It's the distance your vehicle covers in one second. That will vary with your speed. Here's how to measure it.

Wait until the vehicle ahead passes a shadow on the road, a pavement marking, or some other clear landmark. Then count off the seconds like this: "One thousand-and-one, one thousand-and-two," and so on, until you reach the same spot. It takes about a second to say "one thousand-and-one."

		Miles Traveled				
		10	20	50	100	200
Speed	65 MPH	9 minutes	18 minutes	48 minutes	1 hour 32 minutes	3 hours 4 minutes
	60 MPH	10 minutes	20 minutes	50 minutes	1 hour 40 minutes	3 hours 20 minutes
	55 MPH	11 minutes	22 minutes	55 minutes	1 hour 49 minutes	3 hours 38 minutes

Figure 7-12. Driving faster doesn't save much time, but it does increase the chance of a crash.

Compare your count with the rule of one second for every 10 feet of length. If you are driving a 40-foot vehicle and only counted up to two seconds, you're too close. Drop back a little and count again until you have four seconds of following distance.

At speeds over 40 mph, you must add one second for safety. For example, if you are driving a 40-foot vehicle, you should leave four seconds between you and the vehicle ahead. In a 60-foot rig, you'll need six seconds. Over 40 mph, you'd need five seconds for a 40-foot vehicle and seven seconds for a 60-foot vehicle. Add another second for adverse driving conditions. These could be bad weather, poor lighting, or slippery roads. Give yourself more space for night driving. Under the worst conditions, you might need as many as seven seconds of space in front of your 40-foot vehicle at highway speeds. Some states have specific following-distance requirements.

Space Behind

You can't stop others from following you too closely. There are things you can do to make it safer, however.

Stay to the right. Heavy vehicles are often tailgated when they can't keep up with the speed of traffic. This tends to happen when you're going uphill. If a heavy load is keeping your speed low, stay in the right lane if you can. When going uphill, you should not pass another slow vehicle unless you can get around quickly and safely.

Many car drivers follow large vehicles closely during bad weather, especially when it is hard to see the road ahead. It's often hard to see whether a vehicle is close behind you. It helps to be visually alert. Then you might suspect a vehicle is behind you because you didn't see it pass you or exit.

If you find yourself being tailgated, here are some things you can do to reduce the chances of a crash. First, avoid quick changes. If you have to slow or turn, signal early and reduce speed very gradually.

Second, increase your following distance. Opening up room in front of you will help you to avoid having to make sudden speed or direction changes. It also makes it easier for the tailgater to get around you. If in spite of all your efforts the tailgater rear-ends you, at least you won't add a front-end collision with the vehicle in front of you to the problem. Don't increase your speed.

The tailgater will only do likewise. It's safer to be tailgated at a low speed than a high speed. Don't turn on your taillights or flash your brake lights to shake up the tailgater.

Space at the Sides

Commercial vehicles are often wide. They take up most of a lane. Make the best use of what little space you have. Keep your vehicle centered in your lane. Try to have a little clearance on each side.

Avoid driving alongside others. There are two dangers in traveling alongside other vehicles. Another driver may change lanes suddenly and turn into you. Or, you may be trapped when you need to change lanes.

Find an open spot where you aren't near other traffic. When traffic is heavy, it may be hard to find an open spot. If you must travel near other vehicles, try to keep as much space as possible between you and them. Also, drop back or pull forward so that you are sure the other driver can see you.

Strong winds make it hard to stay in your lane, especially for lighter vehicles. This problem can be even worse coming out of tunnels. Don't drive alongside others if you can avoid it.

Space Overhead

Hitting overhead objects is a danger. You learned this in the section on backing. Make sure you always have overhead clearance.

Don't assume that the heights posted at bridges and overpasses are correct. Repaving or packed snow may have raised the height of the road since the signs were posted. This reduces the clearance.

The weight of a cargo van changes its height. An empty van rides higher than a loaded one. Just because you got under a bridge when you were loaded doesn't mean that you can do it when you are empty.

If you doubt you have safe space to pass under an object, go slowly. If you aren't sure you can make it, take another route. Warnings are often posted on low bridges or underpasses, but sometimes they are not.

Some roads can cause a vehicle to tilt. There can be a problem clearing objects along the edge of the road, such as signs or trees. Where this is a problem, drive a little closer to the center of the road.

Before you back into an area, get out and check for overhanging branches or electric wires. It's easy to miss seeing them while you are backing. (Also check for other hazards at the same time.)

Space Below

Don't forget about the space under your vehicle. That space can be very small when a vehicle is heavily loaded. Railroad tracks can stick up several inches. This is often a problem on dirt roads and in unpaved yards where the surface around the tracks can wear away. Don't take a chance on getting hung up halfway across. Drainage channels across roads can cause the end of some vehicles to drag. Cross such depressions carefully.

Space for Turning

You need space around your CMV in order to turn safely. Large vehicles turn widely. Also, the rear wheels don't follow the same path as the front wheels. Remember what this is called? It's "off-tracking." Because of wide turning and off-tracking, large vehicles can hit other vehicles or objects during turns.

Right turns are more difficult than left turns. You can't see what's happening at the right of your vehicle as well as you can see to the left. Turn slowly to give yourself and others more time to avoid problems. If you cannot make the right turn without swinging into another lane, turn wide as you complete the turn (see Figure 7-13). Don't turn wide to the left as you start the turn, as shown in Figure 7-14. A driver following you may think you are turning left. The driver may then try to pass you on the right. You may crash into the other vehicle as you complete your turn.

Figure 7-13. The correct way to make right turns.

Figure 7-14. The WRONG way to make right turns!

Keep the rear of your vehicle close to the curb. This will stop other drivers from passing you on the right.

If you must cross into the oncoming lane to make a turn, watch out for vehicles coming toward you. Give them room to go by or to stop. However, don't back up for them. You might hit someone behind you.

Left turns are a little easier. Make sure you have reached the center of the intersection before you start the left turn. If you turn too soon, the left side of your vehicle may hit another vehicle because of off-tracking.

If there are two turning lanes, always start your left turn from the right-hand turn lane (see Figure 7-15). Don't start in the inside lane. You may just have to swing right to make the turn. Drivers on your right may be hard for you to see. You may crash into them.

Space to Enter or Cross Traffic

Be aware of the size and weight of your vehicle when you cross or enter traffic. Large vehicles accelerate slowly and need extra space. When heavily loaded, they accelerate even slower. You may need a much larger gap to enter traffic than you would in a car.

Before you start across a road, make sure you can get all the way across before traffic reaches you.

NIGHT DRIVING

You are at greater risk when you drive at night. Drivers can't see hazards as soon as in daylight, so they have less time to respond. Drivers caught by surprise are less able to avoid a crash.

Night driving offers a number of vision problems. Also, you're simply not as alert and responsive at night. You must allow for both these handicaps.

Poor Vision

People can't see as sharply at night or in dim light. Also, the eyes need time to adjust to seeing in dim light. Think about walking into a dark movie theater. At first you can hardly see at all. After your eyes adjust to the dim light, you can see a little

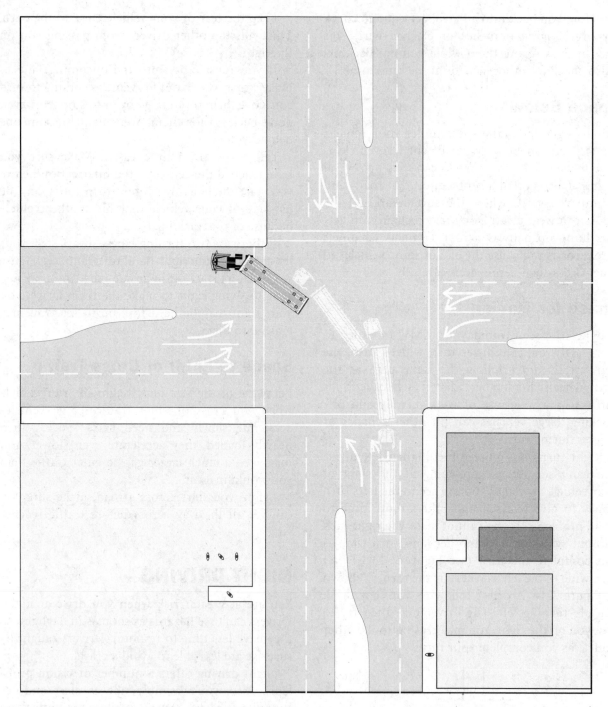

Figure 7-15. Turning left when there are two turning lanes.

better. Of course, you can't see as well as you could in bright light.

Drivers can be blinded for a short time by bright light. It takes time to recover from this blindness. Older drivers are especially bothered by glare. You can probably remember being temporarily blinded by a camera flash or by the high beams of an oncoming vehicle.

It can take several seconds to recover from glare. Even two seconds of glare blindness can be dangerous. A vehicle going 55 mph will travel more than half the distance of a football field during that time. Don't look directly at bright lights when driving. Look at the right side of the road. Watch the sidelines when someone coming toward you has very bright lights.

In the daytime there is usually enough light to see well. This is not true at night. Some areas may have bright street lights, but many areas will have poor lighting. On most roads you will probably have to depend entirely on your headlights.

Less light means you will not be able to see hazards as well as in daytime. Pedestrians, joggers, bicyclists, and animals don't usually have lights or reflectors. They're hard to see. There are many accidents at night involving pedestrians, joggers, bicyclists, and animals.

Even when there are lights, the road scene can be confusing. Traffic signals and hazards can be hard to see against a background of signs, shop windows, and other lights. Reduce your speed, enough to be sure you can stop in the distance you can see ahead.

At night your headlights will usually be the main source of light for you to see and for others to see you. You can't see nearly as much with your headlights as you can see in the daytime. With low beams you can see ahead about 250 feet. With high beams you can see about 350 to 500 feet. Again, adjust your speed to keep your stopping distance within your sight distance. This means going slowly enough to be able to stop within the range of your headlights. Otherwise, by the time you see a hazard, you will not have time to stop.

Night driving can be more dangerous if you have problems with your headlights. Dirty headlights may give only half the light they should. This cuts down your ability to see and makes it harder for others to see you. Make sure your lights are clean and working.

Headlights can be out of adjustment. If they don't point in the right direction, they don't give you a good view and they can blind other drivers. Have a qualified person make sure they are adjusted properly.

In order for you to be seen easily, the following must be clean and working properly:

- reflectors
- marker lights
- clearance lights
- taillights
- identification lights

At night your turn signals and brake lights are vital for telling other drivers what you intend to do. Make sure you have clean, working turn signals and stop lights.

It is also vital to have clean windshields and mirrors. Bright lights at night can cause dirt on your windshield or mirrors to create a glare of its own. This can block your view. Think about times you have driven toward the sun just as it has risen or is about to set. You really find out how clean your windshield is then! Clean your windshield on the inside and outside for safe driving at night.

Poor lighting isn't the only handicap you have at night. Fatigue (being tired) and lack of alertness are bigger problems at night than during the day. The body's need for sleep is beyond a person's control. Most people are less alert at night, especially after midnight. This is particularly true if you have been driving all day.

Drivers may not see hazards as soon or react as quickly, so the chance of a crash is greater. If you are sleepy, the only safe cure is to get off the road and get some sleep. If you don't, you risk your life and the lives of others.

Here's another hazard of night driving. You're more likely to run into drunk drivers. Drunk drivers and drivers under the influence of drugs are a hazard to themselves and to you. Be especially alert around the times bars close. Watch for drivers who have trouble staying in their lane, maintaining speed, or who stop without reason. These are all signs of being under the influence of alcohol or drugs.

Pre-trip Inspection for Night Driving

Make special preparations for driving at night. Pre-trip your vehicle and yourself with reduced visibility in mind. Do a complete pre-trip inspection of your vehicle. Pay special attention to checking all lights and reflectors. Clean all those you can reach.

Check your own fitness to drive. Make sure you are rested and alert. If you are drowsy, sleep before you drive! Even a nap can save your life or the lives of others.

If you wear eyeglasses, make sure they are clean and not scratched. Don't wear sunglasses at night.

PASS BILLBOARD

You may think that if you usually work nights, you'll get used to sleeping during the day. You may figure you'll get used to being awake at night. More and more, studies on what is called "circadian rhythms" are showing this just isn't so. The human body just doesn't like being awake when it's dark.

Also, you have a sort of internal clock that keeps track of how much rest you've had. When you need rest, your body is going to fight your attempts to stay up and about. This battle between your mind and your body adds to your fatigue.

These studies have become very important to long haul drivers. These drivers sleep when their log says it's time to sleep. It doesn't matter if it's day or night. Because of this, they're often at a disadvantage no matter how hard they try to be good drivers.

More and more, magazines for commercial drivers are covering this subject. If you often drive through the night or sleep in short shifts, pay attention to articles on circadian rhythms, inner clocks, and the need for sleep. You'll find helpful hints for overcoming this handicap.

Night Driving Tactics

Avoid blinding others. Glare from your headlights can cause problems for drivers coming toward you. They can also bother drivers going in the same direction you are, when your lights shine in their rear-view mirrors. Dim your lights before they cause glare for other drivers. Dim your lights within 500 feet of an oncoming vehicle and when following another vehicle within 500 feet.

Avoid glare from oncoming vehicles. Remember, don't look directly at lights of oncoming vehicles. Look slightly to the right at a right lane. If other drivers don't put their low beams on, don't try to "get back at them" by putting your own high beams on. This increases glare for oncoming drivers and increases the chance of a crash.

Use your high beams when you can. Some drivers make the mistake of always using low beams. This seriously cuts down on their ability to see ahead. Use high beams when it is safe and legal to do so. Use them when you are not within 500 feet of an oncoming vehicle.

Also, don't let the inside of your cab get too bright. This makes it harder to see outside. Keep the interior light off. Adjust your instrument lights as low as you can and still read the gauges.

Avoid a heavy meal before you drive. Definitely avoid alcohol. Keep your cab slightly on the cool side and well ventilated. Keep your eyes moving.

Stop at least every two hours. Move around or get some rest.

In spite of all this, you'll eventually get sleepy. When you do, stop driving at the nearest safe place. People often don't realize how close they are to falling asleep even when their eyelids are falling shut. This is a very dangerous condition.

When you're this tired, a walk in the cool air won't make you more alert. You'll just be a cold sleepy driver. Coffee and other drinks with caffeine won't make you more alert. They'll only make you a jittery, bug-eyed, sleepy driver. The only sure cure for sleepiness and fatigue is sleep.

DRIVING IN ADVERSE CONDITIONS

Driving under ideal conditions is no easy job. Driving under adverse or unfavorable conditions takes real skill. Any of the following can make your driving job that much harder:

- winter weather
- very hot weather
- mountain terrain

You should prepare for adverse conditions with the same care you prepare for night driving. If conditions change on you from normal to adverse, you must know how to adjust.

DRIVING IN WINTER

Winter weather can present some of the worst driving challenges you'll face. Snow and ice reduce traction. Blowing snow can reduce visibility. Fighting the weather and trying to maintain control over your vehicle can wear you out. So fatigue is a problem.

Slippery Surfaces

Very simply, when the road is slippery, don't make any sudden or abrupt moves. Start gently and slowly. When first starting, get the feel of the road. Don't hurry. Drive slowly and smoothly.

Turn as gently as possible. Don't brake any harder than you need to. Give yourself lots of space and following distance. Then you can brake gently when you need to slow or stop. Don't use the engine brake or speed retarder. They can cause the driving wheels to skid on slippery surfaces.

Avoid passing slower vehicles. Go slow and watch far enough ahead to keep a steady speed. Then you won't have to keep slowing down and speeding up. Take curves at slower speeds. Don't brake while in curves. Be aware that as the temperature rises to the point where ice begins to melt, the road becomes even more slippery. Slow even more.

Don't drive alongside other vehicles. Keep a longer following distance. When you see a traffic jam ahead, reduce your speed or stop to wait for it to clear. Try hard to anticipate stops early and slow gradually.

If it is very slippery, you shouldn't drive at all. Stop at the first safe place.

Pre-trip Inspection for Winter Driving

Before driving in winter weather, make your regular pre-trip inspection. Pay extra attention to the items that are important in cold weather.

Make sure the cooling system is full. There must be enough antifreeze in the system to protect against freezing. You can check this with a special coolant tester.

Make sure the defrosters work. They are needed for safe driving. Make sure the heater is working and that you know how to operate it. You may have other heaters, such as mirror heaters, battery box heaters, or fuel tank heaters on your vehicle. If you plan to use them, make sure they work.

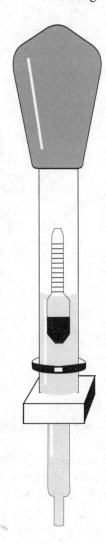

Figure 7-16. A coolant tester.

Make sure the windshield wiper blades are in good condition. The wiper blades should press against the window hard enough to wipe the windshield clean. Otherwise, they may not sweep off snow properly. Make sure the windshield washer works and there is washing fluid in the washer reservoir. Use windshield washer antifreeze to keep the washer liquid from freezing.

If you can't see well enough while driving (for example, if your wipers fail), find a safe place to stop safely. Fix the problem before you go on.

Don't take chances with your tires. Make sure they have enough tread. The drive tires must provide traction to push the rig over wet pavement and through snow. The steering tires must have traction to steer the vehicle. Enough tread is especially important in winter conditions. You should have at least 4/32-inch tread depth in every major

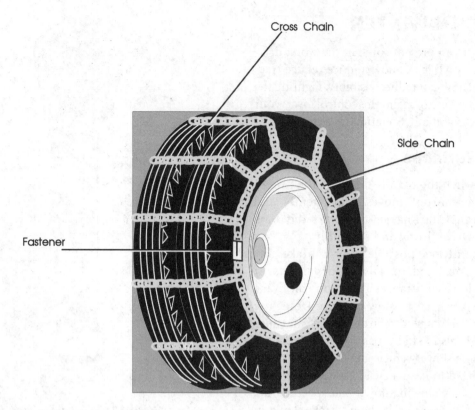

Cross Chain

Side Chain

Fastener

Figure 7-17. Snow chains mounted on a dual tire.

groove on front wheels and at least 2/32-inch tread depth on other wheels. More would be better. Use a gauge to determine if you have enough tread for safe driving.

You may find yourself in conditions where you can't drive without snow chains, even to get to a place of safety. Carry the right number of chains and extra cross links. Make sure they will fit your drive tires. Check the chains for broken hooks, worn or broken cross links, and bent or broken side chains. Learn how to put the chains on before you need to do it in snow and ice.

Make sure the lights and reflectors are clean. Lights and reflectors are especially important during bad weather. Check from time to time during bad weather to make sure they are clean and working right. Remove any ice, snow, and dirt from the lights, reflector, windshield, windows, and mirrors before starting. You may need to use a windshield scraper, snow brush, or windshield defroster.

Remove all ice and snow from handholds, steps, and deck plates that you use to enter the cab or to move about the vehicle. This will reduce the chance of slipping.

Remove ice from the radiator shutters. Make sure the winterfront is not closed too tightly. If the

shutters freeze shut or the winterfront is closed too much, the engine may overheat and stop.

Exhaust system leaks are especially dangerous when you have your cab tightly buttoned up against winter's cold. Loose connections could allow poisonous carbon monoxide to leak into your vehicle. Carbon monoxide gas will make you sleepy. In large enough amounts it can kill you. Check the exhaust system closely for loose parts and for sounds and signs of leaks.

DRIVING IN THE RAIN

Remember that roads are more slippery at the start of a rainstorm than during it. That's because the rain mixes with oil and grease on the road. If the rain continues, it will wash the oil away. So be most careful when it first starts raining. When you drive in heavy rain or deep standing water, your brakes will get wet. Water in the brakes can cause the brakes to be weak, to apply unevenly, or to grab. This can cause a lack of braking power. The wheels can lock up and/or pull to one side. Water in the brakes can also cause a "jackknife" when pulling a trailer. Avoid driving through deep puddles or flowing water.

PASS BILLBOARD
How to Mount Snow Chains

Check the condition of your chains. Look for broken links and fasteners.

Straighten out the chain. Drape the chain over the tire, with the open ends of the cross-chain hooks facing away from the tire. Fasteners can be on the outside of the tire.

Tuck the first cross-chain under the front of the tire. Drive the vehicle forward until the fasteners are at hub-level. Be careful not to drive over the fasteners.

Get out, straighten, and center the chain. There should be an equal amount on either side of the tire. Lift the ends of the side chains and decide which links will be hooked into the fasteners.

If you are chaining duals, fasten the two center chains first. Then do the inside chains, followed by the outside chains.

On singles, fasten the inner chain first. Then fasten the outer chain.

To remove snow chains, unhook the cross chains from the side chain fasteners on the outside. If you have duals, unhook the inside cross chains next. Spread the chains on the ground. Drive the vehicle off the chains. Be careful not to run over the fasteners. Last, remove the chains from the tires.

If you must drive through puddles, reduce your speed. Shift into a low gear. Gently put on the brakes. This presses the linings against brake drums or discs and keeps mud, silt, sand, and water from getting in. Increase the engine rpm and cross the water while you keep light pressure on the brakes.

When you come out of the water, maintain light pressure on the brakes for a short distance. This will heat them up and dry them out.

Make a test stop when safe to do so. Check behind to make sure no one is following. Then apply the brakes to be sure they work right. If not, dry them out some more. Do not apply too much brake pressure and accelerate at the same time or you can overheat brake drums and linings.

DRIVING IN HOT WEATHER

Hot weather will tax your cooling system, engine, and tires. You can't lower the temperature outside. You can keep from making it worse, though. Reduce your speed, enough to prevent overheating. High speeds create more heat for tires and the engine. In desert conditions the heat may build up to the point where it is dangerous. The heat will increase chances of tire failure, or even fire, and engine failure.

Slippery Surfaces

Watch for bleeding tar. Tar in the road pavement often rises to the surface in very hot weather. Spots where tar "bleeds" to the surface are very slippery.

Pre-trip Inspection for Hot Weather

If you'll be running in hot weather, do a normal pre-trip inspection. Pay special attention to the tires, cooling system, and engine.

Check the tire mounting and air pressure. Inspect the tires every two hours or every 100 miles when driving in very hot weather. Air pressure increases with temperature. Don't let air out or the pressure will be too low when the tires cool off.

Tires can get hot enough to burst into flame. If a tire is too hot to touch, remain stopped until the tire cools off. Otherwise the tire may blow out or catch fire. You can cool down a tire with water, of course, or even sand or dirt. If you have neither, you'll just have to give the tire time to cool by itself.

Pay special attention to recapped or retreaded tires. Under high temperatures the tread may separate from the body of the tire.

The engine oil helps keep the engine cool, as well as lubricating it. Make sure there is enough engine oil. If you have an oil temperature gauge, make sure the temperature is within the proper range while you are driving.

Before starting out, make sure the engine cooling system has enough water and antifreeze. Your owner's manual should have some information for you on this. (Antifreeze helps the engine under hot conditions as well as cold conditions.)

While you're driving, check the water temperature or coolant temperature gauge from time to time. Make sure that it remains in the normal range. If the gauge goes above the highest safe temperature, there may be something wrong that could lead to engine failure and possibly fire. Stop driving as soon as safely possible and try to find out what is wrong.

The coolant in your vehicle is held under pressure. Never remove the radiator cap or any part of the pressurized system until the system has cooled. Taking the cap off releases the pressure. Steam and boiling water can spray out and burn you. If you can touch the radiator cap with your bare hand, it is probably cool enough to open.

Some vehicles have sight glasses or see-through coolant overflow containers or coolant recovery containers. These allow you to check the coolant level even while the engine is hot. If the container is not part of the pressurized system, you can safely remove the cap and add coolant even when the engine is at operating temperature.

If you have to add coolant to a system without a recovery tank or overflow tank, follow these steps:

- shut the engine off
- wait until the engine has cooled
- protect your hands (use gloves or a thick cloth)
- turn the radiator cap slowly to the first stop (this releases the pressure seal)
- step back while pressure is released from cooling system
- when all pressure has been released, press down on the cap and turn it further to remove it
- visually check the coolant level, and add more coolant if the level is below "Add" or "Low"
- replace the cap and turn it all the way to the closed position

Learn how to check V-belt tightness on your vehicle by pressing on the belts. Your owner's manual will tell you how slack they should be. Belts that are too loose will not turn the water pump or fan properly. If the fan doesn't cool the water, it will overheat, and so will the engine. Also, check belts for cracking or other signs of wear. Heat just puts more stress on them. That's why they always seem to break when you need them the most.

Make sure coolant hoses are in good condition. Heat can turn a little break into a big one. If a hose breaks while you're driving, it can lead to engine failure and even fire.

MOUNTAIN DRIVING

Gravity plays a major role in mountain driving. If you have a heavy load, you will have to use lower gears and go slower to climb hills. When you head down, gravity will tend to speed you up. You must go slowly enough that your brakes can hold you back without getting too hot. When brakes become too hot, they may start to "fade." This means that you have to apply them harder and harder to get the same stopping power. If you continue to use the brakes hard, they can continue to fade until you can't slow or stop at all.

Pre-trip for Mountain Driving

If you'll be running in the mountains, do a normal pre-trip inspection. Inspect your braking system thoroughly. If your vehicle has an engine retarder, make sure it's working.

When you're driving, highway signs will give you plenty of warning that a long downgrade is coming up. You'll usually be given a chance to "pull out" at the top of the hill. This is an area to the side of the road where you can safely leave the vehicle and check the brakes. Do it. Don't just stomp on the brake pedal and decide that the brakes work. Do a complete brake check, as you would during a pre-trip.

Your vehicle may have manually adjusted brakes. Check the travel of the slack adjusters. If the distance is too great, you'll lose braking power. They should be fixed before you continue. Slack adjustment is discussed in greater detail in Chapter 10.

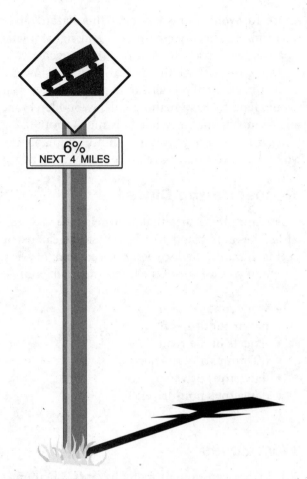

Figure 7-18. Watch for highway signs that warn you of downgrades.

Downhill Tactics

These dangers can be avoided by going downhill slowly. Use the gear (and retarder, if you have it) that allows you to descend safely while using the brakes very little or not at all.

Gear Selection for Downhill Driving

No matter what the size of your vehicle, going down long, steep grades can cause your brakes to fail if you go too fast. Using lower gears will help you keep from going too fast. Lower gears allow engine compression and friction to help slow the vehicle. This is true whether you have an automatic transmission or a manual transmission.

If you do have a large vehicle with a manual transmission, don't wait until you have started down the hill to shift down. You will probably get stuck in neutral. You won't be able to shift into another gear. In neutral, you will have no engine braking at all. You'll be coasting, which is illegal and dangerous. Be in the right gear before starting down the hill.

With older vehicles, a rule for choosing gears was to use the same gear going down a hill that you would need to climb the hill. However, new vehicles have low friction parts and streamlined shapes for fuel economy. They may also have more powerful engines. This means that they can go up hills in higher gears. There is less friction and air drag to hold them back going down hills. For that reason, you may have to use lower gears going down a hill than would be required to go up the hill.

Find out what is right for your vehicle. If you don't know, try this. Stop at the top of the hill. As you restart, upshift to a gear that will set the vehicle's speed at 20 mph while keeping the rpm just under the rated engine speed. Then stay in that gear until you reach the bottom of the hill.

Braking on Downgrades

When going downhill, brakes will always heat up. As you know, the reason brakes work is because the brake shoes or pads rub against the brake drum or disks to slow the vehicle. This creates heat. Brakes are designed to take a lot of heat. However, brakes will eventually fail from too much heat. Excessive heat results from trying to slow from too high a speed too many times or too quickly. Brakes will fade (have less stopping power) when they get very hot. They can get to the point where they will no longer slow the vehicle.

The right way to use your brakes for long downhill grades is to go slowly and use snub braking to stay at a safe speed.

Don't fan the brakes. You may think that applying and releasing the brakes repeatedly will allow them to cool enough so they don't become overheated. Tests have proven this is not true. Brake drums cool very slowly. The amount of cooling between applications is not enough to prevent overheating. Fanning the brakes requires heavier brake pressures than steady application does.

So, select the right gear, go slowly enough, and use snub braking to maintain a safe speed.

Loss of Braking Power

What if, in spite of your best efforts, you lose your brakes? Knowing what to do will not only help you earn your CDL. It will save your life.

Escape ramps have been built on many steep mountain grades. Escape ramps are made to stop runaway vehicles safely without injuring drivers and passengers. Escape ramps lead your vehicle into a long bed of loose soft material, such as pea gravel. The gravel offers enough extra friction to slow a runaway vehicle.

Escape ramps are sometimes built on an upgrade. Your vehicle then has to overcome gravity as well as friction. The two forces together are enough to stop your vehicle.

Know where the escape ramps are on your route. Maps for professional drivers pinpoint escape ramp locations. While you're driving, note the highway signs that tell you a ramp is coming up.

If there is no escape ramp available, take the least hazardous escape route you can. Head for an open field, or a side road that flattens out or turns uphill. Make the move as soon as you know your brakes don't work. The longer you wait, the faster the vehicle will go and the harder it will be to stop.

ROAD HAZARDS

You can't avoid road hazards if you don't see them in the first place. What is a hazard? A hazard is anything or anyone on the road that could be a danger.

Picture this. A car in front of you is headed toward the freeway exit. You think he's going to leave the freeway. Instead, his brake lights come on and he begins braking hard. This could mean that the driver is uncertain about taking the off-ramp. He might suddenly return to the highway. He may come to a complete stop! This car is a hazard. If the driver of the car cuts in front of you, or stops, it is no longer just a hazard. It is an emergency. If you increased your speed, thinking he was going to leave the freeway, you may be in big trouble.

Be Prepared

Seeing hazards before they become accidents helps you prepare. You will have more time to act. In the example above, you should have slowed, not gone faster. Then you could change lanes or slow even more to avoid a crash. Seeing this hazard gives you time to check your mirrors and signal a lane change. Being prepared reduces the danger.

If you did not see the hazard until the slow car pulled back on the highway in front of you, you would have to do something quickly. Sudden braking or an abrupt lane change is much more likely to lead to a crash. If you have not been managing your space, you may have no room to move at all.

Recognize the Clues

There are often clues that will help you see hazards. The more you drive, the better you can get at seeing hazards. Reduce your speed and be very careful if you see any of the following road hazards:

- work zones
- pavement drop off
- objects in the road
- off-ramps and on-ramps
- dangerous drivers
- dangerous non-drivers

Work Zones

Road work can present many hazards. There may be narrower lanes, sharp turns, or uneven surfaces. Other drivers stop paying attention to their driving so they can watch the work. Workers and construction vehicles may get in the way. Drive slowly and carefully near work zones. Use your four-way flashers or brake lights to warn drivers behind you that you are slowing.

Drop Off

Sometimes the pavement drops off sharply near the edge of the road. If you are driving too near the edge, your vehicle can tilt toward the side of the road. It can hit signs, tree limbs, and other objects along the side of the road. Also, it can be hard to steer as you cross the drop off, go off the road, or come back on.

Objects in the Road

Things that have fallen on the road can be hazards. They can damage your tires and wheel rims. They can damage electrical and brake lines. They can get caught between dual tires and cause severe damage.

Some obstacles that appear to be harmless can be very dangerous. For example, cardboard boxes may be empty. They could also contain some solid or heavy material that could cause damage. The same is true of paper and cloth sacks.

Be alert for objects of all sorts. If you see them early enough, you can avoid them without making sudden, unsafe moves.

Off-ramps and On-ramps

Freeway and turnpike exits can be particularly dangerous for commercial vehicles. Off-ramps and on-ramps often have speed limit signs posted. Remember, these speeds may be safe for automobiles, but may not be safe for larger vehicles or heavily loaded vehicles. The safest speed is not the posted one. It's the one that gives you the most control over your vehicle.

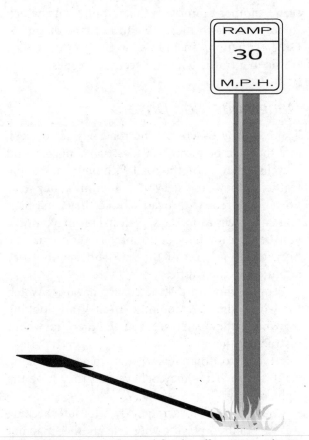

Figure 7-19. The speed posted for the off-ramps may be too high for your CMV.

Exits that go downhill and turn at the same time can be especially dangerous. The downgrade makes it hard to reduce speed. Braking and turning at the same time can be a dangerous practice. Make sure you are going slowly enough before you get on the curved part of an off-ramp or on-ramp.

As you learned earlier, off-ramps can be dangerous for other reasons. Drivers ahead of you may slow in order to exit. They may decide suddenly to move for the exit. Or, they may start to exit and change their minds.

On-ramps mean drivers entering the highway. They may not always be very skilled at doing this. They'll probably assume you can slow to make room for them. If you can't, a crash can result. If you're in the merge lane, manage your space and speed very carefully.

Dangerous Drivers

Other drivers can present still more hazards. Learn how to read the clues.

Confused drivers often change direction suddenly or stop without warning. Confusion is common near freeway or turnpike interchanges and major intersections. Stopping in the middle of a block, changing lanes for no apparent reason, backup lights suddenly going on are all clues to confusion. Hesitation is another clue. So is driving very slowly, using brakes often, or stopping in the middle of an intersection. You may also see drivers who are looking at street signs, maps, and house numbers. These drivers may not be paying attention to you.

Tourists unfamiliar with the area can be very hazardous. Clues to tourists include car-top luggage and out-of-state license plates.

Slow drivers are hazards. Seeing slow-moving vehicles early can prevent a crash. Some vehicles (mopeds, farm machinery, construction machinery, tractors, etc.) by their nature are slow. Some of these will have the "slow moving vehicle" symbol to warn you. This is a red triangle with an orange center. Watch for it. Proceed with caution around slow moving vehicles.

Drivers signaling a turn may be a hazard. They may slow more than expected or stop. If they are making a tight turn into an alley or driveway, they may go very slowly. If they are blocked by pedestrians or other vehicles, they may have to stop on the roadway. Vehicles turning left may have to stop for oncoming vehicles.

Drivers in a hurry may feel your CMV is keeping them from getting where they want to go on time. Such drivers may pass you without a safe gap in the oncoming traffic, cutting too close in front of you. Drivers entering the road may pull in front of you in order to avoid being stuck behind you. Be careful around drivers who are going faster than the flow of traffic and making many lane changes.

Drivers who are sleepy, have had too much to drink, are on drugs, or who are ill are hazards. Be suspicious of drivers who are:

- weaving across the road or drifting from side-to-side
- leaving the road (dropping right wheels onto the shoulder, or bumping across a curb in a turn)
- stopping at the wrong time (stopping at a green light or waiting for too long at a stop)
- driving with the window open in cold weather
- speeding or slowing suddenly, driving too fast or too slow

As you learned earlier, you should be especially alert for drunk and sleepy drivers late at night.

Learn to read the body movements of other drivers. Drivers look in the direction they are going to turn. You may sometimes get a clue from a driver's head and body movements that this driver may be going to make a turn even though the turn signals aren't on. Drivers making over-the-shoulder checks may be going to change lanes. These clues are most easily seen in motorcyclists and bicyclists. Watch other road users and try to tell whether they might do something hazardous.

Be alert for drivers whose vision is blocked. Vans, loaded station wagons, and cars with the rear window blocked are examples. Rental trucks should be watched carefully. The driver's vision to the sides and rear of the truck is limited. A rental truck's driver is likely unfamiliar with the vehicle. In winter, vehicles with frosted, ice-covered, or snow-covered windows are hazards.

Vehicles may be partly hidden by blind intersections or alleys. If you only can see the rear or front end of a vehicle but not the driver, then he or she can't see you. Be alert because the driver may back out or enter into your lane. Always be prepared to stop.

Delivery trucks can present a hazard. The driver's vision is often blocked by packages or vehicle doors. Drivers of step vans, postal vehicles, and local delivery vehicles often are in a hurry. They may suddenly step out of their vehicle or drive their vehicle into the traffic lane.

Parked vehicles can be hazards when the people start to get out. Or, they may suddenly start up and drive into your way. Watch for movement inside the vehicle or movement of the vehicle itself that shows people are inside. Brake or backup lights and exhaust are clues that a driver is about to move.

Be careful of a stopped bus. Passengers may cross in front of or behind the bus. Often, they can't see you.

Some situations are just accidents waiting to happen. Anywhere vehicles meet there is a chance of danger. Be cautious where lanes of traffic merge. This could be at an on-ramp, as you've seen. Another example is the end of a lane, which forces vehicles to move to another lane of traffic. Other situations include slow moving or stalled traffic in a traffic lane and accident scenes.

Dangerous Non-Drivers

There can be others on the road besides drivers, and they can be hazards. Pedestrians, joggers, and bicyclists may be on the road with their back to the traffic, so they can't see you. Sometimes, they wear portable stereos with head sets, so they can't hear you either. On rainy days, pedestrians may not see you because of hats or umbrellas. They may be hurrying to get out of the rain and may not pay attention to the traffic.

People who are distracted are hazards. Watch for where they are looking. If they are looking elsewhere, they can't see you. Be alert to what's distracting them.

Be alert to others even when they are looking right at you. They may believe that they have the right-of-way.

Children tend to act quickly without checking traffic. Children playing with one another may not look for traffic and are a serious hazard. Look for clues that children are in the area. For example, someone selling ice cream is a hazard clue. Children may be nearby and may not see you.

Drivers changing a tire or fixing an engine often do not pay attention to the danger that roadway traffic is to them. They are often careless. Jacked-up wheels or raised hoods are hazard clues.

Accidents are particularly hazardous. People involved in the accident may not look for traffic. Passing drivers tend to look at the accident. People often run across the road without looking. Vehicles may slow or stop suddenly.

People in and around shopping areas are often not watching traffic because they are looking for stores or looking into store windows.

You should always be looking for hazards. They may turn into emergencies. Look for the hazards in order to have time to plan a way out of any emergency. When you see a hazard, think about the emergencies that could develop. Figure out what you would do. Always be prepared to take action based on your plans. That makes you a prepared, defensive driver. You will improve not only your own safety but the safety of all road users.

EMERGENCY MANEUVERS

In spite of your best efforts, you may be faced with an emergency. Traffic emergencies occur when two vehicles are about to collide. Vehicle emergencies occur when tires, brakes, or other critical parts fail. Your chances of avoiding a crash depend upon how well you take action.

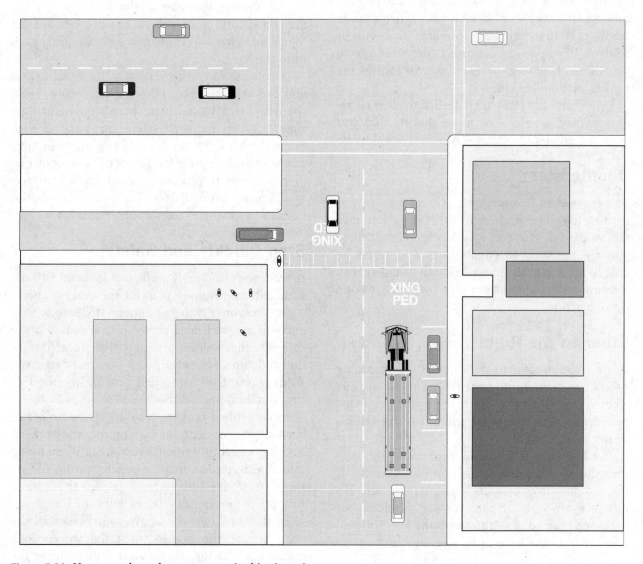

Figure 7-20. How many hazards can you spot in this picture?

Stopping May Not Be Safe

Stopping is not always the safest thing to do in an emergency. When you don't have enough room to stop, you may have to steer away from what's ahead. You can almost always turn to miss an obstacle more quickly than you can stop. Keep in mind, though, that top-heavy vehicles may flip over.

Turn Quickly and Safely

To turn quickly, you must have a firm grip on the steering wheel with both hands. The best way to have both hands on the wheel if there is an emergency is to keep them there all the time. Use the grip described at the beginning of this chapter.

A quick turn can be made safely, if it's done the right way. Don't apply the brake while you are turning. It's very easy to lock your wheels while turning. If that happens, you may skid out of control. If you have been managing your speed, you should be going slowly enough to control the vehicle without braking.

Don't turn any more than needed to clear whatever is in your way. The more sharply that you turn, the greater the chances of a skid or rollover.

Countersteer

Be prepared to "countersteer." Countersteering is turning the wheel back in the other direction once you've passed whatever was in your path. Unless you are prepared to countersteer, you won't be able to do it quickly enough. Think of emergency steering and countersteering as two parts of one driving action.

Steer to the Right

In most cases, steer to the right. If an oncoming driver has drifted into your lane, a move to your right is best. If that driver realizes what has happened, the natural response will be to return to his or her own lane.

If something is blocking your path, the best direction to steer will depend on the situation. If the shoulder is clear, going right may be best. No one is likely to be driving on the shoulder, but someone may be passing you on the left. You will know if you have been using your mirrors.

If you are blocked on both sides, a move to the right may be best. At least you won't force anyone into an opposing traffic lane and a possible head-on collision.

If you have been using your mirrors, you'll know which lane is empty and can be safely used.

Leaving the Road

In some emergencies, you may have to drive off the road. It may be better than colliding with another vehicle. Most shoulders are strong enough to support the weight of a large vehicle.

If you do leave the road, avoid using the brakes until your speed has dropped to about 20 mph. Then brake very gently to prevent skidding on a loose surface. Keep one set of wheels on the pavement if possible. This helps to maintain control.

Stay on the shoulder. If the shoulder is clear, stay on it until your vehicle has come to a stop. Signal and check your mirrors before pulling back onto the road.

If you are forced to return to the road before you can stop, use the following procedure. Hold the wheel tightly and turn sharply enough to get right back on the road safely. Don't try to edge gradually back on the road. If you do, your tires might grab unexpectedly. You could lose control.

When both front tires are on the paved surface, countersteer immediately. The two turns should be made as a single "steer-countersteer" move.

Stop Quickly and Safely

If somebody suddenly pulls out in front of you, your natural response is to hit the brakes. This is a good response if there's enough distance to stop and you use the brakes correctly. You should brake in a way that will keep your vehicle in a straight line and allow you to turn if it becomes necessary. You can use the "controlled braking" method or the "stab braking" method.

In controlled braking, you apply the brakes as hard as you can without locking the wheels. Keep steering wheel movements very small while doing this. If you need to make a larger steering adjustment or if the wheels lock, release the brakes. Reapply the brakes as soon as you can.

In stab braking, you apply your brakes all the way. Release the brakes when the wheels lock up. As soon as the wheels start rolling, apply the brakes fully again. It can take up to one second for the wheels to start rolling after you release the brakes. If you reapply the brakes before the wheels start rolling, the vehicle won't straighten out.

Don't jam on the brakes! Emergency braking does not mean pushing down on the brake pedal as hard as you can. That will only cause the wheels to lock up and skid. If the wheels are skidding, you cannot control the vehicle.

Brake Failure

You've learned about the loss of braking power on downgrades. There, a runaway ramp is your escape route. What if your brakes fail and there is no escape ramp?

First, remember that brakes kept in good condition rarely fail. That's why brake inspection and maintenance is so important.

Most hydraulic brake failures occur for one of two reasons:

- loss of hydraulic pressure
- brake fade on long hills

When the system won't build up pressure, the brake pedal will feel spongy or go to the floor. If this happens, downshift. Putting the vehicle into a lower gear will help to slow the vehicle. You can also pump the brakes. Sometimes pumping the brake pedal will create enough hydraulic pressure to stop the vehicle.

The parking or emergency brake is separate from the hydraulic brake system. Therefore, it can be used to slow the vehicle. However, be sure to press the release button or pull the release lever at the same time you use the emergency brake. That way you can adjust the brake pressure and keep the wheels from locking up.

Find an escape route. While slowing the vehicle, look for an escape route. This could be an open field or side street as well as a designated escape ramp. Turning uphill is a good way to slow and stop the vehicle. Make sure the vehicle does not start rolling backward after you stop. Put it in low gear and apply the parking brake. You may have to roll back into some obstacle that will stop the vehicle.

When your brakes fail on a downgrade, use the tactics described in the "Mountain Driving" section of this chapter.

Tire Failure

There are four important things that safe drivers do to handle a tire failure safely:

- Be aware that a tire has failed.
- Hold the steering wheel firmly.
- Stay off the brake.
- After stopping, check all the tires.

The sooner you realize you have a tire failure, the more time you will have to react. Having just a few seconds to remember what it is you're supposed to do can help you. Recognize the major signs of tire failure.

Sound is your first clue. The loud "bang" of a blowout is an easily recognized sign. Because it can take a few seconds for your vehicle to react to the tire failure, you might think the sound is coming from some other vehicle. To be safe, any time you hear a tire blow, assume it's yours.

If the vehicle thumps or vibrates heavily, it may be a sign that one of the tires has gone flat. With a rear tire, that may be the only sign you get.

If the steering feels "heavy," it is probably a sign that one of the front tires has failed. Sometimes, failure of a rear tire will cause the vehicle to slide back and forth or "fishtail." However, dual rear tires usually prevent this. Any of these signs is a warning of possible tire failure. To maintain control of the vehicle, hold the steering wheel firmly. If a front tire fails and you're holding the wheel loosely, the force can twist the steering wheel out of your hand. You must have a firm grip on the steering wheel with both hands at all times.

Stay off the brake. It's natural to want to brake in an emergency. However, braking when a tire has failed could cause you to lose control of the vehicle. Unless you're about to run into something, stay off the brake until the vehicle has slowed down. Then brake very gently, pull off the road, and stop. After you've come to a stop, get out and check all the tires. Do this even if the vehicle seems to be handling all right. If one of your dual tires goes, the only way you may know it is by getting out and looking at it.

SKID CONTROL AND RECOVERY

Recovering from a skid can really challenge your emergency maneuvering skills.

A skid happens whenever the tires lose their grip on the road. A skid can result from over-braking. That is braking too hard and locking up the wheels. Skids also can occur if you use the engine retarder when the road is slippery.

Over-steering can lead to skidding. Don't turn the wheels more sharply than the vehicle can turn.

Supplying too much power to the drive wheels can cause them to spin. That can lead to a skid.

Driving too fast is a common cause of skidding. Most serious skids result from driving too fast for road conditions. If you manage your speed, you don't have to accelerate suddenly, brake hard, or over-steer to avoid a collision.

Drive-Wheel Skids

By far the most common skid is one in which the rear wheels lose traction. This can result from over-braking or over-acceleration.

Skids caused by acceleration usually happen on ice or snow. They can be easily stopped by taking your foot off the accelerator. If it is very slippery, push the clutch in. Otherwise the engine can keep the wheels from rolling freely and regaining traction.

Rear-wheel braking skids occur when the rear-drive wheels lock. Because locked wheels have less traction than rolling wheels, the rear wheels usually slide sideways in an attempt to "catch up" with the front wheels. In a bus or straight truck, the vehicle will slide sideways in a "spin out." When you're towing a trailer, a drive-wheel skid can let the trailer push the towing vehicle sideways, causing a sudden jackknife.

To correct a drive-wheel braking skid, first let off the brake. This will let the rear wheels roll again and keep the rear wheels from sliding any further. If you are on ice, push in the clutch to let the wheels turn freely.

Turn quickly. When a vehicle begins to slide sideways, quickly steer in the direction you want the vehicle to go—down the road. You must turn the wheel quickly.

As a vehicle turns back on course, it has a tendency to keep right on turning. Unless you turn the steering wheel quickly the other way, you may find yourself skidding in the opposite direction. Your quick turn must be followed by countersteering.

Learning to stay off the brake, turn the steering wheel quickly, push in the clutch, and countersteer in a skid takes a lot of practice. You don't often get to "practice" skid recovery. Usually, when you're trying to get out of a skid, it's the real thing. About the only place to get this practice is on a large driving range or "skid pad."

Front-Wheel Skids

Most front-wheel skids are caused by driving too fast for conditions. They can also result from worn front tires. If your cargo is loaded so that not enough weight is on the front axle, you could skid.

In a front-wheel skid, the front end tends to go in a straight line no matter how much you turn the steering wheel. On a very slippery surface, you may not be able to steer around a curve or turn.

When a front-wheel skid occurs, the only way to stop the skid is to let the vehicle slow. Stop turning or braking so hard. Reduce your speed as quickly as possible without skidding.

COUPLING AND UNCOUPLING

We're almost at the end of the skills CDL drivers are required to have. As you learned in Chapter 4, you must have some knowledge about hazardous materials. Even if you don't plan to haul them, you must be able to recognize a hazardous materials shipment when you see one. The basic information about hazardous materials was included in the FMCSR summary in Chapter 5 and is presented in FMCSR Part 397. Make sure you have all the information under your belt.

CDL drivers must also show they know how to inspect their vehicle and why it's important to do so. Vehicle inspections are covered in detail in Chapter 9.

So there's one last area of information to cover before we leave this chapter. That's the coupling and uncoupling of tractors and trailers. If you plan to drive a straight truck and never to pull a trailer, you may not need these skills. You will have to take your CDL Skills Test in a Group B or

C truck, without a trailer. If you plan to drive a Group A truck, or pull a trailer, you will be tested on your knowledge of coupling and uncoupling a tractor and semitrailer. You'll have to know how to inspect this vehicle combination.

Coupling and uncoupling, and inspecting vehicle combinations, is covered in Chapter 11. Don't skip that chapter if you are going for a Group A CDL.

You've covered a lot of ground in this chapter. The "list" of vital skills is long and detailed. You may already have many of these skills. Or, you may have found quite a few new ones to master. To be a safe, skilled, licensed commercial driver, you must be in command of them all.

You may need several sessions with this chapter and quite a bit of practice with the tactics and techniques described here. Keep at it until safe vehicle control comes naturally to you.

PASS POST-TRIP

Instructions: For each true/false test item, read the statement. Decide whether the statement is true or false. If it is true, circle the letter A. If it is false, circle the letter B. For each multiple-choice test item, choose the answer choice—A, B, C, or D—that correctly completes the statement or answers the question. There is only one correct answer. Use the answer sheet provided on page 175.

1. Rough acceleration is not only likely to damage the vehicle, it can cause you to lose control.
   A. True
   B. False

2. How far ahead of the vehicle should a driver look while driving?
   A. nine to 12 seconds
   B. 18 to 21 seconds
   C. 12 to 15 seconds
   D. 12 to 15 feet

3. When driving 55 mph on dry pavement, allow _____ to bring the vehicle to a stop.
   A. the length of a football field
   B. twice the length of the vehicle
   C. half the length of a football field
   D. the length of the vehicle

4. Hydroplaning _____.
   A. only occurs when there is a lot of water
   B. only occurs at speeds above 55 mph
   C. is more likely if tire pressure is low
   D. is more likely if the road is banked

5. You're driving a 40-foot vehicle and keeping with the flow of traffic at 50 mph. You should keep _____ seconds of distance between you and the vehicle ahead.
   A. four
   B. five
   C. six
   D. seven

6. When there are two turning lanes, you should start a left-hand turn from the right or outer lane.
   A. True
   B. False

7. When you are dead tired, the best way to refresh yourself and get ready to drive again is to _____ .
   A. have a drink
   B. go for a brisk walk
   C. drink some coffee
   D. get some sleep

8. Roads are most slippery and dangerous during _____ a rainstorm.
   A. the first 15 minutes of
   B. the last 15 minutes of
   C. the heaviest part of
   D. a pause in

9. "Fanning the brakes" is a recommended downhill braking tactic.
   A. True
   B. False

10. You may hear a drive tire blow out before you feel it.
    A. True
    B. False

CHAPTER 7 ANSWER SHEET

Test Item	Answer Choices				Self - Scoring Correct	Incorrect
1.	A	B	C	D	☐	☐
2.	A	B	C	D	☐	☐
3.	A	B	C	D	☐	☐
4.	A	B	C	D	☐	☐
5.	A	B	C	D	☐	☐
6.	A	B	C	D	☐	☐
7.	A	B	C	D	☐	☐
8.	A	B	C	D	☐	☐
9.	A	B	C	D	☐	☐
10.	A	B	C	D	☐	☐

Cut Here

Chapter 8

Loading, Securing, and Hauling Loads

In this chapter, you will learn about:

- how natural forces affect loaded vehicles
- size and distribution regulations
- how to load cargo properly
- how to secure and protect loads
- how to inspect loads
- the challenge of driving a loaded vehicle

To complete this chapter you will need:

- a dictionary
- a pencil or pen
- blank paper, a notebook, or an electronic notepad
- colored pencils, pens, markers, or highlighters
- a CDL preparation manual from your state Department of Motor Vehicles, if one is offered
- the Federal Motor Carrier Safety Regulations pocketbook (or access to U.S. Department of Transportation regulations, Parts 383, 391, 392, and 393 of Subchapter B, Chapter 3, Title 49, Code of Federal Regulations)
- the owner's manual for your vehicle, if available

PASS PRE-TRIP

Instructions: Read the statements. Decide whether each statement is true or false. If it is true, circle the letter A. If it is false, circle the letter B.

1. Axle weight is how much an axle weighs.
   A. True
   B. False

2. For stability, the center of gravity should always be in the middle of the cargo compartment.
   A. True
   B. False

3. If you always operate at the legal size limits set by states, you can be certain you will be operating safely.
   A. True
   B. False

Chapters 6 and 7 contained a lot of information about vehicle safety control systems and safe driving. If you don't handle it properly, your CMV can be dangerous to you and to others on the roadway and alongside it. So those chapters, and the skills described in them, are very important.

In a way, though, Chapter 8 is what the CMV driver's job is all about. After all, why drive a CMV in the first place? It's to haul cargo. An empty truck isn't earning anyone any money. Since carriers try to keep deadhead miles—miles without cargo—to a minimum, you will almost always be hauling something.

Like most everything else in CMV driving, there's a right way and a wrong way to transport cargo. As you might expect, the wrong way usually means damage, and sometimes injury and death. Cargo handling is so important that several parts of the FMCSR refer to it.

Subpart G of FMCSR Part 383 lists the principles and procedure for the proper handling of cargo among the specific knowledge elements required to get a CDL.

FMCSR Part 391 states that CMV drivers must:

- be able to tell if cargo is distributed properly
- be able to secure cargo

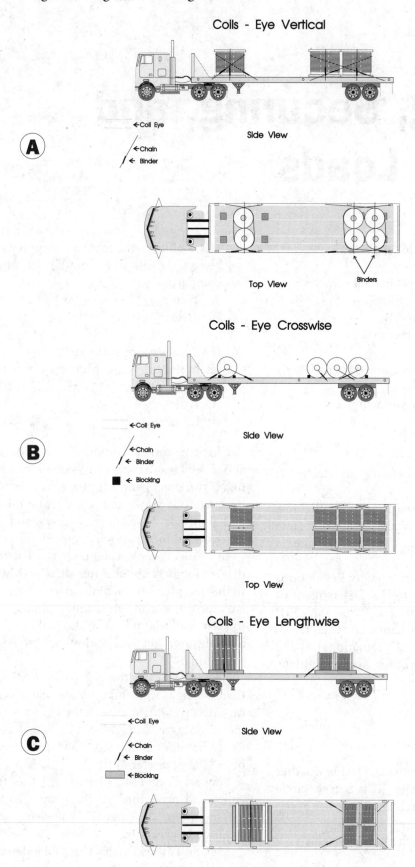

Figure 8-1. Metal coils chained on a flatbed (A) with eyes vertical, (B) with eyes crosswise, and (C) with eyes lengthwise.

FMCSR Part 393 gives some rules about transporting cargo. It describes how to equip trucks and trailers to keep cargo from shifting or falling. This part also describes how to secure coils of metal on a flatbed and how to tie down other metal cargo.

This part of the FMCSR also covers hauling intermodal cargo containers. It states specifications for securement devices.

There is far more to know about hauling cargo than what's in those few pages. This chapter will give you the information you need to comply with the regulations. This will help you earn and keep your CDL.

This chapter will tell you what you need to know to haul most types of cargo in a straight truck or with a single trailer. Note that in some driving jobs, your cargo is people. To transport people, you will need a Passenger Endorsement.

In other driving jobs, your cargo is bulk liquids. You would need a Tank Vehicle Endorsement for this. Hauling bulk liquids is covered in Chapter 12.

The last special type of cargo is hazardous materials. As you know, you'll need a Hazardous Materials Endorsement to haul placarded loads. Read Chapter 13 to prepare for the Hazardous Materials Endorsement tests.

To haul cargo using double or triple trailers, you'll need a Doubles/Triples Endorsement. Be sure to read Chapter 11 for more information on that subject.

FORCES OF NATURE

Even if the only cargo that you have ever hauled was the family's groceries home from the market, you may have noticed something about cargo. It doesn't always stay where you put it. It doesn't matter what the cargo is. It doesn't matter what type of vehicle you put it in. Be it a car or a commercial truck or trailer, cargo shifts. Sometimes nothing bad results. Sometimes the cargo is damaged. If the cargo shifts badly enough, it can overturn your vehicle. It can spill out on the road and cause a major accident.

You may have also noticed that cargo affects the way your vehicle handles. Sometimes the vehicle pulls to one side. It may feel top-heavy. It may feel like it's going to tip. It may be harder to get it rolling. You may have a hard time stopping the vehicle.

This is because, as solid and stable as it may seem, cargo is subject to the forces of nature. These forces are:

- friction
- gravity
- inertia

Friction

Simply put, friction is one surface rubbing against another. You've run into this term before. We mentioned friction in terms of how brakes work. Friction between your tires and the road is what gives you traction and the ability to move. Friction also comes into play in cargo loading and securement.

Gravity

Gravity has to do with the natural attraction between two masses. Like friction, gravity helps to hold your vehicle on the road. Too much gravity will keep your vehicle from moving. Gravity helps keep your cargo securely in place. It can also move your cargo out of place. The pull of gravity on your vehicle and the cargo in it can make the vehicle hard to handle.

Inertia

Inertia is the tendency of an object that is in motion to stay in motion. Objects that are standing still will tend to stay that way. For example, if your truck is moving, it will tend to continue to move. It will take a greater force, like a lot of friction, to overcome the inertia and make the vehicle stop. If it is standing still, the vehicle will tend to remain standing still. You will have to overcome inertia to get the truck to move.

Inertia affects cargo the same way. When you place your cargo in your truck, it's standing still. It will stay that way until a force makes it move. Once it's in motion, it takes another force to get it to stop.

Cargo and the Forces of Nature

The three forces often act on your cargo together.

When you stack one box on top of another, there is friction between the two boxes. This keeps the top box from sliding off the bottom box. There is also gravity pulling on the stack of

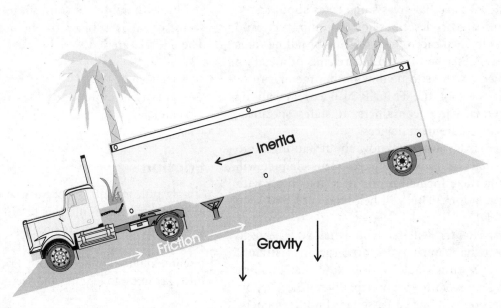

Figure 8-2. The forces of nature can work for or against you, your vehicle, and your load.

boxes. If there were no gravity, they would simply fly around weightlessly. Inertia is also working to keep the boxes in place.

This is all fine, as long as the forces are present in the right amounts. As you have seen, some friction helps keep one box on top of the other. Too much friction creates heat. That can start a fire. Not enough friction, and the boxes don't stay in place. Head uphill, and gravity works on the stack in a different way than it did on level ground. Without enough friction (or anything else) to hold the boxes in place, they begin to slide away from each other. Once the boxes are in motion, inertia works so they continue to move around. Once you had a nice stack of boxes in your vehicle. Now you have loose boxes rolling around inside.

Cargo securement devices help the forces of nature to work for you, not against you. However, they only help if they are in good condition and you use them correctly. Later in this chapter, you'll learn how to do just that.

SIZE AND BALANCE

The size and balance of your vehicle and cargo are important for two reasons. One is that states, counties, and even cities have laws about vehicle size limits. Laws regulate how heavy your vehicle and cargo can be. They also regulate height, width, and length.

Other laws state how the weight should be distributed in the vehicle. The idea is to keep from having too much weight on any one portion of the vehicle. Why? An overloaded or poorly loaded vehicle can damage the road surface. Poor loading affects how the vehicle handles, as you have learned. It affects steering, braking, and speed control. If you lose control of your vehicle, you could have an accident. "Safety" is the second reason why size and balance are important.

Size and Balance Terms

The laws use certain terms in specific ways. Let's define those terms first.

Gross vehicle weight (GVW) is the total weight of a single vehicle, plus its load. Gross combination weight (GCW) is the total weight of the power unit or tractor, plus any trailers, plus the cargo. (This would also apply to a vehicle towing another vehicle as cargo.) Gross vehicle weight rating (GVWR) is a rating given to the vehicle by its manufacturer. It is the top GVW for that single vehicle plus its load. Gross combination weight rating (GCWR) is—that's right! This rating is set by the vehicle manufacturer. And it's the top GCW for the combination of power unit plus trailer or trailers plus cargo. (Again, the cargo could be a vehicle, as in a towaway operation.)

Axle weight is not how much the axle weighs. It is how much weight the axle (or set of axles) transmits to the ground. Axles hold the wheels, which enable the vehicle to move. That's true. Axles also hold the brake assembly. Axles also support the vehicle and its load. Each axle carries part of the total weight of the vehicle and load. That weight is then transmitted to the ground. That's axle weight.

The maximum legal axle weight may be reduced still further by a bridge law. Like roads, bridges can handle just so much weight at any one point. There must never be more than that at any point or the bridge will break. Many states have laws to prevent this. Those are called bridge laws. A bridge formula is used to determine how much weight is put on any point by any group of axles. A tandem axle is an example of a group of axles. If there is more than one set of tandems, the formula takes into account how close the tandems are to each other. The result could be an even lower maximum axle weight for each axle in the group.

Tires, wheels, and suspensions all support the vehicle. They all carry weight. You learned in Chapter 5 that tires, wheels, and suspensions are rated. The manufacturer states how much weight these parts can carry. The most weight a tire can carry safely is called "tire load." As you learned, the rating is taken at a stated inflation pressure. If the tire were underinflated or overinflated, this rating might no longer apply. An underinflated or overinflated tire may not be able to carry safely the same load it could carry at the proper inflation pressure.

The same idea applies to other parts that have a load or weight rating. Examples are coupling devices, such as saddle mounts, tow bars, and chains. Securement devices have load or weight ratings.

Loading a device beyond its rating is illegal and unsafe.

Weight Limits

There are specific laws on weight. States have limits for GVWs, GCWs, and axle weights. Some states have bridge laws. You must know the limits set by each state in which you run. Truckers' map books often have details on specific state size limits and bridge laws. You must stay within those limits.

There are laws about balance, too. As mentioned, the weight must be balanced. That is, it must be distributed so each axle bears an equal amount of the load.

To be able to get the load balanced, you should understand the idea of "the center of gravity." The center of gravity is the point where weight acts as a force. That is, weight itself becomes a matter of concern. The center of gravity affects the vehicle's stability. You should aim to put the center of gravity where it will have the most support. This will not always be the center of the cargo compartment. This is best explained by example.

First, remember that your vehicle's axles provide the support. Not the cargo area floor, the axles. This includes the front axles under the cab. Imagine your empty vehicle on level ground. Picture all the wheels touching the road. Draw a line from wheel to wheel, making a rectangle. When your vehicle's center of gravity is over the center of this rectangle, each axle supports an equal amount of weight. Your vehicle is most stable.

OK, where is the center of gravity? Think about your vehicle. Right now it's empty. Where is the heaviest part of it? It's probably in the engine compartment. Where is your support rectangle? It's mostly under the cargo compartment. The center of gravity is not well supported. The front axles are bearing most of the weight. That's why an empty vehicle can be unstable. If you lose traction under your rear wheels, you could easily go into a rear-wheel skid.

Let's fill the cargo compartment with boxes of books. Where is the center of gravity now? Remember to include the entire vehicle: cab, engine, cargo, and all. The cargo is probably as heavy as the engine. Taken as a whole, the center of gravity is somewhere around the front half of the cargo compartment. Picture a vertical line drawn through the center of gravity. It would drop right through the center of the rectangle, to the center of the earth. Each axle is supporting an equal amount of weight. The vehicle is stable.

Now let's fill that cargo compartment with gold bars. They're very heavy. The engine is heavy, but the cargo is heavier. The center of gravity of the vehicle, cargo and all, is now more toward the rear of the cargo compartment. It's no longer centered over the support rectangle. The vehicle is no longer stable. Most of the weight is being supported by the rear axles.

You may not notice this on dry, level ground. So let's hit a wet spot in the road. Remember, the front wheels do not have much of a load on them. The rear wheels are providing the support. You now lose what little friction there was between your front tires and the road. You go into a front-wheel skid.

This would not have happened had your load been better balanced. You may have even been within weight limits. Still, you should have loaded your cargo a little differently. You should have loaded it a little more toward the front of the cargo compartment. That would have moved the center of gravity. The front axles would have received their fair share of the weight.

Height Limits

States have height as well as weight limits. Again, there are two reasons. Overhead structures, like overpasses, could be damaged by too-tall vehicles or loads trying to go under them. Also, vehicles or loads that are too tall are unstable. Height limits ensure safety.

You saw that you could be carrying the legal amount of weight and still be unsafe. That was a center of gravity problem. You can be within the legal height limits and have a center of gravity problem with height, also. Here's how.

Let's take the load of gold bars again. We'll load them over the support rectangle, in three even layers. Where is the center of gravity? It's centered over the rectangle, yes. It's also low. It's not toward the roof of the cargo compartment. There's only air there. Air doesn't weigh enough to add to the total weight of the vehicle and cargo. The vehicle is stable.

Now let's stack the bars in the center of the cargo compartment. Load them in stacks up to the roof. Still no problem. The center of gravity is over the center of the rectangle. Now, however, it is also high. There is weight near the roof. Remember, think of the total vehicle: the cab, the cargo compartment, the cargo. The "average" of all that weight has a location. It has both a horizontal and a vertical location. In this case, the vertical location is high. This is what is meant by a "high center of gravity."

Still no problem, on a level surface. Now take that vehicle on a banked road. The wheels stay in

contact with the road. The body of the vehicle tilts away from the high side of the bank. Where is the center of gravity? It's tilted, too. It's no longer over the center of the support. If you drop your vertical line through the center of gravity, it doesn't pass through the center of the earth. It's off to one side. The vehicle is unstable. It could tip over. This type of accident is not uncommon. Proper cargo loading could prevent it.

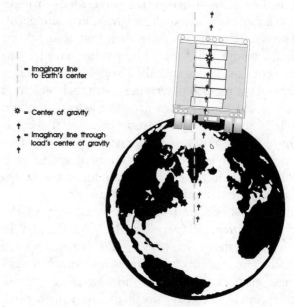

= Imaginary line
  to Earth's center

✳ = Center of gravity

† = Imaginary line through
    load's center of gravity

Figure 8-3. A high center of gravity can cause the vehicle to tip over.

Length and Width Limits

States have length and width laws, too. On today's busy roads, drivers of normal-sized CMVs have a hard enough time managing what little space they have. You can see how oversized vehicles could present problems. You usually need a special permit to take an oversized vehicle on the highway. Your vehicle may only be allowed on the road at certain times. You may even be escorted by police or a pilot car. These are ways states control the use of the roadways by very large or heavy CMVs.

There may not always be specific laws against overloading a part or device. Still, it is illegal to operate in an unsafe condition. An overloaded or badly loaded vehicle is unsafe. So, you could be put out of service simply for being in an unsafe condition.

Keep this in mind as well. Size and balance laws usually assume good driving conditions. Adverse conditions turn small handling problems into big ones. In bad weather, or on bad roads, it may not be safe to operate at the legal limits. You may have to reduce the size of your load. It's for certain you should lower your speed, increase your following distance, and drive with care.

LOADING CARGO

Now you have some awareness of how the forces of nature affect a loaded vehicle. You see that the location of the center of gravity is important. Now look at some specific loading tactics.

Loading Flatbeds

Loads can only be stacked so high on flatbeds. Find out what the height limits are, and stay within them. Keep the center of gravity low or you are more likely to tip.

Loading Vans

You may load a van by hand, stacking the cargo yourself on the floor of the truck. The cargo may also come on pallets. Even if you don't load the truck yourself, remember that you are responsible for it. Everything that happens to the load after you sign for it is your responsibility. Be sure the loading is done properly.

To load stacks of boxes securely, use the tiered stacking method shown in Figure 8-4. This will distribute the weight of a tier of freight equally. By overlapping rows so that the freight in the tier ties together, there's less chance the cargo will shift.

Start on either side on the floor in the nose of the cargo compartment. Load the first carton in the corner against the front wall. Load all the way across the floor, completing one row of cartons in the tier.

Start the next row of the tier on the side opposite the start of the first row. If the cartons in the first row do not exactly fit across the unit, leave the space. As long as the carton being loaded to start the next row does not overhang more than half its width, do not worry. This overlap is good and it starts an overlap all the way across the row. Wedge this first carton tight against the wall and each carton in the row against the loaded cartons.

Continue loading rows across the cargo compartment. Always start the next row on the side on which the previous row was finished. This builds a tier of freight from the floor to the roof.

You can use this method for sacks as well as cartons. The different sizes and shapes of the freight will affect how closely you can follow this tier technique. You will have to make some adjustments for different sizes of cartons. For instance, you could load the length of the freight the long way instead of across. This reduces the chance of the tier shifting backward or forward when you stop and start. The overlapping rows reduce shifting from side-to-side.

Aim to build your tiers as straight and as level as possible. Put the heavier, stable freight in the bottom tiers. Put the lighter freight on top of the tiers. You know why. This makes for a low center of gravity, which is more stable. Always place cartons right-side up.

If you're loading pallets, make sure the pallets don't lean. Each pallet should be placed so it is tight against the one in front of it. Start with rows

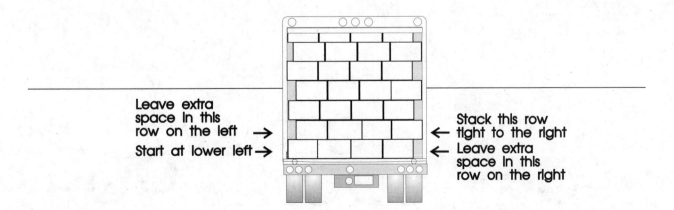

Leave extra space in this row on the left →
Start at lower left →

← Stack this row tight to the right
← Leave extra space in this row on the right

Figure 8-4. Cargo loaded in tiers.

down the length of the cargo compartment. Leave some space between the rows and between the rows and the walls.

Loading Refrigerated Vehicles

Air must circulate all around the load. This helps to maintain an even temperature. It keeps sections of the cargo from drying out. Be especially careful to leave spaces around and between the rows. Keep the load away from the rear door, especially when extreme outdoor temperatures could affect the cargo.

Make sure the cargo is at the proper temperature before you load it. The truck's refrigeration unit isn't supposed to cool down a load. It's supposed to keep a load that's already cool at the proper temperature.

Loading Other Vehicles for Towing

In a towaway operation, your cargo will be one or more other vehicles. You may use a tow bar-pintle hook connection. (Only one vehicle in the towaway operation may be towed this way.) If you use a saddle mount, you may tow three vehicles. You may use no more than three saddle mounts at one time. You may also tow a vehicle by putting it on the towing or towed vehicle. The FMCSR calls this "full mounting."

Using a Tow Bar

You may not connect the tow bar to the bumper of a towed vehicle weighing more than 5,000 pounds. The tow bar should have a safety chain or cable. This will at least keep the towing and towed vehicles connected if something should happen to the tow bar.

There should be enough space between the two vehicles so you can make turns. If there is not enough space between them, the towed vehicle will come up against the towing vehicle in the turn. Too much space, and the tow bar or chains can hang too low.

You also want the towed vehicle to track the towing vehicle. You don't want the towed vehicle swaying, whipping, or fish-tailing behind the towed vehicle.

Make sure that the lights and reflectors have power and that they work. (Later in this chapter you'll find a list of the lights and reflectors you must have when towing vehicles.)

Decking

Loading vehicles as cargo using saddle mounts is called decking. You may have only three saddle mounts in a combination.

The saddle mount is mounted on the frame of the towing vehicle. The saddle mount has a kingpin. The kingpin fits in the fifth wheel of the towing vehicle. Make sure this connection is secure.

The saddle mount has hooks (C-clamps or U-bolts) that connect with the axle of the towed vehicle.

When saddle mounts are used, some of the wheels of the towed vehicles are on the road. The brakes on these wheels must work. So must the lights and reflectors. (This chapter includes a list of the lights and reflectors you must have when towing vehicles.)

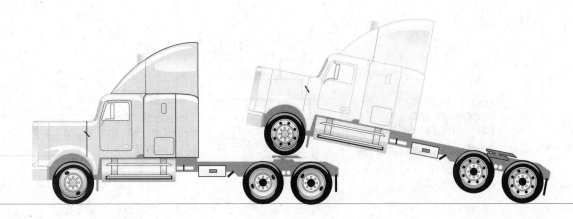

Figure 8-5. Decking.

If the steering isn't locked, the front wheels of the towed vehicle might turn out past the widest part of either the towed vehicle or the towing vehicle. This isn't allowed. So, the movement of the front wheel must be restrained.

The steering mechanism of the towed vehicles must be locked in a straight forward position. If it isn't, the towed vehicle must be towed with the front end on the towing vehicle.

Sometimes, the towed vehicle will not clear the frame of the towing vehicle. You may need to raise up the front end of the towed vehicle. You may use hardwood blocks for this. This will provide the clearance you need.

Full Mounting

Sometimes towed vehicles may be carried on the towing vehicle. This is called "full mounting." Because of the weight of the towed vehicle, you could have a serious center of gravity problem here. It's vital that weight be well-distributed on the towing vehicle.

Loading Livestock

Livestock can move around in the cargo compartment if they are not confined. Their movements can rock the vehicle. This can really give you handling problems. You may have to confine the animals if they don't fill the cargo compartment. Use false bulkheads to bunch them together and limit their movement.

SECURING LOADS

You can be a big help to yourself when you load cargo properly. That way, you don't give yourself any center of gravity problems. Still, proper loading isn't always enough to keep your load in place. You may need to use cargo securement devices. You may also have to take steps to protect the load.

Securing Flatbed Loads

The most common items used to secure loads on flatbeds are:

- cables and winches
- webbing straps and winches
- chains and load binders

Use tiedowns that are strong enough to do the job. The combined strength of all cargo tiedowns must be 1/2 times the weight of the cargo being tied down.

Use enough tiedowns for the amount of cargo. The minimum number of tiedowns required to secure an article or group of articles against movement depends on the length, weight, and bracing of the items. FMCSR Part 393.110 gives further details. The whole purpose of chaining is to hold the load down and prevent its movement sideways or forward or backward. As you can see from Figure 8-6, proper placement and the direction of the chains will prevent such movement.

Illustrations in FMCSR Part 393 tell you in detail how coils of metal must be secured to

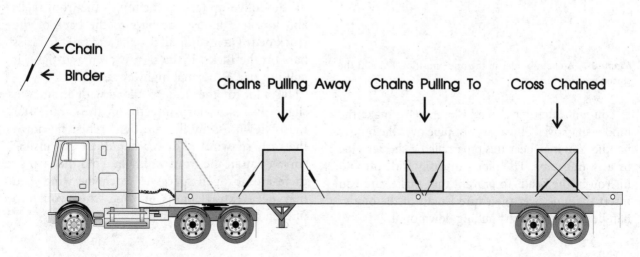

Figure 8-6. Chaining a flatbed load.

restrict movement. The dotted lines represent the tiedown assemblies. (See Figure 8-1.) The instructions in the FMCSR are quite clear. We won't repeat the instructions here. So make sure you read this part of the FMCSR thoroughly.

The steel hooks on the chain are hooked to a steel rail on the sides of the trailer. Or, they can be hooked to rings in the floor. The chain is placed over or around the cargo and stretched tight by hand.

However, no matter how hard you try, when you stretch chain by hand, it will not be tight enough to keep the cargo from shifting. The chain must be tightened even more. That's the purpose of a binder. You attach the binder end hooks to the chain with the binder lever open. Then you pull the binder lever to the closed position to tighten up the chain.

Keep your fingers clear when using this device. The proper way to install the load binder is with the legs open. Pull the handle down or toward you as you attempt to snap it over the center.

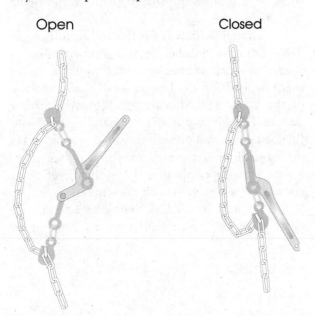

Open   Closed

Figure 8-7. A chain binder in the open position (A) and the closed position (B).

You will often have to add leverage to make the binder snap over. Slip a piece of pipe over the binder handle. You may hear this pipe called a cheater pipe or a swamp pipe. The pipe can easily slide off and hit you in the head. To prevent this, make sure you insert the pipe over the entire length of the binder handle before you start pulling down on it.

Some containerized loads have their own tiedown devices. They may have locks that attach directly to a special frame. Others will be loaded onto flatbeds and tied down just like any other large cargo.

Cables, straps, and chains all help keep the load in place. However, they can crush, bend, or cut the load. So you must take steps to protect the cargo from being damaged by these securement devices. Place "V boards" or wooden blocks between the cable, strap, or chain to protect the edge of the load.

You may have to use blocking or bracing to keep the cargo in place. You can use blocking on the front, back, or sides—or all around. Bracing goes from the upper part of the cargo to the floor or walls of the cargo compartment.

Use blocks that are shaped to fit snugly against the cargo. Then secure them to the floor. If the floor is wood, you can nail the blocks in place. If the trailer floor is metal, you can use chains to hold the bracing in place against the wheels. Make sure the lumber used for the brace is without knots or cracks. Make sure the nails that are used are at least twice as long as the thickness of the board through which they are being driven. They should be nailed straight down or at an angle against the movement of the cargo.

You may have to tarp your load. ("Tarp" is short for tarpaulin.) It covers and protects exposed cargo. It also protects others on or alongside the road from cargo that flies off or spills from the vehicle. Some states have laws that require you to cover any flatbed or open-top vehicle.

Tarps are tied down with rope, fabric webbing, or elastic cords with hooks. To tarp a load, you'll lift the rolled up tarp to the top of the front racks and unroll it across the bars to the back of the truck bed. Then you pull it tight and tie it to cross bars on the racks. If the tarp is even and tight, it will not flap at normal highway speed.

It's a lot tougher to tarp a load with an uneven shape. Place the tarp directly on the cargo after the tiedown assemblies are tight. Then tie down the tarp so wind and weather don't get inside. An overlap in the front will help. Fold the tarp so there are no open spaces to catch the wind. You may need extra lengths of rope when you tarp uneven cargo.

Figure 8-8. Tarping a load.

Water can damage a load. Inspect the bed of the cargo compartment for holes. If the floor is not waterproof, cover it with plastic. Sometimes the top and sides of loads come covered with plastic. This plastic usually does not go around the bottom of the load. Placing plastic on the truck bed and tarp over the plastic used by the shipper will keep the load dry.

You may have to "smoke tarp" a load. This means using a tarp to cover the front part of the load. The tarp prevents the cargo from being discolored by smoke coming from the exhaust stack.

Prepare the truck bed for a large heavy load that will be set in place by forklift or crane by putting dunnage (loose packing material) in place on which the load will rest. This will allow a forklift or sling to get under the load at the receiving destination. Placing a solid heavy load directly on the surface of the trailer would cause major unloading problems.

Front-end Structure

Something else that needs protection is the driver. Front-end structures protect you from your cargo if there is a collision. Front-end structures are called header boards or headache racks. You must

have one, or some structure that meets the same requirements, to block the forward movement of any cargo you carry.

Securing Van Loads

Van loads aren't as hard to secure and protect as flatbed loads are. The van body itself provides stability to the load. Use a load lock if you need extra stability. Load locks are long poles that stretch across the width of the trailer. Once in place they are pushed against the sides of the trailer with a jack-like device. These should always be used in the rear of the load to prevent any boxes from falling. Place one near the top of the load and another about halfway down. Be sure they are firmly in place.

The van body itself also protects the cargo, so you don't need tarps or plastic. Before you load, though, make sure that the inside of the trailer is clean and dry. Check for nails, splinters, or anything else that could damage the cargo.

You don't need a header board. The structure of the van body itself would protect you if the load should move forward.

CARGO INSPECTION

Regulations require that you inspect all cargo, including tarped loads. The only exception is sealed cargo. Of course you can't inspect sealed cargo. You are still responsible if such cargo exceeds gross weight or axle limits, though.

Check the cargo as part of your pre-trip inspection. Once you're on the road, you must check the load often. Make your first inspection within the first 50 miles. Make your second inspection after three hours or 150 miles, whichever comes first. Then inspect every three hours or 150 miles after that. Some unusual loads, such as refrigerated ones, should be checked more often.

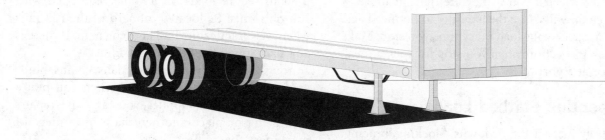

Figure 8-9. A header board blocks the forward movement of cargo.

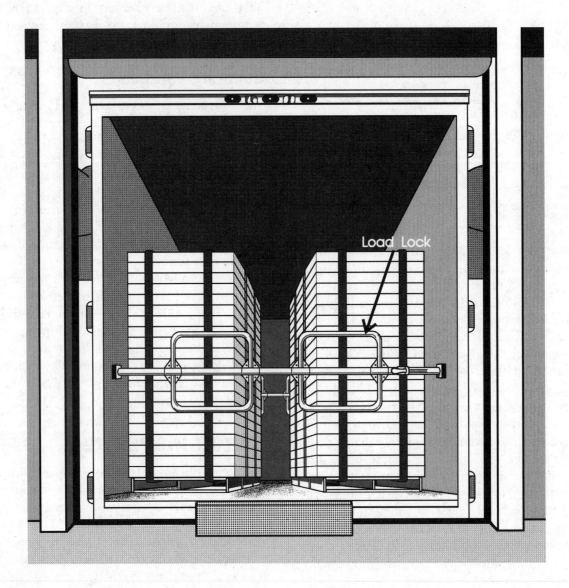

Figure 8-10. A load lock.

When you perform your three-hour checks, you'll walk around the truck to check tires and lights as well as the cargo. At night, you should have a flashlight in one hand and a tire hammer in the other. Lightly tapping the chains with the hammer will tell you if they are tight or loose. A tight chain will cause the hammer to bounce back.

Last, check your load every time you stop. Make a quick inspection when you stop for fuel, a meal, rest, or at a port of entry.

Inspecting Flatbed Loads

Carefully inspect the tiedowns, blocking, bracing, and tarping on flatbed loads. Make sure none of the nails are pulling away. Make sure that each tiedown is attached and secured so that it will not loosen, come unfastened, or either open or release while the vehicle is moving. All tiedowns and other components of a cargo securement system used to secure loads on a trailer equipped with rub rails must be located inboard of the rub rails whenever practicable. Pull at each chain. If there is any slack, open the binder and adjust it.

You should carry a hammer, nails, chains, binders, and a good supply of rope when you pull a flatbed. With these supplies, you can make repairs to bracing or place another chain on a piece of machinery that appears loose.

Inspecting Van Loads

Inspecting a van load is also a fairly simple process. It's mostly just a simple check to make sure nothing has shifted out of place. If you have done a good job of loading and securing the cargo, nothing should.

Towaway Inspection

Make sure that the required lights and reflectors work on both the towing and the towed vehicle. The towing vehicle must have:

- a headlamp on each side of the front
- a turn signal on each side of the front
- a clearance lamp at each side of the front
- a side-marker lamp at each side toward the front
- a taillamp at each side of the rear
- a stop lamp at each side of the rear

The last vehicle in the combination must have:

- a side-marker lamp at each side toward the rear
- a taillamp at each side of the rear

- a stop lamp at each side of the rear
- a turn signal at each side of the rear
- a clearance lamp at each side of the rear
- a reflector at each side of the rear

When towing vehicles with a tow bar, inspect the tow bar connection. Make sure that all parts are secure and undamaged. See that the safety chain or cable is hooked up. Make sure all the pins in the length adjustment are in place. You may want to take a "test drive" before you leave the yard to make sure you can stop, start, and turn with no problems.

In a towaway operation, check your load within the first 50 miles of the trip. Check again whenever you change your duty status. Also inspect the mounting after the vehicle has been driven for three hours, or every 150 miles, whichever occurs first.

DRIVING WITH A LOAD

You have seen in general how the forces of nature and the center of gravity affect how your vehicle handles. Here are some specific handling problems you may run into when carrying a load.

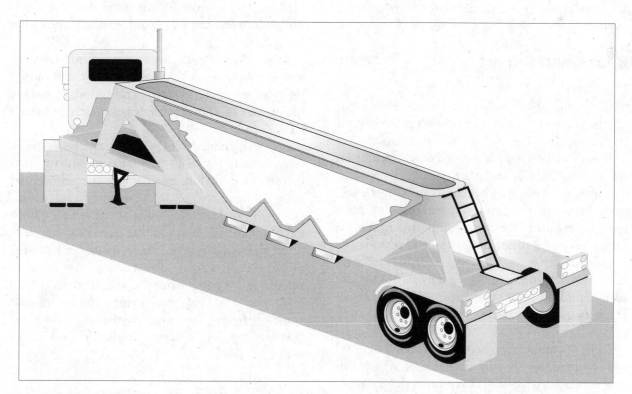

Figure 8-11. The shifting of liquid loads makes curves and sharp turns dangerous.

Starting and Stopping

Adding weight increases the effects of gravity. When you are hauling a light load, the effect of gravity is small. You can move easily.

When hauling a heavy load, gravity pulls harder on your vehicle. The engine has to work harder to overcome the pull of gravity. It takes more effort to put the vehicle in motion.

Your load, and the pull of gravity, affects your ability to stop. A heavy load increases the effect of gravity. That gives you more traction. That means you can stop better, in less time, and in a shorter distance. A light load does not increase the pull of gravity. Since you're moving faster and more easily, it takes more time and distance to stop the vehicle. In short, when your vehicle is empty or lightly loaded, you need more stopping distance.

Poor weight distribution can make axles too light. They lose contact with the road. As you have seen, this makes it easy to skid.

Turning

Loading the vehicle puts extra weight on the axles. This includes the steering axle. The more weight on the steering axle, the harder it is to steer. Too much weight on the rear axles makes the steering axle too light. This also decreases your steering control.

Banks and Curves

If you have a high center of gravity, you will really notice it on banked and curved roads. Poor loading may cause the vehicle to tip over if the bank is steep enough or the load is badly unbalanced.

If your flatbed load is not tied down securely, it can shift to the side or fall off.

Today, most meat is hauled in tubs. You'll still sometimes see it transported as sides of meat hung from rails in the cargo compartment. This is called "swinging meat," and with good reason. The motion of the truck can cause the sides of meat to swing back and forth. Once they start, they can really build up some momentum. This can result in a very unstable vehicle, especially on curves and ramps.

Dry bulk tanks often have a high center of gravity. The load can shift inside the tank. Again, this makes taking curves and sharp turns dangerous.

Upgrades and Downgrades

A loaded vehicle will perform differently from an empty one on upgrades and downgrades. Overloaded trucks pull a hill very slowly. Be aware that other drivers are more prone to tailgate you or try to pass.

On the downgrade, a loaded vehicle will really pick up speed. You must be in the right gear before you head downhill. If you are going too fast, you will have to use your brakes. Trying to slow or stop a heavily loaded vehicle going downhill at high speed sets the stage for brake failure. This is a type of accident that is all too common, and easily prevented.

Take It Slow and Easy

When you transport cargo, adjust your driving to reduce the chance of shifting. Make no sudden moves. Don't swerve and don't stop short. Take turns and curves slowly and carefully. While safe under most conditions, the posted speed limits may be too high for your loaded vehicle.

Keep a safe following distance. You must carefully manage the space around your vehicle. Give yourself time, space, and distance to maneuver.

Do not pull off the roadway onto an uneven surface. Drive more slowly than usual on off-ramps and on-ramps. Avoid parking areas that have sharp inclines.

Most drivers will haul some kind of cargo almost all the time. Vehicles are rarely sent out on the road empty. So all drivers must understand the basic rules of transporting cargo safely to earn a CDL license.

A poorly loaded vehicle is a dangerous vehicle. Cargo can fall off and become a hazard to other drivers. Overloading can damage the vehicle and make it hard to control. Under adverse conditions, an overloaded or unbalanced vehicle can be a rushing missile that can't be steered or stopped.

To prevent this, you have to understand the principles of weight and balance. You have to understand how the forces of nature act on a loaded vehicle. Then you can load and secure your vehicle properly. You can drive the vehicle in a way that leaves you in control, despite the problems a load can create.

PASS POST-TRIP

Instructions: For each true/false test item, read the statement. Decide whether the statement is true or false. If it is true, circle the letter A. If it is false, circle the letter B. For each multiple-choice test item, choose the answer choice—A, B, C, or D—that correctly completes the statement or answers the question. There is only one correct answer. Use the answer sheet provided on page 193.

1. A vehicle that is in motion will tend to stay in motion because of _____.
   A. friction
   B. gravity
   C. inertia
   D. state regulations

2. Loading the vehicle incorrectly can _____.
   A. result in damage to the road surface
   B. make the vehicle hard to steer
   C. make the vehicle hard to stop
   D. all of the above

3. The gross combination weight (GCW) of a vehicle includes _____.
   A. the weight of the entire vehicle but not the cargo
   B. the weight of the entire vehicle and the cargo
   C. the weight of the cargo but not the vehicle
   D. the weight of the power unit only

4. The gross combination weight rating (GCWR) of a vehicle is _____.
   A. the weight of the entire vehicle plus its load
   B. the manufacturer's stated weight for the vehicle
   C. the total maximum axle weight allowed by the state
   D. the maximum weight stated by the manufacturer for a specific vehicle plus the load

5. Axle weight is how much an axle weighs.
   A. True
   B. False

6. Vehicles are rated for the weight they can carry. So are _____.
   A. tires, suspension systems, and coupling devices
   B. engines, brakes, and drive shafts
   C. hoses, belts, and wires
   D. switches, lubricants, and controls

7. For stability, the center of gravity should always be in the middle of the cargo compartment.
   A. True
   B. False

8. If you always operate at the legal size limits set by states, you can be certain you will be operating safely.
   A. True
   B. False

9. Cargo securement devices can help the forces of nature work for you, not against you, if _____.
   A. you use them correctly
   B. they are in good condition
   C. they are rated for the load
   D. all of the above

10. During your trip, you must inspect cargo at least once every _____, whichever comes first.
    A. hour or 50 miles
    B. two hours or 200 miles
    C. three hours or 150 miles
    D. 10 hours or 500 miles

CHAPTER 8 ANSWER SHEET

Test Item	Answer Choices				Self - Scoring Correct	Incorrect
1.	A	B	C	D	☐	☐
2.	A	B	C	D	☐	☐
3.	A	B	C	D	☐	☐
4.	A	B	C	D	☐	☐
5.	A	B	C	D	☐	☐
6.	A	B	C	D	☐	☐
7.	A	B	C	D	☐	☐
8.	A	B	C	D	☐	☐
9.	A	B	C	D	☐	☐
10.	A	B	C	D	☐	☐

Cut Here

Chapter 9

Vehicle Inspection

In this chapter, you will learn about:

- commercial motor vehicle inspection

To complete this chapter you will need:

- a dictionary
- a pencil or pen
- blank paper, a notebook, or an electronic notepad
- colored pencils, pens, markers, or highlighters
- a CDL preparation manual from your state Department of Motor Vehicles, if one is offered
- the Federal Motor Carrier Safety Regulations pocketbook (or access to U.S. Department of Transportation regulations, Parts 383, 393, and 396 of Subchapter B, Chapter 3, Title 49, Code of Federal Regulations)
- the owner's manual for your vehicle, if available

PASS PRE-TRIP

Instructions: Read the statements. Decide whether each statement is true or false. If it is true, circle the letter A. If it is false, circle the letter B.

1. You could be put out of service in one state for a vehicle condition that was acceptable in another state.
   A. True
   B. False

2. Steering wheel free play should not be more than 10 degrees on a 20-inch steering wheel.
   A. True
   B. False

3. A windshield is considered defective if it has even one tiny crack.
   A. True
   B. False

The vehicle inspection is an important part of the CDL test. FMCSR Part 383 states that CMV drivers must be familiar with FMCSR Part 396. As you now know, that includes vehicle inspection.

Look also at the appendix to Part 383, Subpart G, if that's included in your copy of the FMCSR. That's the Required Knowledge and Skills Sample Guidelines. Many states use these guidelines to develop their CDL tests. Item (e) in the guidelines suggests CMV drivers should show they can do a proper inspection. It states that drivers should be able to make certain repairs and do some maintenance. CMV drivers should also know how to tell when problems develop during operation.

Most states will have you perform an inspection as part of the Skills Test. During that time, you may also be asked to describe in some detail the different parts and systems on your vehicle as you inspect them. Or, the examiner may have you stop your inspection from time to time to answer questions about your equipment.

Chapters 4 through 8, which you just finished, should help prepare you for this part of the CDL tests. You know what is required to be on your vehicle. You know how these systems work. You can understand how certain defects keep you from operating your vehicle safely. You know how to load cargo properly. You understand how you, as the driver, are responsible for the safety of your vehicle.

There's one more thing that will help you pass the inspection part of the CDL tests. That's to have an inspection routine. If you have a routine, you won't fail to check the same things every time. You'll be less likely to forget or overlook something. You won't have to stop and make decisions. You won't have to ask yourself "Should I check this?" "Can I skip it this time?" "Did I already inspect that?"

In this chapter we'll focus on straight trucks.

We'll cover inspecting combination vehicles in Chapter 11. You should practice that inspection if you plan to get a Group A CDL, or want the doubles/triples endorsement.

Practice the inspection that fits the vehicle in which you plan to take your tests.

This chapter goes hand-in-hand with Chapters 5 and 6. Chapter 5 described the parts that motor vehicles must have. Chapter 6 described how they work. It described some defects for which you should be on the lookout. That information won't be repeated here. Have a good grasp of Chapters 5 and 6 before you begin Chapter 9.

As you practice your inspection routine, also practice explaining what you are inspecting and why. State why you are looking for certain defects. State why such defects are problems. People may wonder why you are walking around talking to yourself, but you will be well prepared for this part of the CDL tests.

WHY INSPECT

One of the first things to know about vehicle inspection is why you do it. Inspecting your vehicle helps you to know your vehicle is safe. As you know, federal laws require inspection. Your state may also have laws on vehicle inspection.

WHO INSPECTS

The motor carrier has the main inspection responsibility. You may recall that motor carriers must perform a periodic inspection of their vehicles. This is often referred to as the "annual inspection."

Motor carriers usually assign the routine inspections (pre-trip and post-trip) to the CMV drivers.

Federal and state inspectors also inspect commercial vehicles. These inspections may be performed while the vehicle is on the road, in operation. An unsafe vehicle can be put "out of service" at this time. Do you recall how long the vehicle remains out of service? It's until the driver or owner fixes it.

WHEN TO INSPECT

You'll also recall that you do a pre-trip inspection before each trip. You look for problems that could cause a crash or breakdown. These must be repaired before you can take the vehicle out.

You do a post-trip inspection at the end of the trip, day, or duty shift on each vehicle you operated. This may include filling out a vehicle condition report (DVIR). In this report you list any problems you find. The inspection report helps the vehicle owner know when to fix something.

You should also ensure your safety and the safety of others by inspecting your vehicle during a trip. You should:

Figure 9-1. State police may inspect your vehicle while on the road.

- watch gauges for signs of trouble
- use your senses (look, listen, smell, and feel) to check for problems

Check critical items when you stop. Critical items are those necessary for safe operation. Remember what they are? The list includes:

- lights and wiring
- brakes
- windows
- fuel and fuel system
- coupling devices on combination vehicles
- tires
- windshield wipers
- defrosters
- rear-view mirrors
- horn
- speedometer
- floor
- rear bumper
- flags on projecting loads
- seat belts
- emergency equipment
- cargo securement devices
- frame
- cab
- wheels and rims
- steering
- suspension

Most vehicles will have at least those parts and systems.

State laws may require vehicles to have still more equipment. We will cover some of the items commonly required by state regulations. You should inform yourself about the requirements of your home state, and every state through which you will pass. You could be put out of service in one state for a vehicle condition that was acceptable in another state.

WHAT TO INSPECT

Lights and Reflectors

In your inspection, make sure you have all the required lights and reflectors. (Review Chapter 6 to refresh your memory about the requirements.) Make sure they all work. Make sure they are clean. Dirt on the lights and reflectors can cut down the amount of light they give. You can't see, or be seen, well with dirty lights and reflectors. Refer to pages 72–81 for diagrams showing all the required lights and reflectors.

Electrical System

Look for wiring problems. Open and short circuits could result from the following:

- broken or loose wires
- worn insulation that exposes the wire
- bare wires that touch each other
- corroded connections

Loose wires can be reattached. Wires that are broken should be replaced. So should wires with worn insulation. Corrosion can be cleaned off with a wire brush.

If your vehicle uses fuses, check them. Replace any that have blown.

Inspect the battery. (Review Chapter 6 for a description of a complete battery check.)

Two gauges on your dashboard, the ammeter and the voltmeter, help you judge the battery's condition. Readings outside the normal range should alert you to defects. Your vehicle may also have a charging circuit warning light. This light would come on if your battery wasn't charging during operation.

Brake System

Your vehicle may have hydraulic or vacuum brakes, or you may have air brakes. If you have air brakes, you must make a special inspection of that system. You'll learn those special steps in Chapter 10. Here you'll learn about inspecting brake parts that all vehicles have. Plus, we'll look at inspecting hydraulic and vacuum brake system parts.

Remember, if you plan to drive a vehicle with air brakes, you must answer special air brake questions on the CDL Knowledge Test. You'll have to show you can inspect the air brake system, too. Be sure to read Chapter 10 if you don't want an air brake restriction on your CDL license.

Most brake systems have brake shoes and brake drums. Some will have disc brakes on the front axle. For the most part, you have to remove the wheel to inspect a disc brake. That's beyond the scope of a pre-trip inspection. Short of doing that, check the hub for cracks and leaking fluid. Some disc brakes have a device that will warn you when the pads get too thin. You'll hear a squealing noise when braking if this is the case.

At each wheel, check the brake drums and shoes. Look for cracks in the brake drum. Cracked drums must be replaced. Look for oil, grease, and brake fluid on the shoes or pads. Any of these will prevent the brakes from working properly. Brake fluid on the shoes or pads signals problems in the brake lines as well. Look for missing or broken brake shoes. These must be replaced or repaired. Brake pads don't usually wear evenly. But at their thinnest point, they should be no thinner than 1/4 inch.

Look at the brake lines and hoses. There should not be any leaks or breaks. Look for worn or weakened spots. Make sure all the lines and hoses are connected properly. There shouldn't be any kinks or twists in the lines.

Check each wheel cylinder. Leaks and loose connections are signs of trouble.

You should check the level of hydraulic fluid in the master cylinder. You'll do this when you inspect the engine area. Different manufacturers provide different means for doing this. In some cases, you can see through the tank. Or, a sight glass may be provided. Check the manual for your vehicle to see what's available to you. In all cases, there will be some kind of fluid level marking. Make sure the fluid level comes up to that mark.

As you might expect, leaks around the master cylinder mean trouble.

When you're in the cab, you'll check the operation of the brake system. Here's how to check hydraulic brakes. The engine should be on, and the transmission should be in neutral. Pump the brake pedal three times. Then press firmly on the brake pedal. Keep pressing for no less than five seconds (some manufacturers recommend 30 seconds). The pedal should not move.

If the pedal is not firm, you may have air in the lines. If the pedal sinks slowly to the floor, that's a sign there's a leak. These problems must be fixed before the vehicle may be driven.

Time is critical for tests you perform during vehicle inspection. You don't need a stopwatch to count seconds, though. Instead, say to yourself: "One thousand one." It takes about one second to say that to yourself. To count five seconds, say "One thousand one, one thousand two, one thousand three, one thousand four, one thousand five."

Test vacuum brakes. If you have to push hard on the brake pedal to get the brakes to work, you may have defects in the vacuum system. Another sign of problems would be brake fade.

While in the cab, you'll also test the parking brakes. Put your seat belt on. Put the vehicle in gear and allow it to move forward at a slow speed. Apply the parking brake. It should stop the vehicle. If it doesn't, it must be repaired before the vehicle goes anywhere.

An alternative way to test the parking brakes is to put the parking brake on and gently pull against it in low gear to test that the parking brakes will hold.

To test the service brake, go forward at about five mph. Push the brake pedal firmly. If the vehicle pulls to the left or the right, you could have brake troubles. Any pause before the brakes catch is another sign of brake problems. You may also "feel" the brake pedal doesn't work right. Perhaps it travels too far before the brakes apply or it takes too much effort. This too could mean brake problems. Have the service brake checked and any problems repaired before driving.

Windows and Glass

Inspect the window glass and windshield. You must have a clear view in order to drive safely. The glass must not be dirty or discolored. Glass that has been factory-tinted to reduce glare is allowed (but sun screen film you apply yourself is not).

A certain area of your windshield must be free of decals and stickers (see Figure 9-2). You should not have stickers at all on the cab side windows. The only stickers allowed are those required by law. Place these at the bottom of the windshield. They should extend upwards no farther than 4 inches into the viewing area.

The glass may have some minor cracks and "dings." Single cracks no larger than 3/4 inch in diameter—if not closer than 3 inches to other similar damage—are permitted, as long as they don't connect with other cracks. A whole spider web of cracks is not allowed.

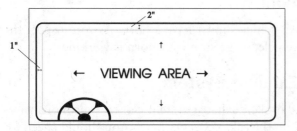

Figure 9-2. Your viewing area must be clean and free of "illegal" stickers.

Fuel System

Check the fuel tank or tanks. See that they are securely mounted, not damaged or leaking. Make sure the fuel crossover line is secure. It should not be hanging so low that it would be exposed to high grade crossings or objects bouncing up from the road. The filler caps must be on firmly. Of course, see that the tanks contain enough fuel. Tanks should not be more than 95 percent full. Fuel expands as it warms up. Filling your tanks slightly less than full leaves some room for this expansion.

Coupling Devices

If you are following along in your FMCSR, you'll see the next item of required equipment covered is coupling devices. This includes the saddle mounts, tow bars, pintle hooks, and safety chains used in towaway operations. Check all the parts that join vehicles together. They should not be bent or warped. The safety chains should not have any broken or twisted links. Make certain the connections are all secure. Check that the lights, reflectors, steering, and brakes work as required on the towed vehicles as well as the towing vehicle. (Review Chapters 5 and 8 if you don't recall the requirements.)

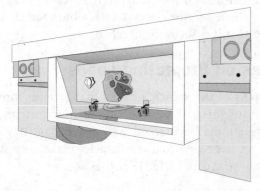

Figure 9-3. Check for bent or warped coupling devices.

If you are inspecting a vehicle combination as part of your CDL test, then you are going for a Group A CDL or the doubles/triples endorsement. You will find combination vehicle inspection covered in Chapter 11.

Tires

It is simply dangerous to drive with bad tires. Look for problems such as body ply or belt material showing through the tread or sidewall. Neither the tread nor the sidewall should be separating from the tire body. Look for cuts that are so deep that the ply or belt shows through. Look for cut or cracked valve stems or missing valve stem caps. Naturally, a very low or flat tire would have to be repaired. Don't drive on tires with any of these problems.

Listen for air leaks that could result in a flat. Look for bulges in the rubber that could lead to blowouts. Check the inflation pressure. Some drivers become expert enough to do this accurately by just thumping the tire with a tire billy. You'll never go wrong using a tire gauge, though. It would not be a bad idea to use a tire gauge during the CDL test inspection. Simply kicking the tires is definitely not good enough.

Check the tires for wear. You need at least 4/32-inch tread depth in every major groove on front wheels. You need 2/32-inch tread depth on other wheels.

Other problems are dual tires that come in contact with each other or parts of the vehicle. You should not have two tires of different sizes on the same axle. Radial and bias-ply tires should not be used together on the same axle, either.

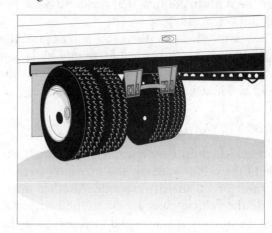

Figure 9-4. Don't mix tire sizes on the same axle.

Last, remember that you may not have regrooved tires with load-carrying capacity equal to or greater than 4,920 pounds on the front wheel of a tractor.

Sleeper Berths and Heaters

Sleeper berths and heaters are listed among miscellaneous parts and accessories. However, most of their specifications are of concern to manufacturers. There's little for the driver to inspect, and they're not included in the inspection report form. For your own safety, note if anything blocks easy access to the sleeper berth entrance or exit. Be alert to leaks that could allow exhaust or fuel fumes to enter the sleeper berth. Be aware, too, of safety threats presented by the heater, such as leaks, loose mounting, frayed wires, worn hoses, and so forth.

Wipers and Defrosters

Your vehicle must have two windshield wiper blades, one on each side of the centerline of the windshield. They must operate automatically, not by manual control. They must clean the windshield of rain, snow, and other moisture so you have clear vision. (A single blade is acceptable if it performs to the specifications in FMCSR Part 393.78.)

As part of your inspection, make sure the wipers work. Make sure the rubber blades do a good job of cleaning. Blades that are stiff, crumbly, or loose can't clear your windshield well.

If your vehicle has a windshield, it must also have a windshield defroster. Turn the defroster on and off as part of your inspection. Put your hand over the vents to confirm that the defroster is putting out warm air.

Test windshield washers to make sure they work. If they don't, you could simply be lacking fluid in the washer reservoir. Or, the fluid lines could be kinked, broken, or leaking.

Rear-view Mirrors

Your vehicle must have a rear-view mirror at each side of the cab. (Some vehicles may have one outside mirror on the driver's side. Another mirror inside the vehicle gives a view to the rear.) Make sure the mirrors are clean and undamaged. Adjust them so that when you are in the driver's seat, you have a view of the highway to the rear along both sides of the vehicle.

Horn

All vehicles must have a horn. As you inspect the vehicle, make sure your horn is working.

Exhaust System

A broken exhaust system can let poisonous gases into the cab or sleeper berth. You should check for loose, broken, or missing exhaust pipes, mufflers, tail pipes, or stacks. Loose, broken, or missing mounting brackets, clamps, bolts, or nuts can lead to problems. Look for exhaust system parts that rub against fuel system parts, tires, or other moving parts of the vehicle. Look for hoses, lines, and wires that are so close to exhaust system parts that they could be damaged by the heat.

Check for exhaust system parts that are leaking. Hold your hand close to the exhaust manifold. You'll be able to feel any leaks. (Don't put your hand too close, though. It's hot, and you could burn yourself.)

Defects in the exhaust system may not be repaired with wraps or patches. If you see such repair work, note it as a defect.

Floor

The floor of your vehicle should be free of holes and openings. Remember, such holes could allow fumes and exhaust gases into your vehicle, which could make you ill. Also make sure that the floor is clean, free of oil or grease. Slippery floors could lead to accidents.

Rear Bumper

Vehicles that are higher than 30 inches off the ground (when empty) must have rear bumpers. If your vehicle requires or has such a bumper, check to see that it is securely attached.

Flags on Projecting Loads

Any part of the vehicle or load that extends more than four inches from the side or four feet from the rear must be marked with a red flag. If you have an oversized load or vehicle, make sure it's marked with a red flag, 12 inches square.

Seat Belts

Most trucks and buses must have a seat belt for the driver. When you inspect your vehicle, make sure the seat belt is firmly attached and in good condition.

Don't forget to put the seat belt on when you take the driving portion of the CDL tests. Nearly all states require the driver (and often the passenger) to wear seat belts.

Emergency Equipment

You must carry a fire extinguisher in all but lightweight vehicles. Most vehicles are required to have one with a 5 B:C rating, or two with a 4 B:C rating. (Fire extinguishers on vehicles hauling hazmat must have a 10 B:C-rated fire extinguisher.) Include the fire extinguisher in your inspection. Check the gauge that tells you the fire extinguisher is fully charged. Check the nozzle to make sure it's clear. Check the ring pin to make sure the tip is intact. Check the pressure gauge. The needle should be in the green area. The extinguisher must be securely mounted to the vehicle, not rolling around loose. It must be stored where you can get to it easily.

Know how to use the fire extinguisher. Do you recall the proper method? If not, review Chapter 6.

If your truck was made on or after January 1, 1974, you must carry three emergency triangles that reflect from both sides. There are other devices you may carry to supplement the triangles, but you must carry the triangles. (If you're driving a truck that was built in 1974 or earlier, there are other warning devices you may use instead of the triangles. If your vehicle is this old, check Section 393.95 of the FMCSR for a full description of those devices.)

Figure 9-5. Make sure you have the required warning devices and emergency equipment.

Unless your vehicle uses circuit breakers, make sure you have spare fuses. You should know how to install them.

Cargo Securement

Cargo must be safely and securely loaded. We covered cargo loading in more detail in Chapter 8. Briefly, though, inspect the following:

- tailgate
- doors
- tarps
- spare tire
- binders
- chains
- winches
- braces and supports

All must be in good working order. Nothing should be flapping or hanging loose.

Tarps must be tied down so they don't flap or billow. If they are too slack, they could fill with air and balloon out. This could block your view in your rear-view mirrors.

Cargo must be loaded so it won't shift or fall off the vehicle.

If your vehicle doesn't have a cab guard, your load may call for a header board or similar device. Make sure it's strong, free of damage, and securely in place.

Cargo that could shift sideways must be held in place by sideboards or stakes. Again, make sure these are in good condition and securely in place. The cargo must not block the view you have while driving. Cargo must not block you in so that you can't move your arms or legs freely. You must be able to get around the cargo easily so you can reach emergency equipment or exit the vehicle in an emergency.

Curbside doors should be securely closed, latched, and locked. Otherwise they could swing open and hit someone on the sidewalk.

If you're hauling a sealed load, you should have security seals in place on the doors.

If you are hauling hazardous materials, you must have the correct papers and placards to go with them. You might also need a Hazardous Materials Endorsement. Review Chapter 5 if you don't exactly recall the rules about hauling hazmat.

Frame

As you inspect your vehicle, check the condition of the frame. Look for loose, cracked, sagging, bent, or broken parts. The frame should not be missing any bolts. Make sure all bolts are in good condition and tightened down. Anything attached to the frame must be bolted or riveted. Check these fastenings to make sure they are tight.

Cab

Check the condition of the doors. They should open easily and close securely. Doors may not be wired shut or held closed in a way that makes them hard to open again. Look for any loose, cracked, sagging, bent, or broken parts. The hood must be securely fastened. The seats must not be loose and wobbly. Your vehicle must have a front bumper. It must be firmly attached to the vehicle. It must not stick out so far that it becomes a hazard.

The cab should be clean and neat. Papers, cups, and other loose items could become hazards. If you stop short or turn sharply, they could fly around and hit you or get in your way.

Wheel and Rim Problems

Look for cracked or broken wheels or rims. Missing clamps, spacers, studs, and lugs could cause problems. Mismatched, bent, or cracked lock rings are dangerous. Wheels or rims should not have been repaired by welding.

Rust around wheel nuts may mean the nuts are loose. Use a wrench to check tightness. Examine the stud or bolt holes on the rims. If they are egg-shaped (out-of-round), that signals a problem. Check the supply of oil in the hub. See that there are no leaks.

Suspension System

Broken suspension parts can be extremely dangerous. Look for cracked or broken spring hangers. Note as defects torque rod or arm, U-bolts, spring hangers, or other axle positioning parts that are cracked, damaged, or missing. Loose or bent spring hangers will allow the axle to move out of its proper position. That could result in problems with the steering or driveline. The axle must be in the proper alignment and secure. Power axles should not be leaking lube or gear oil.

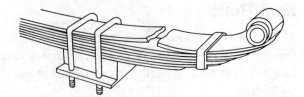

Figure 9-6. Check your suspension for broken parts.

If your vehicle has sliding tandems, make a close inspection of the locking pins. They must not be missing or out of place. The tandems should be securely locked in position.

Another problem to look for is missing or broken leaves in a leaf spring. In many states, if one fourth or more are missing, it will put the vehicle out of service. However, cracks, breaks, or missing leaves are prohibited. Leaves that have shifted might hit and damage a tire or other vehicle part. If your vehicle has coil springs, check the coils for cracks or breaks.

Look for leaking shock absorbers. Air suspension systems that are damaged or leaking are considered defects. Make sure the air pressure regulator valve works properly. First check to see that the vehicle's air pressure gauge shows normal operating pressure. Then check to see that the air suspension doesn't leak any more than three psi in five minutes, if at all.

Steering System

You'll include the steering system in your vehicle inspection. The steering wheel must not be loose. It must not have any missing or cracked spokes.

Remember what steering wheel lash is? That's the number of turns the steering wheel must make before the wheels begin to move. Another term for this is "steering wheel free play." FMCSR Part 393 sets limits on steering wheel lash. It should not exceed the limits set in the regulations for steering wheels of various sizes, manual or power. Anything more than that can make it hard to steer. The steering wheel should move easily both to the right and left.

Look at Figure 9-7. Note the steering wheel lash limit for the steering wheel in your vehicle.

Steering wheel diameter	Manual steering system	Power steering system
16" or less	2" +	4 1/4 " +
18"	2 1/4 " +	4 3/4 " +
20"	2 1/2 " +	5 1/4 " +
22"	2 3/4 " +	5 3/4 " +

Figure 9-7. Steering wheel lash limits.

The steering column should be securely fastened.

Check both the ball-and-socket and U-joints for signs of wear, flaws, or welding. (Damaged U-joints may not be repaired by welding.) There should be no cracks in the steering gear box. All its bolts and brackets should be secure. The Pitman arm must not be loose.

If your vehicle has power steering, all the parts must be in good working order. Look for loose or broken parts. Frayed, cracked, or slipping belts will have to be replaced or adjusted. Look for leaks in the power steering fluid tank and lines. Make sure there is enough power steering fluid. Check for missing nuts, bolts, cotter keys, or other parts. A bent, loose, or broken steering column, steering gear box, or tie rod is a defect.

STATE AND LOCAL REQUIREMENTS

There may be state and local vehicle laws on top of the federal ones we just covered. These may require vehicles to have still more equipment. Remember, you should inform yourself about the requirements of your home state and every state through which you will pass. You could be put out of service in one state for a vehicle condition that was acceptable in another state.

Here are some common state requirements you should know about. You may have to check for these items during your vehicle inspection as well as all those we already covered.

Mud Flaps

Federal law does not require you to have mud flaps on your vehicle. However, some states do. They may have mounting guidelines with which you must comply. It's your responsibility to find out what those guidelines are.

If there are no regulations on mounting mud flaps, use the following as a guide. These suggestions come from the Society of Automotive Engineers.

Measure from the ground to the bottom edge of your vehicle's mud flap. This measurement should be no more than six inches (measured when the vehicle is loaded). This will allow the mud flap to clear snow, curbs, or road hazards.

Your mud flaps should be at least as wide as the tires they cover.

Mud flaps should be mounted as far as possible to the rear from the wheel. They work best at that position.

Remember, check state regulations to see if mud flaps are required, and how they should be mounted.

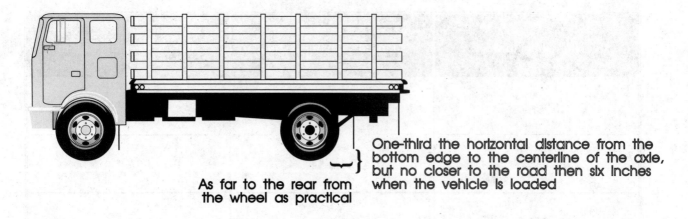

One-third the horizontal distance from the bottom edge to the centerline of the axle, but no closer to the road then six inches when the vehicle is loaded

As far to the rear from the wheel as practical

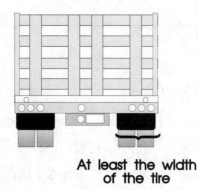

At least the width of the tire

Figure 9-8. Some states may require mud flaps.

Tire Chains

Many states require tire chains during snowy weather. This is especially true in the West. Often, you'll see highway signs telling you to put chains on your tires.

Again, check the regulations of your home state and every state on your route. Include tire chains in your stock of equipment. Practice mounting and removing them. That way you'll be prepared if you're asked to put them on during your CDL tests.

Other Equipment

It is often recommended that you carry the following items as emergency equipment, in addition to flares and reflectors:

- tire changing kit
- accident notification kit
- list of emergency phone numbers

Read the CDL manual your state offers carefully. It may mention these items and others besides. You may see them referred to as "suggested" or "recommended." Take that as a very strong recommendation. Have those items on your vehicle when you take your CDL tests. Know what they are for and how to use them.

INSPECTION ROUTINES

Now you know what equipment you must have on your vehicle. You know what condition is considered safe. You know what defects to include in your inspection. You know why they threaten your safety and the safety of others on the road.

All you need is a system for checking these items. Follow the routine outlined on the next pages. This routine is more or less the same whether you inspect a straight truck or vehicle combination. Pair the checklist with the illustration that matches your vehicle. Stop and perform each check described in the routine at the location marked on the diagram that follows.

Usually, you may have the checklist with you when you take your test. Check with your state to see if for some reason this is not permitted. Otherwise, feel free to use the checklist. You'll be less likely to overlook something if you check off items as you go.

As you inspect your vehicle, tell the examiner what you are inspecting and why. Describe what you find. You're not likely to take a defective vehicle to the test. However, the examiner will want to know that you would recognize a defect if you saw one. So as you inspect your vehicle, mention the defects typically found at the different sites. Explain how those defects impair safety.

There are seven main steps in this inspection routine. They are:

- approach the vehicle
- raise the hood or tilt the cab and check the engine compartment
- start the engine and inspect inside the cab
- check the lights
- walk all around the vehicle, inspecting as you go
- check the signal lights
- check the brakes

Step One: Vehicle Overview

As you walk toward the vehicle, notice its general condition. Look for damage. Note if the vehicle is leaning to one side. This could mean a flat tire. Cargo may be overloaded or may have shifted. Or, it could be a suspension problem. Look under the vehicle for fresh oil, coolant, grease, or fuel leaks. Check the area around the vehicle for people, other vehicles, objects, low hanging wires, or limbs. These could become hazards once the vehicle begins to move.

Look over the most recent vehicle inspection report. Drivers typically have to make a vehicle inspection report in writing each day. The vehicle owner then should see that any items in the report that affect safety are repaired. If the mechanic repaired the defects, or determined no repairs were needed, there will be a certification for that on the form.

You should look at the last report to find out what was the matter, if anything. Inspect the vehicle to find out whether problems were fixed, or whether repairs were not needed.

Step Two: Engine Compartment Check

Check that the parking brakes are on and/or wheels chocked. You may have to raise the hood or open the engine compartment door. If you have a tilt cab, secure loose items in the cab before you tilt it. Then they won't fall and break something.

At this site, check the engine oil level. The oil should be above the "Low" or "Add" mark on the dipstick. See that the level of coolant is above the "Low" level line on the reservoir. Check the condition of coolant-carrying hoses.

Check the radiator shutters and winterfront, if your vehicle has one. Remove ice from the radiator shutters. Make sure the winterfront is not closed too tightly. If the shutters freeze shut or the winterfront is closed too much, the engine may overheat and stop. Inspect the fan. Make sure the blades are sound and not likely to catch on hanging wires or hoses.

If your vehicle has power steering, check the fluid level and the condition of the hoses. Check the dipstick on the oil tank. The fluid level should be above the "Low" or "Add" mark.

Check the fluid level for the windshield washers.

If your battery is located in the engine compartment, perform the battery check described earlier here.

If you have an automatic transmission, check the fluid level. You may have to do this with the engine running. The owner's manual will tell you if that is the case.

Check drive belts for tightness and excessive wear. Press down on the center of the belt. The owner's manual will tell you how much slack there should be. Also, if you can easily slide a belt over a pulley, it is definitely too loose or too worn. Look for leaks in the engine compartment. These could be fuel, coolant, oil, power steering fluid, hydraulic fluid, or battery fluid.

Look for cracked or worn insulation on electrical wiring.

DRIVER'S DAILY VEHICLE INSPECTION REPORT

DRIVER'S NAME: _____ **DATE:** _____

Check items which are defective (X) and supply details about the defect in the "Remarks" section below. Use (4) if inspection was satisfactory.

TRUCK/TRACTOR NO.	DRIVER	MECHANIC	TRUCK/TRACTOR NO.	DRIVER	MECHANIC
Air Lines			Brakes		
Brakes (Emergency)			Brake Connections		
Brakes (Parking)			Bumper		
Body			Coupling Device		
Clutch			Doors		
Cooling System			Hitch		
Coupling Chains			Kingpin		
Coupling Devices			Landing Gear		
Defroster			Lights		
Drive Line			Reflectors		
Engine			Securement Systems		
Exhaust System			Suspension System		
Frame			Tires		
Fuel Tanks			Wheels and Rims		
Heaters			Other		
Horn			**SAFETY/EMERGENCY EQUIPMENT**		
Leaks			Fire Extinguisher		
Lights			Flags		
Oil Pressure			Fuses and Flares		
Rear Vision Mirrors			Reflective Tape		
Reflectors			Reflective Triangles		
Speedometer			Seatbelts		
Steering System			Other		
Suspension System			**REMARKS**		
Tires					
Wheels and Rims					
Windows					
Windshield Wipers/Washers					
Other					

❏ Condition of above vehicle is satisfactory

_____
DRIVER'S SIGNATURE

❏ Above Defects Corrected

❏ Above Defects Need Not Be Corrected for Safe
Operation of Vehicle

_____
MECHANIC'S SIGNATURE

I certify that I am satisfied that this vehicle is in safe operating condition and I have reviewed the last Vehicle Inspection report and verified that required repairs have been completed.

Figure 9-9. A typical Driver's Inspection Report.

Lower and secure the hood, cab, or engine compartment door.

Next, check any handholds, steps, or deck plates. Remove all water, ice, and snow, or grease from handholds, steps, and deck plates that you must use to enter the cab or to move about the vehicle. This will reduce the danger of slipping.

Step Three: Inside the Cab

Get in the cab. Inspect inside the cab. Make sure the parking brake is on. Shift into neutral, "Park" if your transmission is automatic. Start the engine and listen for unusual noises.

Check the gauge readings. The oil pressure should come up to normal within seconds after the engine is started. The ammeter and/or voltmeter should give normal readings. Coolant temperature should start at "Cold" (the low end of the temperature range). It should rise gradually until it reaches normal operating range. The temperature of the engine oil should also rise slowly to normal operating range.

Oil, coolant, and charging circuit warning lights will come on at first. That tells you the lights are working. They should go out right away unless there's a problem.

Make sure your controls work. Check all of the following for looseness, sticking, damage, or improper setting:

- steering wheel
- accelerator ("gas pedal")
- foot brake
- trailer brake (if your vehicle has one)
- parking brake
- retarder controls (if your vehicle has them)
- transmission controls
- interaxle differential lock (if your vehicle has one)
- horn(s)
- windshield wiper and washers
- headlights
- dimmer switch
- turn signal
- four-way flashers
- clearance, identification, and marker light switch(es).

If your vehicle has a clutch, test it now. Depress the clutch until you feel a slight resistance. One to two inches of travel before you feel resistance is normal. More or less than that signals a problem.

Check your mirrors and windshield for the defects described earlier.

Check that you have the required emergency equipment and that it is in good operating condition.

Check for optional items, such as a tire changing kit, and items required by state and local laws, such as mud flaps.

Step Four: Check Lights

Next, check to see that the lights are working. Make sure the parking brake is set, turn off the engine, and take the key with you. Turn on the headlights (on low beams) and the four-way flashers, and get out. Go to the front of the vehicle. Check that the low beams are on and both of the four-way flashers are working. Push the dimmer switch and check that the high beams work. Turn off the headlights and four-way hazard warning flashers. Turn on the parking, clearance, side-marker, and identification lights. Turn on the right-turn signal, and start the "walk-around" part of the inspection.

Step Five: Walk-around Inspection

Walk all around the vehicle, inspecting as you go. Start at the cab on the driver's side. Cover the front of the vehicle. Work down the passenger side to the rear. Cover the rear of the vehicle. Work up the driver's side back to the starting position. Do it this way every time.

Clean all the lights, reflectors, and glass as you go along.

Left Front Side

The driver's door glass should be clean. Door latches or locks should work properly.

Perform a tire, wheel, and rim check at the left front wheel. Perform a suspension system and braking system check. Check the steering parts.

Front

Check the condition of the front axle. Perform a check of the steering system parts. You should actually take hold of steering system parts and shake them to say for sure that they are not loose.

Check the windshield. Pull on the windshield wiper arms to check for proper spring tension. Check the wiper blades for the defects described earlier.

Check the lights and reflectors (parking, clearance, and identification and turn signal lights) at this site. Make sure you have all the required ones as described earlier. They should be clean and working.

Right Front Side

The right front has parts similar to the left front. Perform the same checks. The passenger door glass should be clean. Door latches or locks should work properly. Perform the tire, wheel, and rim check at the right front wheel. Perform a suspension system and braking system check. Check the steering parts.

For cabover (COE) trucks, check the primary and safety cab locks here.

Right Side

The fuel tank is located here. Check to see it is mounted securely, not damaged or leaking. The fuel crossover line must be secure, and not hanging so low that it would be exposed to road hazards. The filler caps must be on firmly. Check the fuel supply. Perform these checks for each fuel tank your vehicle has.

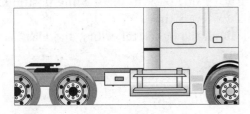

Figure 9-10. Check the fuel tank.

Check all parts you can see at this site. See that the rear of the engine is not leaking. Check for leaks from the transmission. Check the exhaust system parts for defects as described earlier.

Inspect the frame and cross members for bends and cracks.

Electrical wiring should be secured against snagging, rubbing, and wearing.

If your vehicle has a spare tire carrier or rack, check it for damage. It should be securely mounted in the rack. Make sure you have the correct size spare tire and wheel. The spare tire must be in good condition and properly inflated.

Cargo Securement

Check the cargo securement at the right front of the vehicle. Cargo must be properly blocked, braced, tied, chained, and so forth. If you need a header board, make sure it is strong enough and secure. If you have sideboards, check them. The stakes must be strong enough for the load, free of damage, and properly set in place.

Make sure you haven't created handling problems for yourself by overloading the front axles.

Any canvas or tarp must be properly secured to prevent tearing and billowing.

If you are hauling an oversized load, have the necessary signs safely and properly mounted. Make sure you have the permits you need to haul this load.

Check that the curbside cargo compartment doors are securely closed, latched, and locked. See that the required security seals are in place.

Right Rear

At this location, check the wheels, rims, tires, brake system, and suspension parts for the defects described in the first part of this chapter. Check that all parts are sound, properly mounted and secure, and not leaking. Check the axle. Powered axles should not leak.

Check the lights and reflectors required at this location. They should be clean and working.

Rear

Make sure you have a license plate. It should be clean and securely mounted.

If mud flaps are required, check them now. They should be free of damage and fastened according to whatever regulations apply. If there are no legal guidelines, make sure they are at least well-fastened, not dragging on the ground or rubbing the tires.

Check the lights and reflectors required at this location. They should be clean and working.

Cargo Securement

Check the cargo securement at the right rear of the vehicle. The cargo must be properly blocked, braced, tied, chained, and so forth. Check that the tailboard is strong enough and secure. If you have end gates, check them. The stakes must be strong enough, free of damage, and properly set in the stake pockets.

Any canvas or tarp must be properly secured to prevent tearing and billowing. Make sure the tarp doesn't cover up your rear lights.

Check to make sure you haven't distributed the load so as to overload the rear axles.

If you are hauling an oversized load, have the necessary signs safely and properly mounted. Have any lights or flags required for projecting loads. Make sure you have the permits you need to haul this load.

Check that the rear doors are securely closed, latched, and locked.

Left Rear and Front Side

At this location, check the wheels, rims, tires, brake system, and suspension parts just as you did for the right rear and front side. Check that all parts are sound, properly mounted and secure, and not leaking. Check the axle. Powered axles should not leak.

Check the lights and reflectors required at this location. They should be clean and working.

Your vehicle's battery may be mounted here, rather than in the engine compartment. If so, perform the battery check now.

Step Six: Check Signal Lights

Get in the cab and turn off all the lights. Then, turn on the stop lights (apply the trailer hand brake). Or, have a helper step on the brake pedal while you watch to see if the light goes on.

Now turn on the left-turn signal lights. Then get out and check the left side lights, rear and front.

Step Seven: Check Brakes

Now you'll perform some tests with the engine running and the parking brakes off. Get back in the vehicle. Turn off any lights you don't need for driving. Check to see that you have all the required papers, trip manifests, permits, and so forth. Secure all loose articles in the cab.

Start the engine. Test for leaks in the hydraulic brake system. Do you recall the test? Pump the brake pedal three times. Then apply firm pressure to the pedal and hold for five seconds. The pedal should not move.

If your vehicle has air brakes, you would test them now. You'll find the air brake test procedure outlined in Chapter 10.

Put the vehicle in gear, and slowly release the clutch pedal just enough to check if the trailer brakes are working. If you feel a restriction, quickly engage the clutch again. You have now successfully tested your trailer brakes. If you did not have a restriction and the trailer rolls freely, you have faulty trailer brakes that need to be serviced by a licensed air-brake mechanic.

Test how well the service brake works. Remember how? Go about five mph. Push the brake pedal firmly. The brakes should apply evenly all around the vehicle. They should not pull to one side or the other. The brake pedal should not travel all the way to the floor before the brakes apply. Nor should the pedal give you a lot of resistance, or take great effort to apply.

This completes the pre-trip inspection. However, this is not the end of your vehicle inspection responsibilities. You should check your vehicle regularly. You should check your instruments and gauges often. Every time you leave or return to your vehicle, take the long way around. Check the mirrors, tires, lights, and cargo as you go. If you see, hear, smell, or feel anything that might mean trouble, look into it.

You know from Chapter 5 that drivers of trucks and truck tractors must inspect their vehicles during the trip. They should perform an inspection within the first 50 miles of a trip and every 150 miles or every three hours (whichever comes first) afterward. This inspection should include:

- tire pressure
- brakes
- coupling devices

Make sure neither the tires nor the brakes have overheated. You can do this by putting the back of your hand near (not on) the tire or brake drum. (Be careful. When these parts heat up, they get very hot.)

Step 1: Vehicle overview for general condition

Step 2: Engine compartment checks
Fluid levels and leaks
Hoses and belts
Battery
Windshield washer
Wiring

Step 3: Inside the cab checks
Parking brake on
Gauge readings
Warning lights
Controls
Emergency equipment
Optional equipment

Step 4: Check lights
Headlights (low and high beams)
Four-way flashers
Parking lights
Clearance lights
Side marker lights
Identification lights
Right-turn signal

Step 5: Walk-around inspection
A. Left front side: wheel and tire, suspension,
brakes, axle, steering, side marker lamp
and reflector, door glass, latches and locks,
mirrors
B. Front of cab: axle, steering system,
windshield, lights and reflectors
C. Right front side: door glass, latches and
locks, wheels and tires, suspension, brakes,
axle, steering, condition of bed or body, side
marker lamp and reflector
D. Right side: fuel tank, exhaust system, frame
and cross members, wiring, spare tire, lights
and reflectors, mirrors
E. Cargo securement, curbside doors
F. Right rear: wheels and tires, suspension,
brakes, axle, lights and reflectors
G. Rear: lights and reflectors, license plate,
mudflaps, condition of bed or body, rear
doors
H. Cargo securement
I. Left rear and left front: wheels and tires,
suspension, brakes, steering, axle, lights and
reflectors, battery, condition of bed or body

Step 6: Check signal lights

Step 7: Brake check

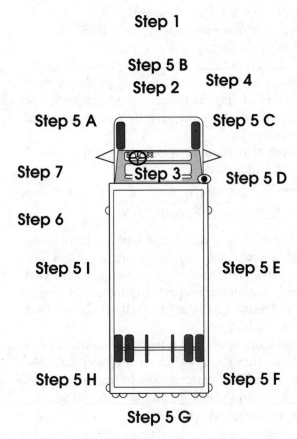

Figure 9-11. Straight Truck Inspection Aid.

As you know, you may have to make a written report each day on the condition of the vehicle you drove. Report anything that could affect safety or lead to a breakdown.

The vehicle inspection report tells the vehicle owner about problems that may need fixing. Keep a copy of your report in the vehicle for one day. That way, the next driver can learn about any problems you have found.

Different driving conditions call for special preparation. Examples are night driving, hot weather, cold weather, and mountain driving. You'll add some steps to your vehicle inspection to make sure you are prepared for these special conditions. Special preparations for driving at night, in extreme temperatures, and in the mountains were detailed in Chapter 7. You may want to go over those sections one more time before going on to the next chapter.

That completes a multi-chapter summary of the knowledge and skills you need to get a "basic" CDL with an air brake restriction and no endorsements. The next four chapters are for those who want to drive vehicles with air brakes, combination vehicles, and vehicles with double or triple trailers. Drivers who plan to haul liquid loads or hazardous materials requiring placards will also need these chapters.

Remember, you can transport hazardous materials not requiring placards with a "basic" CDL. Therefore, all licensed CMV drivers must have some knowledge of hazardous materials. That part of Chapter 5 dealing with FMCSR Part 397 covers this aspect of transporting hazardous materials.

PASS POST-TRIP

Instructions: For each true/false test item, read the statement. Decide whether the statement is true or false. If it is true, circle the letter A. If it is false, circle the letter B. For each multiple-choice test item, choose the answer choice—A, B, C, or D—that correctly completes the statement or answers the question. There is only one correct answer. Use the answer sheet provided on page 213.

1. You could be put out of service in one state for a vehicle condition that was acceptable in another state.
   A. True
   B. False

2. Which of these defects can prevent the brakes from working properly?
   A. cracks in the brake drum
   B. oil, grease, and brake fluid on the shoes or pads
   C. brake pads that are 1/8-inch thick
   D. all of the above

3. A windshield is considered defective if it has even one tiny crack.
   A. True
   B. False

4. The proper level for fuel in your vehicle's fuel tanks is _____.
   A. at "Full"
   B. nearly full
   C. at "Add"
   D. at "Low"

5. The best way to check tire inflation pressure is _____.
   A. use a tire billy
   B. kick the tire
   C. visually inspect the tire to see if it is low
   D. use a tire gauge

6. Hoses and wires that hang close to exhaust system parts don't burn because they are protected by insulation.
   A. True
   B. False

7. Optional equipment includes _____ .
   A. rear-view mirrors
   B. a tire-changing kit
   C. windshield wipers
   D. treaded tires

8. Your emergency equipment must include three reflective devices and _____, if your vehicle uses them.
   A. circuit breakers
   B. spare fuses
   C. pot torches
   D. three accident notification kits

9. Suspension system defects _____.
   A. can allow the axle to move out of its proper position
   B. are too complex to spot during a pre-trip inspection
   C. are present only on older trucks
   D. do not occur on vehicles with air suspensions

10. Steering wheel free play should not be more than $2\frac{1}{2}$ inches on a 20-inch steering wheel.
    A. True
    B. False

CHAPTER 9 ANSWER SHEET

Test Item	Answer Choices				Self - Scoring	
					Correct	Incorrect
1.	A	B	C	D	☐	☐
2.	A	B	C	D	☐	☐
3.	A	B	C	D	☐	☐
4.	A	B	C	D	☐	☐
5.	A	B	C	D	☐	☐
6.	A	B	C	D	☐	☐
7.	A	B	C	D	☐	☐
8.	A	B	C	D	☐	☐
9.	A	B	C	D	☐	☐
10.	A	B	C	D	☐	☐

Cut Here

Chapter 10

Air Brakes

In this chapter, you will learn about:

- how air brakes work
- the parts of the air brake system
- the causes of air brake failure
- how to respond to loss of service brakes
- how to inspect and maintain air brakes
- driving with air brakes

To complete this chapter you will need:

- a dictionary
- a pencil or pen
- blank paper, a notebook, or an electronic notepad
- colored pencils, pens, markers, or highlighters
- a CDL preparation manual from your state Department of Motor Vehicles, if one is offered
- the Federal Motor Carrier Safety Regulations pocketbook (or access to U.S. Department of Transportation regulations, Parts 383, 393, and 396 of Subchapter B, Chapter 3, Title 49, Code of Federal Regulations)
- the owner's manual for your vehicle, if available

PASS PRE-TRIP

Instructions: Read the statements. Decide whether each statement is true or false. If it is true, circle the letter A. If it is false, circle the letter B.

1. The compressed air in the brake chamber transmits the force exerted by the driver's foot on the brake pedal to the foundation brakes.
   A. True
   B. False

2. The emergency brake system is a completely separate system and is completely under the driver's control.
   A. True
   B. False

3. If you have an alcohol evaporator, you do not have to drain the air tanks.
   A. True
   B. False

Many large CMVs have air brakes. Air brakes are a safe way of stopping large vehicles, but only if the brakes are well maintained and used right.

If you want to drive a vehicle with air brakes, you have to show that you know how they work. You must show that you understand what causes them to fail. You must show that you recognize problems and defects in the air brake system. You must show that you know how to operate air brakes. You'll show that you have this knowledge and skill by taking two tests just on air brakes. You'll take a special Knowledge Test and a special Skills Test.

If you can't do these things, you can still get a CDL. It will be a restricted license. You will be restricted to driving vehicles without air brakes.

BRAKING WITH COMPRESSED AIR

Recall the discussion of hydraulic fluid in Chapter 6? There you learned that hydraulic fluid transmits braking force to the mechanical parts in the brake chambers. Stepping harder on the brake pedal puts more pressure on the fluid. That puts more pressure on the mechanical parts.

The air in air brakes works differently. Unlike fluid, air can be compressed. It can be made more compact and stored in a small space. Compressing air creates energy. Keeping air under pressure stores that energy. Releasing the compressed air releases energy. Compressed air actually exerts force on the mechanical parts in the brake chamber.

Air is compressed by the compressor. Stepping on the brake pedal simply opens a valve. This allows the compressed air to leave the reservoir where it's stored. Stepping on the brake pedal harder or repeatedly does not increase the pressure.

Put another way, you can press the pedal down part way. This partially opens the valve. That allows some air pressure to flow to the brake chamber. The result is light braking. Putting the pedal to the floor opens the valve all the way. That allows a full application of air pressure to flow to the brake chamber. Pressing the pedal again gives you another application, but it does not give you greater pressure.

You'll understand this idea better after tracing the flow of air through the system. It's an important idea to understand. The idea explains why pumping or fanning air brakes results in a decrease, rather than an increase, in braking power.

You'll see how compressed air works in the brake chambers. Then you'll see how the compressed air gets to the brake chambers.

THE BRAKE CHAMBER

Air pressure moves the brake shoes and brake pads into contact with the brake drums. This creates friction. The friction ultimately causes the wheels to stop turning.

The Brake Drums

Brake drums are made of iron or steel. They are bolted to the wheels, so the wheel and the drum rotate together. The inside surface of a brake drum must be smooth and uniform. If there are scores or ridges cut into the surface, the brake shoes may not make complete contact with the drum. That could result in poor brake performance.

The braking mechanism, which consists of the brake shoes, the brake linings, and the foundation brakes, is found inside the drum. It is the action of the brake shoes pushing the brake lining against the brake drum surface that produces friction and stops the vehicle.

Brake Shoes and Lining

Each brake drum contains two brake shoes with attached linings that are made of metallic mineral fiber.

Foundation Brakes

Foundation brakes are used at each wheel. Foundation brakes are the parts of the brake that

don't rotate. The most common type is the S-cam drum brake. There are also wedge brakes and disc brakes. Wedge brakes and disc brakes are less common than S-cam brakes.

S-cam Brakes

When you push the brake pedal, air is let into each brake chamber. Air pressure pushes against a diaphragm attached to the push rod. This pushes the push rod out. This moves the slack adjuster, which twists the brake camshaft. The brake camshaft turns the S-cam (so called because it is shaped like the letter "S"). The S-cam forces the brake shoes away from one another and presses them against the inside of the brake drum.

When you release the brake pedal, the S-cam twists back. The return spring pulls the brake shoes away from the drum. That lets the wheels roll freely again.

Locate all the parts mentioned so far in Figure 10-1.

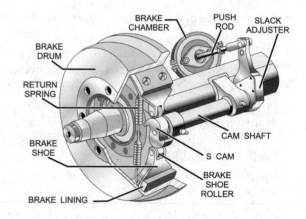

Figure 10-1. S-cam air brake.

Wedge Brakes

In wedge brakes, the brake chamber push rod pushes a wedge directly between the ends of two brake shoes. This shoves them apart and against the inside of the brake drum. Wedge brakes may have a single brake chamber, or two brake chambers, pushing wedges in at both ends of the brake shoes. Wedge type brakes may be self-adjusting or may require manual adjustment.

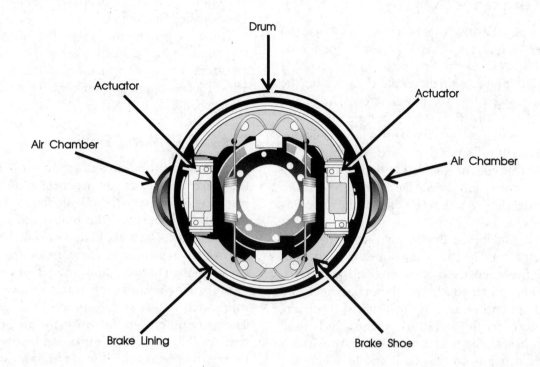

Figure 10-2. Wedge brake.

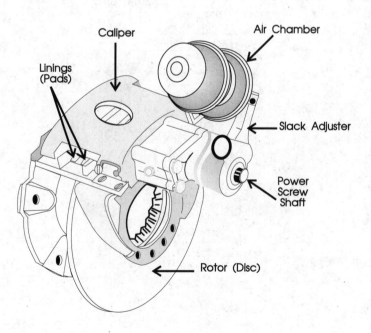

Figure 10-3. Disc brake.

Disc Brakes

In air-operated disc brakes, air pressure acts on a brake chamber and slack adjuster, like S-cam brakes. Instead of the S-cam, a "power screw" is used. The pressure of the brake chamber on the slack adjuster turns the power screw. The power screw clamps the disc or rotor between the brake lining pads of a caliper, similar to a large C-clamp.

Figure 10-3 shows a disc brake assembly.

All three types of foundation brakes have brake drums, linings, and shoes. (Disc brakes have disks, not shoes. However, they work about the same way as shoes.)

You'll find brake drums on each end of the vehicle's axles. The wheels are bolted to the drums. As you have seen, the braking mechanism is inside the drum. When you brake, the brake shoes and linings are pushed against the inside of the drum. This causes friction, which slows the vehicle (and creates heat). The heat a drum can take without damage depends on how hard and how long the brakes are used.

Brake Fade

Too much heat can make the brakes stop working. The drum expands. The brake shoes with their linings have to travel farther to contact the drum.

They don't contact the drum with as much force. This is called "brake fade."

The more you use the brakes, the more the heat will build up. Individual brakes in the system that are doing the work will develop the most heat. They'll lose stopping power first. That will affect vehicle handling.

Loss of Braking Power

The drums can heat up and expand so much that the brake shoes can't contact them at all. That's a complete loss of service braking power.

Continued abuse of the brakes can cause the metal to crack. The brake linings can burn. The seals can soften, expand, and allow grease to leak out. Create enough heat, and the grease itself can ignite.

Brake fade and loss of braking power due to overheated brakes is a common brake problem. There are other ways in which the air brake system can fail. Since it's air pressure that works the brakes, problems with the air supply can mean brake failure too.

Next you'll learn about the other parts of the air brake system. You'll trace the flow of air from the compressor to the brake chamber. Then you'll learn about places in the system where problems can occur.

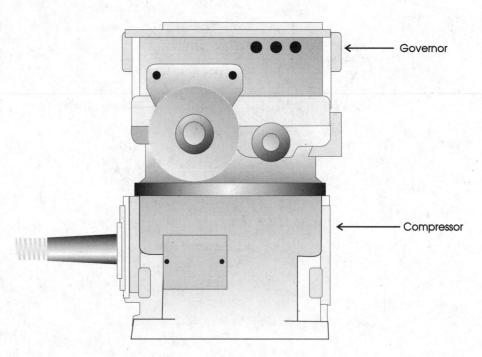

Figure 10-4. Air compressor and governor.

THE BRAKE SYSTEM

Compressor

The compressor is a machine that draws in the air around it. It pumps that air into a smaller space to increase its pressure. Then it pumps the air into the air reservoir system where it is stored in air tanks until it is needed. The engine provides the power for the compressor, so it is usually mounted on the side of the engine.

The compressor may be air-cooled. Or, it may be cooled by the engine's cooling system. It may have its own oil supply, or it may be lubricated by engine oil.

Air Governor

Located on the compressor, the air governor controls the compressor. It regulates the amount of air pressure in the system. When air tank pressure rises to the "cut-out" level, the governor stops the compressor from pumping air. The cut-out level ranges between 120 and 130 psi. The mid-range at 125 psi is common. When the tank pressure falls to the "cut-in" pressure, the governor allows the compressor to start pumping again. Cut-in pressure is around 100 psi.

Air Dryer

When air is compressed, it heats. As air cools off, any moisture in it condenses. Also, small amounts of oil from the compressor are vaporized and travel out of the compressor with the air. When the compressed air cools, this oil also condenses. The result will be a sludge. This can clog and corrode valves if it's not removed from the system.

In freezing weather, this sludge can freeze in the lines and the valves. An alcohol evaporator can help prevent this.

When the air leaves the compressor, it flows through the air dryer. This cleans and removes moisture and vaporized oil from the compressed air. The air dryer does a pretty good job of removing the oil and moisture. The dryer doesn't get it all, though. You'll need to finish the job yourself. You'll learn how in the Inspection and Maintenance section.

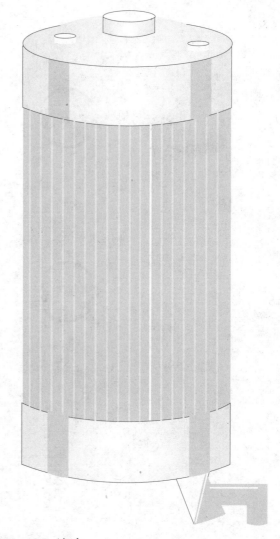

Figure 10-5. Air dryer.

The Air Reservoir System

From the air dryer, the compressed air goes to the reservoir system. Reservoirs (or tanks) store the compressed air until it is needed. Three types of tanks are used in air reservoir systems: single compartment, baffled, and multi-compartment.

The number and size of tanks varies with the vehicle. Figure 10-6 illustrates a dual circuit air brake system with three tanks. A dual circuit system features separate air tanks for the front and rear axles. The first reservoir tank in a dual circuit air brake system is called the main supply tank, or the wet tank. One of the dry air tanks on the tractor supplies air to the brakes on one tractor axle. Another supplies air to the other axles(s) on the tractor. One tank is usually called the primary

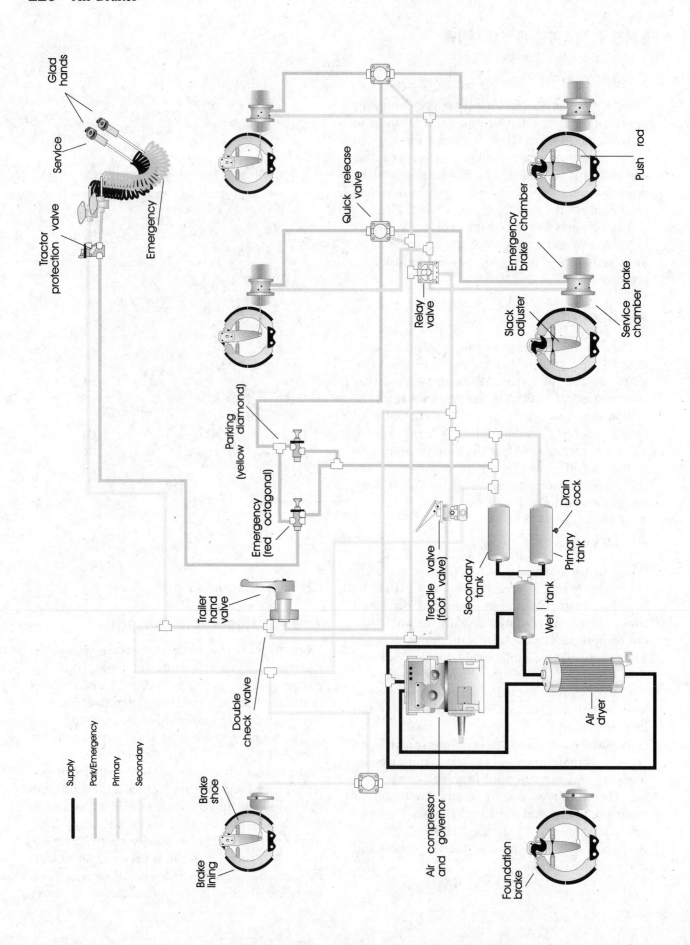

Glad hands

Service

Emergency

Tractor protection valve

Quick release valve

Push rod

Emergency brake chamber

Relay valve

Slack adjuster

Service brake chamber

Parking (yellow diamond)

Emergency (red octagonal)

Drain cock

Primary tank

Secondary tank

Wet tank

Treadle valve (foot valve)

Traller hand valve

Air dryer

Double check valve

Supply
Park/Emergency
Primary
Secondary

Brake shoe

Brake lining

Air compressor and governor

Foundation brake

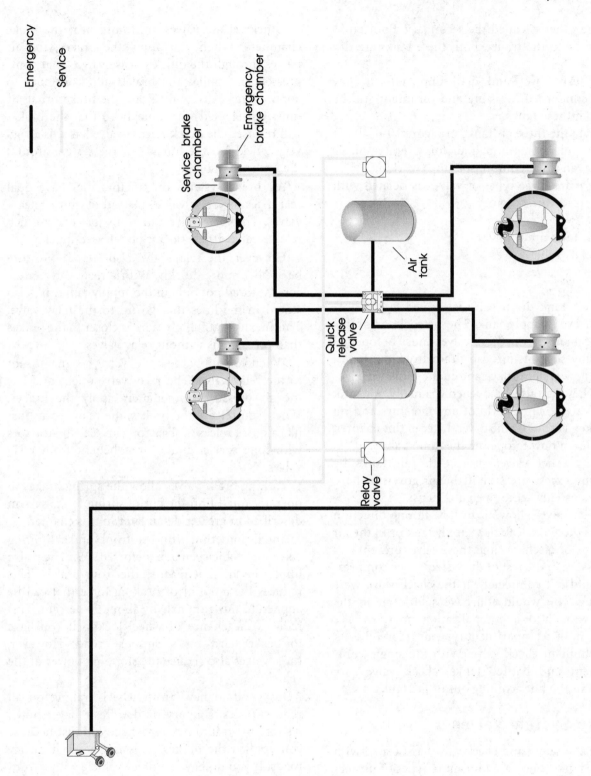

Figure 10-6. Dual circuit air brake system.

and the other is called the secondary. Both tanks supply air to the trailer. Both these tanks are dry tanks.

The terms "wet" and "dry" simply refer to how much condensed moisture and oil might still be found in these tanks.

Locate the three air tanks in Figure 10-6.

The tanks hold enough air for several applications, even if the compressor stops working.

There are three types of valves associated with the air reservoir system:

- air tank drain valves
- safety relief valves
- check valves

An air tank drain valve is located at the bottom of each supply tank. The petcock, or draining mechanism, on these valves must be opened manually so moisture can drain from the tanks. Many new systems have automatic moisture ejectors. They can also be opened manually. Air tank drain valves help get rid of any moisture and oil not taken care of by the dryer. Later in this chapter, you'll learn how to operate the drain valves.

Safety relief valves protect the air tanks by releasing excess pressure if the air governor fails. They're usually set to open at 150 psi.

Check valves allow air to flow in one direction only. If there is a leak in a supply tank or in the air compressor discharge line, these valves prevent loss of pressure in the rest of the system. Your compressor could fail completely. If the check valve were working, you would still have a little air in the tanks, enough for a few applications anyway.

The multi-compartment type air reservoir tank has a built-in check valve. With the single compartment and baffled tanks, check valves are placed in the lines coming out of the tanks.

Brake System Valves

Beside the safety and check valves associated with the air tanks, there are four more types of protective valves:

- quick release valves
- relay valves
- spring brake valve
- front brake limiting valve

Quick release valves are found near the brake chambers. When you apply the brakes, the air passes through the quick release valve. That compressed air applies the brakes and continues to apply them. Once you release the brakes, that air must be released very quickly so that the brakes will release. The quick release valve lets this air go very quickly to the outside of the system around the brake chamber.

Air brake systems on tractors that have dual rear axles use relay valves instead of quick release valves. The relay valve functions just like a quick release valve. It lets the air be released quickly.

However, the relay valve also lets air pressure be applied more quickly. With a relay valve, pressure is stored not only in the supply tanks. It's also stored in the lines that go up to the relay valve. This means that full pressure is closer to the brakes than on systems without relay valves.

When you apply the brake valve, a signal is sent to the relay valve. The relay valve opens and lets the air pressure immediately apply the brakes. When you release the brakes, the relay works just like a quick release valve. You can see why tractors with long service brake lines benefit from relay valves.

Parking brake valves allow the spring brakes to operate. You'll find the entire spring brake system described in greater detail later in this chapter.

You'll sometimes find a front brake limiting valve on vehicles made before 1975. The valve limits the air that reaches the front brakes. This reduces the amount of braking force that can be applied to the front axle by about 50 percent. This reduces the chance of wheel lockup if you have to brake hard on a slippery road. However, these valves also reduce the stopping power of the vehicle.

It's good to have front wheel braking on all road surfaces, slippery or dry. Tests have shown it's rare to go into a front-wheel skid just because you applied the brakes. This is true even on icy roads. If you do have an older vehicle with a front brake limiting valve, you should never have it on when the road is dry. That is unsafe and illegal. Regulations allow you to use this valve if you use it correctly. That is, you should have it on only when roads are slippery.

Many vehicles have automatic front wheel limiting valves. They reduce the air pressure to the front brakes except when the brakes are put on very hard. That would be 60 psi or more application pressure. These valves cannot be controlled manually.

AIR BRAKE SYSTEM OPERATION

Now you know enough to take a close look at how the service brakes work. The compressor draws in surrounding air, compresses it, and pumps it to the air dryer. The air dryer removes moisture from the compressed air, which then flows into the air reservoir system.

You press on the treadle. This sends a signal to the relay valve, if your vehicle has one. The relay valve opens and air enters the brake chamber through the air inlet. The pressurized air pushes the diaphragm. The diaphragm pushes the push rod. The push rod pushes the slack adjuster. The slack adjuster twists. This twisting action turns the brake camshaft. The brake camshaft moves the S-cam, wedge, or disc. This pushes the brake shoes and linings against the brake drum. This creates friction, which slows and stops the turning of the brake drum. Because the brake drum is attached to the wheel, the wheel also stops turning.

When you release the treadle, the signal to the relay valve stops. The relay valve closes the air inlet and quickly releases the air inside the brake chamber.

Air brake systems are really three braking systems combined:

- the service brake system
- the parking brake system
- the emergency brake system

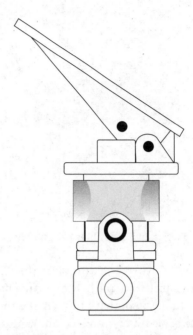

Figure 10-7. Treadle valve.

The service brake system applies and releases the brakes when you use the brake pedal during normal driving. The parking brake system applies and releases the parking brakes when you use the parking brake control. The emergency brake system uses parts of the service and parking brake systems to stop the vehicle in the event of a brake system failure.

The spring brake provides parking and emergency braking. When you apply the spring brake to park, it's the parking brake. When the spring brake comes on automatically in response to a loss of air pressure, it's the emergency brake.

You've seen how compressed air works in the service brake. Now let's see how compressed air works in the spring brake.

COMPRESSED AIR IN THE SPRING BRAKE

The spring brake chamber "piggybacks" on the service brake chamber. A heavy duty spring in the spring brake chamber is held in a compressed position by air pressure. This releases the parking brake. When the air pressure drops, the spring expands. This allows a piston in the spring brake chamber to move. The piston drives the diaphragm and push rod in the service brake chamber. That puts the foundation brakes into operation. The brakes are applied.

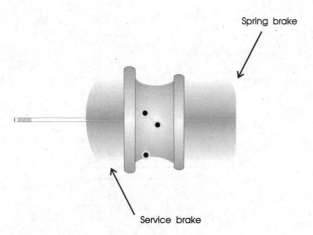

Figure 10-8. Service and spring brake chambers.

This heavy-duty spring will stay in this expanded position by itself. To push the spring back and release the brakes, you have to recharge the spring brake chamber with compressed air. When the system recharges, the spring is pushed back and held back from the brakes. The brakes are released, and you can move the vehicle again.

Recall what FMCSR Part 393 requires of parking brakes? Air pressure may be used to apply the parking brakes. It must not be used to keep the parking brakes applied. Also, when you release the brakes, the system must be such that you can immediately reapply them.

Parking Brake

When you operate this spring brake system on purpose, it's a parking brake. You operate the parking brake control in the cab. Pull the yellow, diamond-shaped knob out. This works a valve that exhausts the compressed air from the spring brake chamber. The spring brakes go into action. The brakes remain on until you recharge the spring brake chamber. You do this by pushing the control knob back in.

See how the parking brake system we just described fills the bill? The air-pressure assisted spring brake uses air only to keep the brakes off. Once on, the strength of the spring itself keeps the brakes on. You cannot take them off again unless you have enough air pressure to keep them off.

Don't use the parking brakes if the brakes are very hot. (They probably are if you've just come down a steep grade.) You can damage them. Let hot brakes cool before you use them. If you use them in cold, wet weather, they can freeze. Then

you won't be able to move the vehicle. Use wheel chocks instead.

If the brakes are wet, you can dry them while driving. Apply the brakes gently while you are in low gear. The heat that's created will dry the brakes.

Never push the brake pedal down when the spring brakes are on. This is called "compounding." The combined forces of air pressure and the springs could damage the brakes. Some vehicles are designed to prevent this. You may not be sure if yours is or not. Also, the system may not always work. So it's still a good idea to remember never to push the brake pedal down while the spring brakes are on.

Above all, never leave your vehicle unattended without either putting the parking brakes on or chocking the wheels. Your vehicle could roll away. You can imagine the damage that would do.

Emergency Brakes

As you know, your vehicle must have emergency brakes. The spring brake serves as an emergency brake as well as a parking brake. You've seen how reducing the air pressure in the spring brake chamber will make the spring brakes come on.

What would cause a problem with your service brakes? A loss of pressure in the system would be one cause. If you lose air pressure in the system, the spring brake chamber will be affected. Lose enough pressure, and there is no longer enough to hold the spring back. The spring brakes come on.

"Enough" in this case is a drop of air pressure to a range of 20 to 45 psi. Should your vehicle's air pressure drop this low, the emergency brakes will come on.

This meets FMCSR requirements very well. The emergency brakes work in response to a loss of brake air pressure. They will stay applied in spite of a loss of air pressure, because they work mechanically.

You don't have to wait for the emergency brakes to come on. You should have some warning there's a problem long before it gets that bad. Your warning devices should alert you when your air pressure drops below normal. That's the time to stop, while you still have some braking power. Get the problem fixed immediately.

How well the spring brakes work depends on how well the brakes are adjusted. You've seen how the service brakes can fail if they are out of adjustment. The spring brakes push out the push rod just like the service brakes do. So if the service brakes are out of adjustment, you won't have good emergency or parking braking power, either.

OPERATIONAL CONTROL VALVES

In the cab are different pedals and knobs you use to control the many operations of the brake system. These are often called valves. Some of them truly are valves. Others are really controls that operate a valve elsewhere in the system.

One of these controls is the brake pedal.

Brake Pedal

The brake pedal is also called a foot or treadle valve. It operates a valve that supplies air pressure to the braking system. When you press on the treadle, air pressure is sent through the air lines to the brake chambers.

If your vehicle uses relay valves, then the air is already at the brake chambers being held back by the relay valves. When you press on the treadle, the relay valve opens. When you release the treadle, the air exhausts through the quick release valve and the brakes are off.

Parking Brake Valve

Another control is the parking brake valve. On late model vehicles, this is a diamond-shaped, yellow, push-pull knob. You pull the knob out to apply the spring brakes for parking. You push it in to release them. (On older vehicles, the parking brakes may be controlled by a lever. On still other vehicles, the parking brake is a round, blue knob.)

Modulating Control Valve

Your vehicle may have a control on the dashboard that regulates the spring brakes. This is called a modulating control valve. You use it to apply the spring brakes gradually, instead of all at once.

If your vehicle is so equipped, check the owner's manual for operating instructions. Or, ask your supervisor about the proper use of this control.

Dual Parking Control Valves

When the main air pressure is lost, the spring brakes come on automatically. To move the vehicle again, you have to release the spring brakes. For this, you need a fresh supply of air pressure. Some vehicles have a separate air tank for the spring brakes. It has a supply of air large enough to release the spring brakes. Then you can move the vehicle off the roadway, out of danger, or to someplace where you can repair the problem in the service braking system that caused the emergency brakes to come on.

This dual control consists of two knobs. One is a push-pull type you use to apply the spring brakes for parking. The other valve is spring loaded in the "Out" position. When you push the control in, air from the separate air tank releases the spring brakes. Then you can move the vehicle. When you release the button, the spring brakes come on again.

There is only enough air in the separate tank to do this a few times. So you must plan your next move carefully. You clearly can't rely on this system to complete your run. It's just enough to move the vehicle away from danger or to a service station.

If you have an older vehicle with a front brake limiting valve, you will find a control for it in the cab. The control is usually marked "Normal" and "Slippery." Switching it to "Slippery" reduces the air pressure at the front wheels. The "Normal" setting gives you full braking power. As you read earlier, front wheel braking has been shown to be good for all road conditions. Most experts recommend you leave the front brake limiting valve on "Normal" at all times.

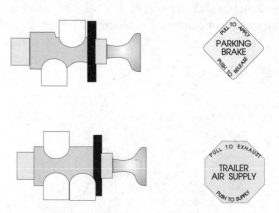

Figure 10-9. Parking and emergency cab control valves.

GAUGES AND WARNING DEVICES

Your vehicle has instruments that inform you of the braking system's operating condition. Warning devices alert you to problems that have occurred.

Air Pressure Gauge

The air pressure gauge, also known as the supply pressure gauge, is marked with the word "Air" on the lower portion of its face. The air gauge shows a reading in pounds per square inch (psi). It indicates psi available in the reservoirs for braking power. The normal gauge reading is 100 to 120 psi (up to 125 in some vehicles). The pressure will vary in this range because of compressor operation and brake application.

If your vehicle has a dual air brake system, there will be a gauge for each half of the system. You may have one gauge with two needles.

Don't drive a vehicle with air brakes until the gauge reads at least 90 pounds psi. Should the pressure drop below 90 pounds when you are driving, stop immediately. If you don't, you may lose all braking power.

Application Pressure Gauge

Some vehicles have an application pressure gauge. This gauge shows the amount of pressure applied to the brakes. Naturally, a heavy brake application will cause a higher reading than a light brake application. When the brakes are released, the gauge pointer should return to "Zero."

When going down steep grades, you may see the brake pressure going up, while your speed remains the same. This is a sign of brake fade. It can also mean your brakes are out of adjustment, leaking, or have other mechanical problems.

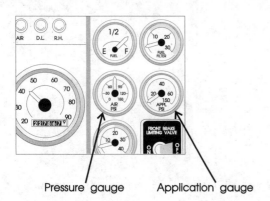

Pressure gauge   Application gauge

Figure 10-10. Air pressure and application gauges.

Warning Devices

Two devices warn you of low air pressure: the warning light and warning buzzer. If you have air brakes, your vehicle must have at least one of them. They are located in the cab. If the pressure in any one of your service air tanks drops below 60 psi while the ignition is on, these devices will alert you. If a low air pressure warning device comes on, pull off the road as soon as it's safe to do so. Do not resume driving until the air pressure problem has been corrected.

Some vehicles with air brakes also use a wigwag low pressure indicator. This is mounted above the windshield. If the air pressure drops below 60 psi, the signal arm will drop down across the windshield. You can't reset it (push it back out of view) until air pressure is brought above 60 psi. Don't operate the vehicle with the signal arm lever down.

One last device connected with the air brake system is the stop light. In an air brake system, air pressure operates an electric switch. This switch turns on the brake lights when you put on the brakes. This tells drivers behind you that you are slowing or stopping.

AIR BRAKES IN TRACTOR-TRAILER COMBINATIONS

As you have seen, when you are pulling a trailer, some of the air supply must go to the trailer service brakes. The air brake systems in tractors designed to pull trailers have more parts. These control the trailer brakes.

Tractor Protection Valve

If your vehicle is designed to pull a trailer, you will have a tractor protection valve. The tractor protection valve has two other names: the Emergency Trailer Air Supply Control Valve and the Breakaway Valve. It's controlled by a push-pull, eight-sided red knob located on your dashboard (pictured in Figure 10-9). Like the spring brake valve, this valve also works automatically.

This valve's job is to protect the tractor air tanks in case of air pressure loss. The valve itself is located at the point where the flexible air lines that go to the trailer are connected. The valve separates the tractor air supply from the trailer air supply.

Wigwag

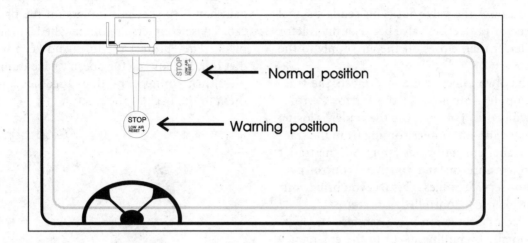

Warning Light

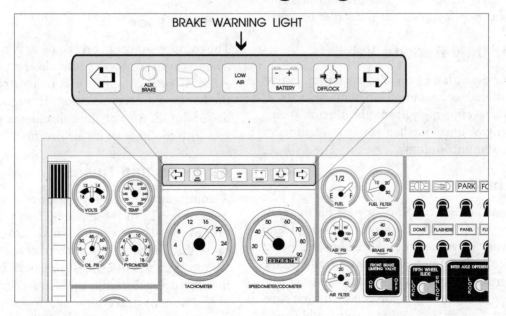

Figure 10-11. Low-air warning devices.

If anything goes wrong with the trailer system that causes it to lose air pressure below 20 to 45 psi, a spring in the tractor protection valve on the dashboard pops the valve out. This sends a signal to the tractor protection valve between the tractor and trailer. It then closes off the air supply to the trailer.

This has two effects. One, it protects the tractor air supply from loss, so the tractor's service brakes will work. Two, because the trailer is losing air pressure and no more is coming from the tractor, the trailer spring brakes apply. So, the trailer spring brakes are on and you have control over the tractor service brakes. This lets you bring your tractor-trailer to a controlled, safe stop.

The tractor protection valve can also be operated manually by pulling it out to the emergency position. When you do that, you close off the trailer brake lines from the tractor. To reconnect the trailer brakes to the tractor, simply push the tractor protection valve in.

Trailer Brake Control

This same type tractor may have a trailer brake control. This lever is called the trailer hand brake, or sometimes a trolley valve or Johnson bar. It operates the trailer brakes only, letting you control the amount of air directed to the trailer brakes. This brake must never be used as a parking brake. It can be used to lock the trailer brakes when coupling or uncoupling or to test the trailer brakes.

Tractor Only Parking Valve

Your vehicle may have a tractor only parking valve. This is offered as an option on some vehicles. It exhausts the air supply from the tractor spring brakes only. You use it when you're bobtailing, or coupling and uncoupling.

Glad Hands

Glad hands are the coupling devices on the ends of the air hoses on the back of your tractor and on the front of your trailer. These hoses connect the service and spring brakes of your trailer to the tractor air supply system. They must be connected properly. Often they are color-coded. In that case, the service brake glad hands are colored blue and the emergency brake glad hands are colored red.

The coupling device is a push, snap-lock type, similar to a radiator cap.

When you're bobtailing, you can connect the glad hands to special couplers on the back of the cab. These couplers are often called "dummy couplers." They protect the lines and keep water and dirt out. If your tractor doesn't have dummy couplers, just connect the lines together and secure them to the back of the tractor.

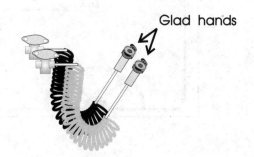

Figure 10-12. Service and emergency brake connections (glad hands).

BRAKE SYSTEM PROBLEMS

Air brake systems do a good job of stopping large vehicles, when the system is well-maintained and used correctly. If it's not, problems can develop.

Problems almost always come up because of a drop in or loss of air pressure. Here's how this can happen.

Compressor

The compressor could break down. What looks like a compressor problem could also be a faulty governor. If your compressor is belt-driven, though, the belt could need attention. It might be broken or too loose. Any of these problems could leave you with too little or no air pressure.

Air Lines and Hoses

A common problem for air lines and hoses are leaks or breaks in the air lines. The normal vibration of the vehicle can loosen air hoses. Heat can cause the rubber to dry out and crack. Hoses that are too close to hot exhaust parts can burn. Hoses that rub up against other parts can get cut.

If a service brake air line breaks, the air will escape from the service line if you apply the service brakes. This will cause a pressure drop.

If you continue to apply the service brakes and pressure falls to between 20 and 45 psi, the spring brakes automatically apply and bring the vehicle to a stop. If the emergency brake air line ruptures, there will be an immediate and rapid loss of pressure in the emergency brake lines. When air pressure falls to between 20 and 45 psi, the tractor protection valve causes the trailer air supply valve to pop out. This will cause the trailer spring brakes to come on.

If the line from the compressor to the main supply tank ruptures, there would be a loss of air from this tank. One-way check valves prevent this. One-way check valves prevent the loss of air from the primary or secondary tanks. If air pressure in either the main or dry air tank drops below 60 psi, one of the warning devices will come on. There should be enough air pressure left in the tank to bring the vehicle to a stop. There will be enough for a limited number of brake applications.

Brake Linings

Over time, friction causes the brake linings to wear and become thinner. This results in reduced contact with the brake drums, and reduced braking power.

It's not only unsafe, but illegal to have brake linings that are too thin. Do you recall what the recommended limit is? You learned in Chapter 6 that you should replace your brake linings when they get thinner than 1/4 inch (1/8 inch for disc brakes).

Brake linings can also become loose, or get soaked with oil or grease. Any of these problems reduces braking power.

Moisture

Moisture in the air system can turn to ice in cold weather. Valves and other parts can freeze.

Other Brake System Parts

You've learned that even the thick metal parts in the brake assembly can crack from the heat. That would result in a loss of braking power. Brake drums must not have cracks longer than one half

the width of the area the brake linings (friction pads) contact.

It is also possible for the gauges, controls, and warning devices to break. In that case, you would not know the true condition of the braking system, or get proper warning of system failure.

MAINTAINING AND INSPECTING AIR BRAKES

Fortunately, most air brake problems can be prevented with a little maintenance. Problems large and small can be spotted in time if you do your pre-trip inspection thoroughly.

Compressor

If the compressor has its own oil supply, making sure it has a good supply of oil will prevent problems. Some compressors are belt-driven. Make sure the belt is in good condition and tightened properly.

Lines and Hoses

Prevent little brake problems from becoming big ones by repairing or replacing brake lines and hoses that look worn or cracked. Check to see that they're not rubbing against each other or other parts. Make sure the lines are straight, not knotted or kinked. Move hoses away from sharp edges that could cut them or hot surfaces that could burn them.

Brake Linings

When linings become too thin, they must be replaced. Up until the time you replace them, you can make up for lining wear with a slack adjustment.

Slack Adjustment

Slack adjusters adjust the brakes to make up for brake lining wear. A slack adjuster is a lever arm attached to the push rod of the brake chamber at the clevis assembly. You can see the slack adjuster in Figures 10-1 and 10-6. Its job is to adjust the position of the S-cam, which then adjusts the distance of the brake shoe from the brake drum.

Slack adjusters can be manually adjusted or automatic. Hand-adjusted slack adjusters have an adjusting nut. Automatic slack adjusters make an adjustment whenever the brakes are applied. They sense the distance the push rod travels each time and keep the brakes in constant adjustment. They can be manually adjusted, if necessary.

To make brake adjustments, you must have special training and be certified as a qualified brake inspector. You must know federal regulations regarding brake maintenance, service, and repair. Knowledge of these regulations is not required to obtain a CDL.

However, you don't need special training or certification to inspect the brakes and determine if the slack adjusters need adjustment. You must know how to do this in order to pass the CDL knowledge and skills tests and avoid an air brake restriction on your license. Here's how:

Make sure you are parked on level ground. Chock the wheels so the vehicle won't move. Turn off the service and parking brake. Otherwise, you won't be able to move the slack adjuster. Wear gloves to protect your hands.

Pull hard on the slack adjuster. It should not move more than about one inch where the push rod attaches to it. If it does, it needs adjustment.

Any needed adjustments should be done by someone qualified. Procedures vary with the type of adjuster.

Moisture

An air dryer can help remove some moisture. Even the best air drying system won't get rid of all the water and oil that condenses out of the compressed air. It tends to collect on the bottom of the air tank. Therefore, each air tank has a drain valve in the bottom. There are three types:

- manual
- automatic
- spit valve

You can open a manual drain valve by turning it one quarter turn. You may also be able to open it by pulling a cable. You should drain this type of tank at the end of each driving day or shift.

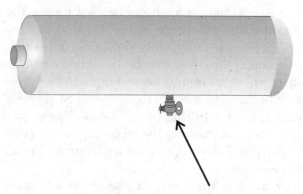

Figure 10-13. Air tank with drain valve.

An automatic drain valve automatically expels oil and water from the tank. These valves can be opened manually as well. You may have to do this if the automatic system fails.

A spit valve is also automatic. It expels water and oil from the tank each time the governor cycles.

Both automatic valves have electric heaters. The heater keeps the valve from freezing in cold weather.

If your vehicle does not drain automatically, daily manual draining is a must.

An alcohol evaporator will help prevent moisture in the system from freezing. Make sure the container has a good supply of alcohol.

Safety relief valves protect the air tanks by releasing excess pressure if the air governor fails. They're usually set to open at 150 psi. If the safety valve releases air, something is wrong. It should be fixed by a mechanic.

Other Brake System Parts

When you first start the electrical system, you should get some response from your gauges and warning devices. This tells you the devices themselves are working. Warning lights should come on for a moment, then go out. The needles on gauges should move, then establish a reading.

If you don't get a response, there could be a defect in the electrical system. You may have a blown fuse, a broken or loose wire, or some other electrical problem. If you can't find and fix the problem, have a mechanic look into it. You must be able to rely on your gauges, controls, and warning devices in order to operate safely.

Bent, cracked, or worn metal parts must be repaired or replaced. If you can't do it yourself, get someone who can to give your brake system some attention. You should not drive off with any part of your brake system in less than top condition. It's simply not safe.

AIR BRAKE SYSTEM INSPECTION

If your vehicle has air brakes, you must add some steps to your normal vehicle inspection routine. You'll learn what they are in detail first. Then you'll find an entire seven-step pre-trip inspection, including the air brake checks. This seven-step routine was first described in Chapter 9, but did not include inspecting the air brakes. We repeat it here, with the air brake inspection included, for your convenience.

Air Brake System Inspection

When you are at the engine compartment, you should check the compressor. Check the oil supply and belt as described in the Maintenance section earlier.

During the "walk-around" part of the inspection, check the hoses for leaks and breaks. At each wheel, check the brake assembly. Look for parts that need replacement or repair. You might find a cracked drum or a loose or worn brake lining. Check the slack adjuster. Have it adjusted if necessary.

Drain all air tanks that must be drained manually. You could find air tanks at the right front, left front, or center of the vehicle. It depends on the manufacturer. Don't forget any air tanks on the trailer.

When you're in the cab, your brake system check should include the following tests. Some of them are performed with the parking brakes off. Make sure the vehicle wheels are chocked so the vehicle doesn't roll while you perform these tests with the parking brakes off.

Test One: Air Pressure Build-Up

This procedure tests pressure build-up time, the low pressure warning indicator, and the air governor.

Open the petcock. Drain the wet tank first, then drain the dry air tanks until the gauges read zero, and close the petcock. Start the engine and run it at a fast idle (600 to 900 rpm). The compressor will start to fill the tanks.

Watch the low air pressure warning mechanism. It should come on and stay on until the pressure builds to more than 60 psi. If the compressor cuts out between 70 and 90 psi, it needs to be adjusted.

In a vehicle with a single air system, the pressure should go from 50 psi to 90 psi within three minutes. In dual air systems, pressure should build from 85 to 100 psi within 45 seconds. If the build-up takes longer, the compressor needs attention.

Some vehicles with larger air tanks can take longer. The owner's manual gives the best advice on this.

Keep filling the tanks until the governor stops the compressor. If it stops below 100 psi or above 125 psi, the governor needs to be adjusted.

The owner's manual may give slightly different upper and lower limits. Run the engine at a fast idle. The air governor should cut out the air compressor at the psi stated in the manual. The air pressure should stop rising.

With the engine still idling, release the spring brakes and step on and off the brake, reducing the air pressure.

 TIP   *Chock your wheels before you begin this test.*

The compressor should cut in at the psi stated in the manual. The air pressure should begin to rise. If the governor cuts in or out too early or too late, it requires adjustment.

If the warning light doesn't go out once you have reached 60 psi, there could be a problem with the light. More seriously, it could be that one part of the dual braking system isn't working.

Test Two: Air Leakage Rate

With these simple steps, you can test the brake system's ability to hold air pressure. With the pressure fully built up (about 125 psi), turn off the engine. Release the parking brakes. Press the service brake pedal firmly. Watch the reading on the pressure gauge and start timing. The pressure should not drop more than 2 psi in one minute for single vehicles. If the pressure drop is greater, something is wrong. Find out what is wrong and see that it's fixed before you drive.

Test Three: Air Leakage Rate

Here's another way to test the combination system's ability to hold air pressure. With the engine turned off, press on the brake pedal. Bring the pressure to 90 psi on the application gauge. Watch the air pressure gauge. The pressure drop should not be more than 3 psi per minute.

Test Four: Warning Device and Emergency Brake

These two steps test your warning device and your spring brake emergency application. With air pressure at about 90 pounds and the engine off, push and release the foot brake, reducing the air pressure, until the low air pressure warning comes on. If the warning indicator fails to come on when the pressure reaches 60 psi or below, get it adjusted before you drive. The warning device should come on before the spring brakes are automatically applied.

 Remember the air pressure at which the spring brakes will automatically come on? It's usually in a range between 20 and 40 psi.

Continue pushing and releasing the foot brake until the emergency brakes apply. The spring brakes apply automatically. If they apply above 45 psi or below 20 psi, something is wrong.

Test Five: Parking Brake

Begin the test from a dead standstill. You'll be moving the vehicle, so remove any wheel chocks you may have put in place. Get in the vehicle. Put the parking brake on. Put the vehicle in a low gear and try to move forward. The parking brake should hold you back. If it doesn't, it needs attention.

Test Six: Service Brake

Again, you'll be moving the vehicle. If you have wheel chocks in place, move them.

Build the air pressure to normal. Release the parking brake. Move the vehicle forward at about 5 mph. Apply the brakes firmly with the brake pedal. The vehicle should not pull to one side or the other. It should stop firmly, promptly, and evenly. If it doesn't, the brakes need adjustment.

If you have a trailer hand valve, repeat the procedure and apply the trailer hand valve to test the trailer service brakes.

SEVEN-STEP PRE-TRIP INSPECTION

As you recall, there are seven main steps in this inspection routine. They are:

- approach the vehicle
- raise the hood or tilt the cab and check the engine compartment
- start the engine and inspect inside the cab
- check the lights
- walk all around the vehicle, inspecting as you go
- check the signal lights
- check the brakes

Step One: Vehicle Overview

As you walk towards the vehicle, notice its general condition. Look for damage. Note if the vehicle is leaning to one side. This could mean a flat tire. Cargo may be overloaded or may have shifted. Or, it could be a suspension problem.

Look under the vehicle for fresh oil, coolant, grease, or fuel leaks. Check the area around the vehicle for people, other vehicles, objects, low hanging wires, or limbs. These could become hazards once the vehicle begins to move.

Look over the most recent vehicle inspection report. Drivers typically have to make a vehicle inspection report in writing each day. The vehicle owner then should see that any items in the report that affect safety are repaired. If the mechanic repaired the defects, or determined no repairs were needed, there will be a certification for that on the form.

You should look at the last report to find out what was the matter, if anything. Inspect the vehicle to find out whether problems were fixed, or whether repairs were not needed.

Step Two: Engine Compartment Check

Check that the parking brakes are on and/or wheels chocked. You may have to raise the hood or open the engine compartment door. If you have a tilt cab, secure loose items in the cab before you tilt it. Then they won't fall and break something.

At this site, check the engine oil level. The oil should be above the "Low" or "Add" mark on the dipstick. See that the level of coolant is above the "Low" level line on the reservoir. Check the condition of coolant-carrying hoses.

Check the radiator shutters and winterfront, if your vehicle has one. In cold weather, remove ice from the radiator shutters. Make sure the winterfront is not closed too tightly. If the shutters freeze shut or the winterfront is closed too much, the engine may overheat and stop. Inspect the fan. Make sure the blades are sound and not likely to catch on hanging wires or hoses.

If your vehicle has power steering, check the fluid level and the condition of the hoses. Check the dipstick on the oil tank. The fluid level should be above the "Low" or "Add" mark.

Here's where you check the windshield washer fluid level.

If your battery is located in the engine compartment, perform the battery check now.

If you have an automatic transmission, check the fluid level. You may have to do this with the engine running. Your owner's manual will tell you if that is the case.

Check drive belts for tightness and excessive wear. Press on the center of the belt. The owner's manual will tell you how much slack there should be. Also, if you can easily slide a belt over a pulley, it is definitely too loose or too worn. Look for leaks in the engine compartment. These could be fuel, coolant, oil, power steering fluid, hydraulic fluid, or battery fluid.

Look for cracked or worn insulation on electrical wiring.

Check the compressor for the air brakes. Check the oil supply. If it's belt-driven, check the belt. It should be in good condition. Check belt tightness.

Lower and secure the hood, cab, or engine compartment door.

Next, check any handholds, steps, or deck plates. Remove all water, ice and snow, or grease from handholds, steps, and deck plates that you must use to enter the cab or to move about the vehicle. This will reduce the danger of slipping.

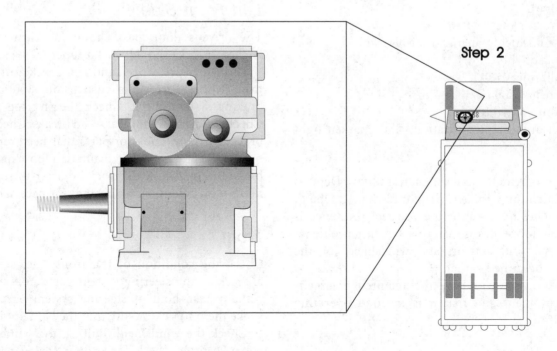

Figure 10-14. Check the air compressor when you're at the engine compartment.

Step Three: Inside the Cab

Get in the cab. Inspect inside the cab. Start the engine. Make sure the parking brake is on. Shift into neutral, or "Park" if your transmission is automatic. Start the engine and listen for unusual noises.

Check the gauge readings. The oil pressure should come up to normal within seconds after the engine is started. The ammeter and/or voltmeter should give normal readings. Coolant temperature should rise gradually until it reaches normal operating range. The engine oil temperature should also rise slowly to normal operating range.

Oil, coolant, and charging circuit warning lights will come on at first. That tells you the lights are working. They should go out right away unless there's a problem.

Make sure your controls work. Check all of the following for looseness, sticking, damage, or improper setting:

- steering wheel
- accelerator ("gas pedal")
- foot brake
- trailer brake (if your vehicle has one)
- parking brake
- retarder controls (if your vehicle has them)
- transmission controls
- interaxle differential lock (if your vehicle has one)
- horn(s)
- windshield wiper and washers
- headlights
- dimmer switch
- turn signal
- four-way flashers
- clearance, identification, and marker light switch(es).

If your vehicle has a clutch, test it now. Depress the clutch until you feel a slight resistance. One to two inches of travel before you feel resistance is normal. More or less than that signals a problem.

Check your mirrors and windshield for the defects described earlier.

Check that you have the required emergency equipment and that it is in good operating condition.

Check for optional items, such as a tire changing kit, and items required by state and local laws, such as mud flaps.

Step Four: Check Lights

Next, check to see that the lights are working. Make sure the parking brake is set, turn off the engine, and take the key with you. Turn on the headlights (on low beams) and the four-way flashers, and get out. Go to the front of the vehicle. Check that the low beams are on and both of the four-way flashers are working. Push the dimmer switch and check that the high beams work. Turn off the headlights and four-way hazard warning flashers. Turn on the parking, clearance, side-marker, and identification lights. Turn on the right-turn signal, and start the "walk-around" part of the inspection.

Step Five: Walk-around Inspection

Walk all around the vehicle, inspecting as you go. Start at the cab on the driver's side. Cover the front of the vehicle. Work down the passenger side to the rear. Cover the rear of the vehicle. Work up the driver's side back to the starting position. Do it this way every time.

Clean all the lights, reflectors, and glass as you go along.

Left Front Side

The driver's door glass should be clean. Door latches or locks should work properly.

Perform a tire, wheel, and rim check at the left front wheel. Perform a suspension system and braking system check. Check the steering parts. Check the air hoses for leaks and breaks. Check the brake assembly. Look for parts that need replacement or repair: a cracked drum or a loose or worn brake lining. Check the slack adjuster to see if it needs adjustment. Drain air tanks if necessary.

Front

Check the condition of the front axle. Perform a check of the steering system parts. You should actually take hold of steering system parts and shake them to say for sure that they are not loose.

Check the windshield. Pull on the windshield wiper arms to check for proper spring tension.

Check the wiper blades for the defects described earlier.

Check the lights and reflectors (parking, clearance and identification, and turn signal lights) at this site. Make sure you have all the required ones as described earlier. They should be clean and working.

Right Front Side

The right front has parts similar to the left front. Perform the same checks. The passenger door glass should be clean. Door latches or locks should work properly. Check the tire, wheel, and rim at the right front wheel. Perform a suspension system and braking system check. Check the steering parts. Check the air hoses for leaks and breaks. Check the brake assembly. Look for parts that need replacement or repair: a cracked drum or a loose or worn brake lining. Check the slack adjuster. Have it adjusted if necessary. Drain air tanks as needed.

For COE trucks, check the primary and safety cab locks here.

Right Side

The fuel tank is located here. Check to see that it is mounted securely, not damaged or leaking. The fuel crossover line must be secure, and not hanging so low that it would be exposed to road hazards. The filler caps must be on firmly. Check the fuel supply. Perform these checks for each fuel tank your vehicle has.

Check all parts you can see at this site. See that the rear of the engine is not leaking. Check for leaks from the transmission. Check the exhaust system parts for defects as described earlier.

Inspect the frame and cross members for bends and cracks.

Electrical wiring should be secured against snagging, rubbing, and wearing.

If your vehicle has a spare tire carrier or rack, check it for damage. It should be securely mounted in the rack. Make sure you have the correct size spare tire and wheel. The spare tire must be in good condition and properly inflated.

Cargo Securement

Check the cargo securement at the right front of the vehicle. It must be properly blocked, braced, tied, chained, and so forth. If you need a header board, make sure it is strong enough and secure. If you have side boards, check them. The stakes must be strong enough for the load, free of damage, and properly set in place.

Make sure you haven't created handling problems for yourself by overloading the front axles.

Any canvas or tarp must be properly secured to prevent tearing and billowing.

If you are hauling an oversized load, have the necessary signs safely and properly mounted. Make sure you have the permits you need to haul this load.

Check that the curbside cargo compartment doors are securely closed, latched, and locked. See that the required security seals are in place.

Right Rear

At this location, check the wheels, rim, tires, brake system, and suspension parts as you did at the right front. Check that all parts are sound, properly mounted and secure, and not leaking. Check the axle. Powered axles should not leak.

Check the lights and reflectors required at this location. They should be clean and working.

Rear

Check the lights and reflectors required at this location. They should be clean and working.

Make sure you have a license plate. It should be clean and securely mounted.

If mud flaps are required, check them now. They should be free of damage and fastened according to whatever regulations apply. If there are no legal guidelines, make sure they are at least well-fastened, not dragging on the ground or rubbing the tires.

Cargo Securement

Check the cargo securement at the right rear of the vehicle. It must be properly blocked, braced, tied, chained, and so forth. Check that the tailboard is

Step 5: Walk-around inspection

    A.   Left front side: wheel and tire, suspension, brakes, axle, steering, side marker lamp and reflector, door glass, latches and locks, mirrors

    B.   Front of cab: axle, steering system, windshield, lights and reflectors

    C.   Right front side: door glass, latches and locks, wheels and tires, suspension, brakes, axle, steering, condition of bed or body, side marker lamp and reflector

    D.   Right side: fuel tank, exhaust system, frame and cross members, wiring, spare tire, lights and reflectors, mirrors

    E.   Cargo securement, curbside doors

    F.   Right rear: wheels and tires, suspension, brakes, axle, lights and reflectors

    G.   Rear: lights and reflectors, license plate, mudflaps, condition of bed or body, rear doors

    H.   Cargo securement

    I.   Left rear and left front: wheels and tires, suspension, brakes, steering, axle, lights and reflectors, battery, condition of bed or body

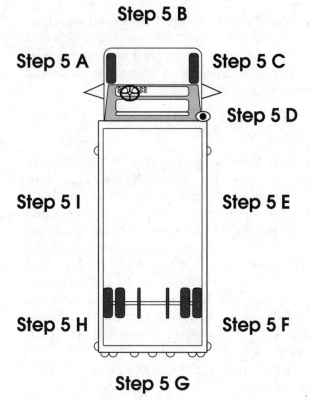

Figure 10-15. Check the air brake assembly when you inspect the wheel area.

strong enough and secure. If you have end gates, check them. The stakes must be strong enough, free of damage, and properly set in the stake pockets.

Any canvas or tarp must be properly secured to prevent tearing and billowing. Make sure the tarp doesn't cover up your rear lights.

Check to make sure that you haven't distributed the load so as to overload the rear axles.

If you are hauling an oversized load, have the necessary signs safely and properly mounted. Have any lights or flags required for projecting loads. Make sure you have the permits you need to haul this load.

Check that the rear doors are securely closed, latched, and locked.

Left Rear and Front Side

At this location, check the wheels, rim, tires, brake system, and suspension parts just as you did for the right rear and front side. Check that all parts are sound, properly mounted and secure, and not leaking. Check the axle. Powered axles should not leak.

Check the lights and reflectors required at this location. They should be clean and working.

Your vehicle's battery may be mounted here, rather than in the engine compartment. If so, perform the battery check now.

Step Six: Check Signal Lights

Get in the cab and turn off all the lights. Then, turn on the stop lights. Or, have a helper step on the brake pedal while you watch to see if the light goes on.

Now turn on the left-turn signal lights. Then get out and check the left side lights, rear and front.

Step Seven: Check Brakes

Now you'll perform some tests with the engine running and the parking brakes off. Make sure you're on a level surface. Chock the wheels.

Get back in the vehicle. Turn off any lights you don't need for driving. Check to see that you have all the required papers, trip manifests, permits, and so forth. Secure all loose articles in the cab. If

Step 1: Vehicle overview for general condition

Step 2: Engine compartment checks
Fluid levels and leaks
Hoses and belts
Battery
Compressor
Windshield washer
Wiring

Step 3: Inside the cab checks
Parking brake on
Gauge readings
Warning lights
Controls
Emergency equipment
Optional equipment

Step 4: Check lights
Headlights (low and high beams)
Four-way flashers
Parking lights
Clearance lights
Side marker lights
Identification lights
Right-turn signal

Step 5: Walk-around inspection
A.   Left front side: wheel and tire, suspension, brakes, axle, steering, side marker lamp and reflector, door glass, latches and locks, mirrors
B.   Front of cab: axle, steering system, windshield, lights and reflectors
C.   Right front side: door glass, latches and locks, wheels and tires, suspension, brakes, axle, steering, condition of bed or body, side marker lamp and reflector
D.   Right side: fuel tank, exhaust system, frame and cross members, wiring, spare tire, lights and reflectors, mirrors
E.   Cargo securement, curbside doors
F.   Right rear: wheels and tires, suspension, brakes, axle, lights and reflectors
G.   Rear: lights and reflectors, license plate, mudflaps, condition of bed or body, rear doors
H.   Cargo securement
I.   Left rear and left front: wheels and tires, suspension, brakes, steering, axle, lights and reflectors, battery, condition of bed or body

Step 6: Check signal lights

Step 7: Brake check

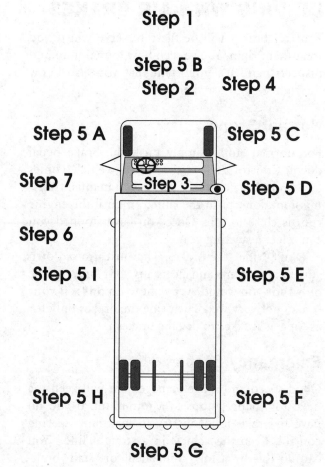

Figure 10-16. Vehicle Inspection Aid for straight vehicles with air brakes.

they are free to roll around, they could hinder your operation of the controls, or hit you in a crash.

Start the engine. Perform air brake tests, numbers one through four. Then stop the vehicle. Remove the chocks. Perform brake tests five and six.

DRIVING WITH AIR BRAKES

Your air brakes will be there to serve you if you treat them right. You've seen how to maintain and inspect them. It's important that you also know the right way to use them.

Normal Stops

For normal stops, simply push the brake pedal down. Control the pressure so the vehicle comes to a smooth, safe stop. If you have a manual transmission, don't push the clutch in until the engine rpm is close to idle. Once you have stopped you can select a starting gear.

Don't "ride" the brakes. That just draws down your air pressure and heats up your brakes. The only time you would ever want to do this is if your brakes got very wet. Then ride the brakes only for as long as it takes to dry the brakes.

Emergency Stops

Good speed and space management will keep the need for "panic" stops to a minimum. If you do have to brake suddenly, try to maintain steering control. Keep the vehicle in a straight line. You can do this by using "controlled" or "stab" braking. These braking techniques were described in Chapter 7. Remember them?

In controlled braking, you apply the brakes as hard as you can without locking the wheels. Don't turn the steering wheel while you do this. If you need to make a larger steering adjustment or if the wheels lock, release the brakes. Reapply the brakes as soon as you have traction again.

In stab braking, you apply your brakes all the way. Release the brakes when the wheels lock up. As soon as the wheels start rolling, apply the brakes fully again. It can take up to one second for the wheels to start rolling after you release the brakes. If you reapply the brakes before the wheels start rolling, the vehicle won't straighten out.

Air Brakes and Stopping Distance

You need increased stopping distance with air brakes. It takes time for air to flow to the brakes after you have stepped on the pedal. The braking response is not as immediate as it is with hydraulic brakes.

Remember the stopping distance formula? It's

$$\begin{array}{r} \text{perception distance} \\ + \;\; \text{reaction distance} \\ + \;\; \text{braking distance} \\ \hline = \;\; \text{total stopping distance} \end{array}$$

You must add the distance taken up by "brake lag" to this formula. At 55 mph on a dry surface, you can travel about 32 feet before the brakes respond to your stepping on the pedal. That would make your total stopping distance over 300 feet—longer than a football field.

Braking on a Downgrade

Brake fade and brake loss are real dangers on a downgrade. It's easy to lose your brakes on a hill, and that's when you really need them, so it's vital that you use them correctly. Start down the hill in a low gear that will give you a lot of engine braking. That way you can use snub braking. Apply the brakes just hard enough to feel the vehicle slow. When the speed has been reduced to about five mph below your safe speed, release the brakes. If your speed rises above the safe level, repeat the procedure.

Above all, don't fan the brakes. Using them hard builds up tremendous heat. Releasing them briefly does not give them enough time to cool off. When you apply them again, you just increase the heat. Brakes won't take too much of this treatment before they expand beyond usefulness, or even burn up.

So again, use a low gear and light steady pressure.

If your brakes are out of adjustment, you'll really notice it on a downgrade. Most long grades offer a "pull-out" at the top of the hill. This is a safe area to the side of the road. Stop, check, and have your brakes adjusted if necessary before heading down. Don't just step on the brakes and decide they're OK. Get out of the vehicle and check them thoroughly.

Low Air Pressure

Heed your low air pressure warning. It means you still have a little air pressure left, and can probably still make a safe stop. Stop and park the vehicle the first chance you get to do so safely. Don't wait until the emergency brakes come on. It's harder to control the vehicle when the spring brakes come on.

As you can see, the idea behind air brakes is not a complicated one. Air brakes are fairly easy to use, maintain, and inspect. The system as a whole is not that complex. It is large, however, and has many parts. So there are many chances for problems to develop. A single problem could cause the whole system to fail. Therefore, if you plan to drive a vehicle with air brakes, you must understand each and every part. You must commit yourself to proper maintenance and thorough inspection. If you do, your air brakes should work dependably and reliably.

PASS POST-TRIP

Instructions: For each true/false test item, read the statement. Decide whether the statement is true or false. If it is true, circle the letter A. If it is false, circle the letter B. For each multiple-choice test item, choose the answer choice—A, B, C, or D—that correctly completes the statement or answers the question. There is only one correct answer. Use the answer sheet provided on page 241.

1. The compressed air in the brake chamber transmits the force exerted by the driver's foot on the brake pedal to the foundation brakes.
   A. True
   B. False

2. When you use brakes hard, the first thing that happens is _____.
   A. the linings become worn
   B. they get out of adjustment
   C. they get hot
   D. they crack

3. If you have an alcohol evaporator, you do not have to drain the air tanks.
   A. True
   B. False

4. If your air compressor fails, _____.
   A. you will have no brakes
   B. the brakes will immediately lock up
   C. gradually decreasing air pressure will continue to be available from the air tanks for a while
   D. air pressure will not be affected at all

5. Front wheel braking is good to have on _____.
   A. dry roads
   B. wet roads
   C. icy roads
   D. all of the above

6. Pushing the brake pedal down when the spring brakes are on will _____.
   A. set the parking brakes
   B. release the parking brakes
   C. damage the brakes
   D. adjust the slack

7. The emergency brake system is a completely separate system and is completely under the driver's control.
   A. True
   B. False

8. The low air pressure warning device should come on before air pressure drops below _____.
   A. 20 psi
   B. 45 psi
   C. 60 psi
   D. 125 psi

9. At 55 mph on a dry surface, the average driver will need more than _____ feet to stop a CMV with air brakes.
   A. 100
   B. 200
   C. 300
   D. 400

10. The best braking technique to use on a downgrade is _____.
    A. snub braking
    B. controlled braking
    C. stab braking
    D. hard braking

CHAPTER 10 ANSWER SHEET

Test Item	Answer Choices				Self - Scoring	
					Correct	Incorrect
1.	A	B	C	D	☐	☐
2.	A	B	C	D	☐	☐
3.	A	B	C	D	☐	☐
4.	A	B	C	D	☐	☐
5.	A	B	C	D	☐	☐
6.	A	B	C	D	☐	☐
7.	A	B	C	D	☐	☐
8.	A	B	C	D	☐	☐
9.	A	B	C	D	☐	☐
10.	A	B	C	D	☐	☐

Cut Here

Chapter 11

Vehicle Combinations

In this chapter, you will learn about:

- coupling and uncoupling a single semitrailer
- coupling and uncoupling doubles
- inspecting vehicle combinations
- pulling trailers safely

To complete this chapter you will need:

- a dictionary
- a pencil or pen
- blank paper, a notebook, or an electronic notepad
- colored pencils, pens, markers, or highlighters
- a CDL preparation manual from your state Department of Motor Vehicles, if one is offered
- the Federal Motor Carrier Safety Regulations pocketbook (or access to U.S. Department of Transportation regulations, Parts 383, 393, and 396 of Subchapter B, Chapter 3, Title 49, Code of Federal Regulations)
- the owner's manual for your vehicle, if available

PASS PRE-TRIP

Instructions: Read the statements. Decide whether each statement is true or false. If it is true, circle the letter A. If it is false, circle the letter B.

1. In a triple combination, the heaviest trailer should be in the middle.
   A. True
   B. False

2. The pre-trip inspection of a double or triple combination should include draining the converter dolly air reservoir.
   A. True
   B. False

3. A fully loaded tractor-trailer is 10 times more likely to roll over in a crash than an empty one.
   A. True
   B. False

If you plan to pull one or more trailers with your vehicle, you will have to take your CDL test in that vehicle. You'll be tested on your knowledge of coupling and uncoupling. You'll have to show you can inspect the combination. This is true whether you're pulling a single trailer, semitrailer, doubles, or triples.

You'll have to pass the Knowledge Test all CDL applicants must take. You must show you have additional knowledge on combination vehicles. Then, once you pass the Skills Test in a "representative vehicle," you'll have a CDL with an "A" code. You'll be licensed to drive Group A ("combination") vehicles.

If you want to pull doubles and triples, you're not done yet. You'll need the doubles/triples endorsement. You'll not only have to take the Skills Test while pulling a trailer. You'll also have to answer additional knowledge questions about pulling doubles and triples.

In this chapter, we'll cover the material drivers need to get a CDL for combination vehicles. If you plan to haul only a single trailer, that's the only part of this chapter you will need. If you want the doubles/triples endorsement, you should complete the whole chapter.

Remember, FMCSR Part 383 states that drivers of tractor-semitrailer combinations must know the correct way to couple and uncouple a tractor and semitrailer. They must know why and how to inspect the vehicle combination.

This chapter begins with coupling and uncoupling a single semitrailer. Toward the end of this you'll find the inspection routine for a vehicle combination.

COUPLING A SINGLE SEMITRAILER

Coupling and uncoupling aren't hard procedures, but for safety's sake, they must be done right.

What's involved in coupling? To put it very simply, you're going to back the tractor up to the trailer so the coupling assemblies connect. Then you must supply electricity to the trailer so the lights will work. If you have air brakes, you must supply the trailer with air. Also, you must make sure you can control the trailer from the tractor.

Slow and steady wins the race here. Follow the procedures step by step every time. Use the rear-view mirrors. Know your equipment. You have to know the width of the tractor as compared to the width of the trailer. Then remember that the center of the fifth wheel is always in the center of the tractor frame and the kingpin is always in the center of the front of the trailer.

Inspect Area

Before you begin any coupling procedure, walk the area around the trailer and tractor. Look for anything in your path that could damage the tractor or trailer. Boards lying on the ground can fly up when popped from the ground by a tire. Nails, glass, or other objects can damage a tire. Make sure the way is clear before you begin the first stages of alignment.

Inspect Vehicles

Check the coupling devices on the tractor and the trailer. Look for damaged or missing parts. Inspect the tractor frame and the fifth wheel mounting. There should be no cracked or bent parts. The mounting should be secure. Make sure the trailer kingpin isn't bent or broken.

Lubricate the fifth wheel. Lubrication reduces friction between the fifth wheel and the trailer. That gives you better steering control.

Pre-position the fifth wheel. It should be tilted down (toward the rear of the tractor). The locking jaws should be open.

Stabilize Vehicles

Work on the most level ground you can find if you have a choice about it. Uneven ground will make

your task just that much harder. Apply the spring brakes. As you move the tractor under the trailer, it can move the trailer. Having the brakes on will keep the trailer from rolling backward.

Some older trailers don't have spring brakes. If yours doesn't, put chocks at the rear of your trailer tires. Perhaps you're not sure if you have spring brakes or not. You can tell just by looking at the brakes. Find the service brake chamber. Look for the spring brake chamber "piggybacking" on the service brake chamber. Or, look at the number of air lines going into the brake chamber. Spring brakes will have two air lines, service brakes have only one. Either way, if you don't have spring brakes, use chocks to keep your vehicle from moving.

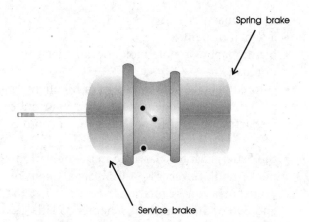

Figure 11-1. Look for the spring brake chamber mounted on the service brake chamber.

Align Tractor

A proper coupling requires you to center the kingpin in the fifth wheel within a small margin of error. Before you back, you must align the tractor and trailer precisely. If they don't, you'll have to move the tractor or trailer before you can complete the coupling.

Place the tractor so the trailer kingpin is as near to the center of the fifth wheel V-slot as possible. Locate the center of the trailer by using its sides as a gauge in your rear-view mirrors. As you straighten the tractor in front of the trailer, the corners of the trailer will appear evenly in the mirrors. If there is an unequal amount of the trailer in one mirror, you are too far to the other side.

Don't try to back under the trailer at an angle. You could damage the landing gear or push the trailer sideways.

Back Slowly

If you back slowly you can steer the tractor in the desired direction to align the fifth wheel with the kingpin. Keep backing until the fifth wheel just touches the trailer. Don't hit the trailer!

Align Trailer

Put on the parking brake and put the transmission in neutral. Go out and compare the level of the fifth wheel with the height of the kingpin. If the trailer is too far below the fifth-wheel level, the tractor may strike and damage the nose of the trailer. If the trailer is higher than the fifth-wheel level, the trailer will come over the fifth wheel when you try to back under it. You could damage the rear of the tractor that way. The trailer should be low enough that it is raised slightly by the tractor when the tractor backs under it.

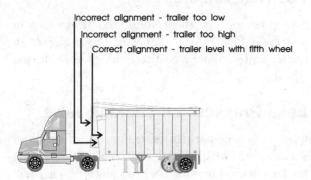

Incorrect alignment - trailer too low

Incorrect alignment - trailer too high

Correct alignment - trailer level with fifth wheel

Figure 11-2. Aim for a good vertical alignment.

If you are either too high or two low, you will have to use the landing gear to raise or lower the trailer to the fifth-wheel level. It isn't all that easy to roll up a loaded trailer.

The landing gear crank is hinged and swings under the trailer to a latch. The latch secures it to the frame while the trailer is in motion. If the crank handle were allowed to swing freely while the trailer is moving, it could damage a nearby vehicle or passing pedestrian. Always secure the crank handle when you're done using it.

Connect Air Lines

The first step of the actual coupling of the tractor and trailer is to connect the air supply lines. First, check the rubber seals on both connections. If one is worn or cracked, have it replaced immediately.

Otherwise, you might end up with a service line air leak you wouldn't notice until you put on the brakes.

There are two air supply lines: the service brake line and the emergency brake line. These are almost always colored red for emergency and blue for service. The connections at the front of the trailer will usually be painted the color of the air hose that should be connected to them. Secure each air supply line to the color-matched trailer connection.

If you cross the air lines on a new trailer with spring brakes, you will not be able to release the trailer brakes. If you cross the air lines on an old trailer without spring brakes, you could drive off and not have any braking power on the trailer.

Lock Trailer Brakes

After you return to the tractor, charge the trailer air supply by pushing in the red trailer air supply knob. When pressure is normal, apply the trailer parking brakes by pulling out the trailer air supply knob. This puts on the trailer brakes to keep it from moving. Release the tractor brakes by pushing in the yellow tractor parking knob.

Back Under Trailer

Put the transmission in the lowest reverse gear and back—slowly. Keep backing until the tractor comes into contact with the trailer. You will feel a definite bump. Don't bump the kingpin too hard. Continue backing slowly until progress is stopped by the kingpin locking into the fifth wheel.

Check Connection

Raise the landing gear slightly. Check the connection by pulling forward very slowly. If the tractor will not move, the connection is complete. If it does move, stop immediately and back again.

Put the transmission in neutral. Put the parking brakes on. Shut off the engine. Take the key with you so someone else can't move the truck while you are under it.

Get out of the tractor and check to see that the fifth wheel locking jaw is closed around the trailer kingpin. To do this you must look under the front of the trailer at the fifth wheel. Use a flashlight if it's too dark to see clearly. If the jaw is closed you will see it locked securely around the back of the kingpin. There should be no space between the upper and lower fifth wheel. If there is space, the kingpin may be on top of the closed fifth wheel jaws. The trailer could come loose very easily.

If the connection is good, check that the locking lever is in the "lock" position. Make sure the safety catch is in position over the locking lever. You may have to put it in place by hand. (Some fifth wheels don't have this lever.)

If the connection is poor, don't drive the coupled unit. Get it fixed.

Connect Electrical Power and Check Air Supply

Connect the electric supply cable to the trailer. Fasten the safety catch. Inspect the air and electrical lines for signs of damage. Make sure air and electrical lines will not hit any of the vehicle's moving parts.

Return to the tractor and turn on the emergency flashers and trailer lights. Walk around the trailer to make sure the lights are working. Check the clearance lights and side-marker lights. Walk to the back of the trailer and check the turn signals. Turning on the emergency flashers at the tractor will make the turn signals flash.

While you're checking out the trailer tires, gauge the tire inflation if you didn't do this in your routine inspection. Also, check for worn hoses or loose connections on the trailer's brake system. The hiss of escaping air may be heard coming from these areas. Rusty connections should be closely inspected for cracks. Should any leaks exist, have them repaired immediately. A ruptured air hose or broken connection is sure trouble while the vehicle is in motion.

Raise Landing Gear

The trailer is finally connected to the tractor and ready to roll. First, though, you must remove any tire chocks. If they are yours, put them in the tractor storage compartment. If they belong in the yard, find out where they should be stored and put them there. Roll up the landing gear. Use low gear to begin raising the landing gear. Once the weight is fully supported by the tractor, you can switch to high gear.

Roll the gear as high as it will go. Never drive with the landing gear up only part way. It may catch on railroad tracks and objects in the road.

Check for enough clearance between the rear of the tractor frame and the landing gear. When the tractor turns sharply, it must not hit the landing gear.

Check that there is enough clearance between the top of the tractor tires and the nose of the trailer.

UNCOUPLING A SINGLE SEMITRAILER

Before you start, make sure the surface that the landing gear will be sitting on when lowered can support a heavily loaded trailer. A trailer can sink into hot asphalt or loose dirt. On such surfaces you should always place something—a wide plank or pad—under the landing gear plate. Cement is virtually the only surface that will support a loaded trailer without allowing it to sink.

Position Vehicles

Straighten out so the tractor and trailer are in a straight line. If you park with the trailer at an angle, you can damage the landing gear during uncoupling.

Ease Pressure on Locking Jaws

Once you have parked the trailer on a firm, level surface, shut off the trailer air supply to lock the trailer parking brakes. Back up slightly. This will take some of the pressure off the fifth wheel locking jaws. It will be easier to release the fifth wheel locking lever. Put the tractor parking brakes on while the tractor is pushing against the kingpin. This will hold the vehicle combination with the pressure off the locking jaws.

Chock Trailer Wheels

Get out of the tractor. If the trailer doesn't have spring brakes, chock the trailer wheels. Once the trailer tires are chocked, lower the landing gear.

Lower Landing Gear

If the trailer is empty, you can lower the landing gear until it makes contact with the ground. If the trailer is loaded, lower the landing gear to the ground. Then turn the crank in low gear a few extra turns. This will lift some of the weight off the tractor. You'll find it easier to unlatch the fifth wheel. Also, it will be easier to couple the next time. Don't lift the trailer off the fifth wheel, though.

Disconnect Air Lines and Electrical Cable

Next you should remove the air supply lines and electrical cable from the trailer. Stow each in its proper position at the rear of the tractor. Connect the air line glad hands to the dummy couplers at the back of the cab. Or, you could couple them together. Hang the electrical cable with the plug end down. This prevents moisture from getting into the cable. Secure the lines so they won't be damaged by moving the tractor.

Unlock Fifth Wheel

Procedures for opening the release handle lock vary based on the model of the fifth wheel. Follow the procedure outlined in your owner's manual for moving the release handle to the "open" position. Keep your legs and feet clear of the rear tractor wheels. If the vehicle were to move, you could be injured.

Pull Tractor Partially Clear of Trailer

Pull the tractor forward until the fifth wheel comes out from under the trailer. Stop with the tractor frame under the trailer. This prevents the trailer from falling to the ground if the landing gear collapses or sinks.

Secure Tractor

Apply the parking brake. Shift into neutral.

Inspect the Trailer Supports

Before you put all the trailer weight on the landing gear, make sure the ground is supporting the trailer. Make sure the landing gear is not damaged and can still support the trailer.

Pull Tractor Clear of Trailer

Release the parking brakes. Check the area and drive the tractor clear.

Once the tractor is completely uncoupled from the trailer, pull away from the trailer slowly. You will know when you are free of the trailer. The rear of the tractor will rise as the shock absorbers extend with the release of the trailer weight.

That covers coupling and uncoupling a semi-trailer. To get a CDL for combination vehicles, you also need to know why and how to inspect this combination. You'll find the inspection routine after the following description of coupling and uncoupling doubles and triples.

If you want the doubles/triples endorsement, you will have to take a knowledge test on:

- coupling and uncoupling multiple trailers
- where to put the heaviest trailer
- driving with multiple trailers
- sharing the road with other motorists

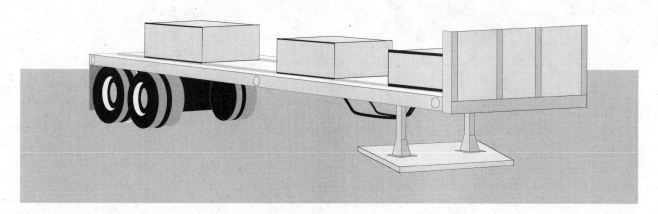

Figure 11-3. You may need to put a plank under the landing gear if the ground is soft.

While we're on the subject of coupling and uncoupling, let's cover the procedure for hooking and unhooking doubles and triples.

COUPLING DOUBLES

Secure Second (Rear) Trailer

If the second trailer doesn't have spring brakes, drive the tractor close to the trailer. Connect the emergency line, charge the trailer air tank, and disconnect the emergency line. This will set the trailer emergency brakes (if the slack adjusters are correctly adjusted). Chock the wheels.

Couple the tractor and the first semitrailer as just described. For safe handling on the road, the more heavily loaded semitrailer must always be in first position behind the tractor. The lighter trailer should be in the rear. This is extremely important. Perhaps you think this has something to do with the center of gravity? You're right. You'll find this covered in greater detail in the "Pulling Trailers Safely" section of this chapter.

Position Converter Dolly

Position the converter dolly in front of the second (rear) trailer. Release the dolly brakes by opening the air tank petcock. If the dolly has spring brakes, use the dolly parking brake control.

If you don't have to move it too far, wheel the dolly into position by hand so it is in line with the kingpin. Otherwise, use the tractor and first semitrailer to pick up the converter dolly. Here's how.

Position the combination as close as possible to the converter dolly. Move the dolly to the rear of the first semitrailer and couple it to the trailer. Lock the pintle hook. Secure the dolly support in the raised position. Pull the dolly into position as close as possible to the nose of the second semitrailer. Lower the dolly support. Unhook the dolly from the first trailer. Wheel the dolly into position in front of the second trailer in line with the kingpin.

Connect Converter Dolly to Front Trailer

Back the first semitrailer into position in front of the dolly tongue (drawbar). Hook the dolly to the front trailer. Lock the pintle hook. Secure the converter gear support in the raised position.

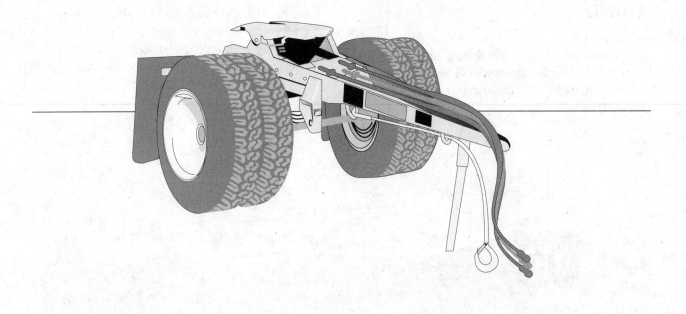

Figure 11-4. A converter dolly.

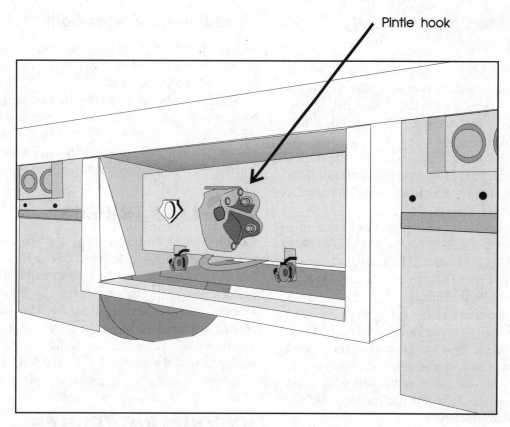

Pintle hook

Figure 11-5. Pintle hook at the rear of the trailer.

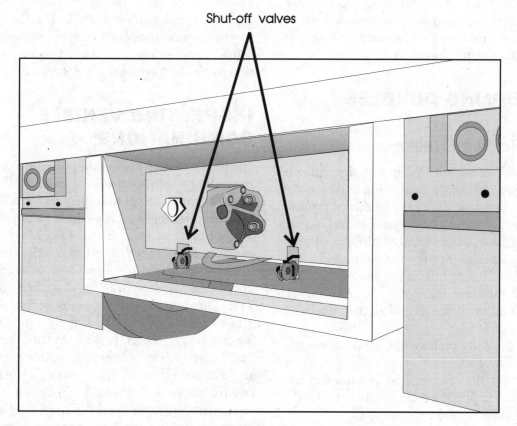

Shut-off valves

Figure 11-6. Shut-off valves.

Connect the Converter Dolly to Rear Trailer

Make sure the trailer brakes are locked or the wheels are chocked. Make sure the trailer is slightly lower than the center of the fifth wheel. That way the trailer will be raised slightly when the dolly is pushed under.

Back the converter dolly under the rear trailer. Raise the landing gear slightly off the ground. This will keep the landing gear from being damaged if the trailer moves.

Test the coupling by pulling against the kingpin of the number two semitrailer. Make a visual check of the coupling. There should be no space between the upper and lower fifth wheel. The locking jaws must be closed on the kingpin.

Connect the safety chains, air hoses, and light cords. Close the converter dolly air tank petcock. Close the shut-off valves at the rear of the second trailer (the service and emergency shut-offs).

Open the shut-off valves at the rear of the first trailer (and on the dolly if it has them). Raise the landing gear completely.

Charge the trailers (push in the "Trailer Air Supply" knob). Check for air at the rear of the second trailer by opening the emergency line shut-off. If air pressure isn't there, something is wrong and the brakes won't work.

UNCOUPLING DOUBLES

Uncouple Rear Trailer

Park the vehicle combination in a straight line on firm level ground. Apply the parking brakes so the vehicles won't move. Chock the wheels of the second trailer if it doesn't have spring brakes. Lower the landing gear of the second semitrailer enough to remove some weight from the dolly.

Close the air shut-offs at the rear of the first semitrailer (and on the dolly if it has them). Disconnect all the dolly air and electric lines and secure them.

Release the dolly brakes. Release the converter dolly fifth-wheel latch.

Slowly pull the tractor, first semitrailer, and dolly forward to pull the dolly out from under the rear semitrailer.

Uncouple Converter Dolly

Lower the dolly landing gear. Disconnect the safety chains. Apply the converter dolly spring brakes or chock the wheels.

Release the pintle hook on the first semitrailer. Slowly pull clear of the dolly. Never unlock the pintle hook with the dolly still under the rear trailer. The dolly tow bar may fly up. You could get hurt. And, it could be very difficult to recouple.

COUPLING TRIPLES

Couple the second and third trailers using the method for coupling doubles. Uncouple the tractor and pull away from the second and third trailers. Then couple the tractor to the first trailer. Use the method already described for coupling tractor-semitrailers. Move the converter dolly into position and couple the first trailer to the second trailer using the method for coupling doubles. The triples combination is now complete.

UNCOUPLING TRIPLES

Uncouple the third trailer by pulling the dolly out. Then unhitch the dolly, using the method for uncoupling doubles. Uncouple the rest of the combination as you would any double combination using the method we've described.

INSPECTING VEHICLE COMBINATIONS

Use the seven-step inspection procedure described in Chapter 9 to inspect your vehicle combination. There are more things to inspect on a vehicle combination than on a single vehicle. Many of these are just more of what are on a single vehicle (for example, tires, wheels, lights, and reflectors).

However, there are also some new things to check. Those steps will be described in detail first. Then you'll find the entire seven-step pre-trip inspection, including stopping to check the coupling and trailer brakes. As first described in Chapter 9, this seven-step routine was suited more to inspecting straight trucks, or trucks pulling trailers on a tow bar. You'll find it repeated here, with the steps for inspecting tractors with air brakes pulling one, two, or three trailers (semitrailers and full trailers).

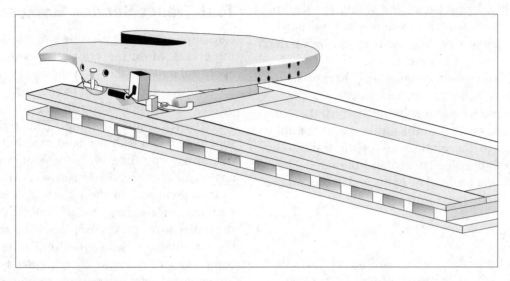

Figure 11-7. A sliding fifth wheel.

Lower Fifth Wheel

Check that the lower fifth wheel is securely mounted to the frame. Look for missing or damaged parts. There should be no visible space between the upper and lower fifth wheel.

The locking jaws should be around the shank, not the head of the kingpin. The release arm should be properly seated and the safety latch or lock engaged.

Upper Fifth Wheel

See that the glide plate is securely mounted to the trailer frame. Check the kingpin for damage.

Air Lines to Trailer

The air lines should be properly connected to the glad hands, with enough slack for turns. All lines must be free from damage and air leaks.

Sliding Fifth Wheel

Check the slider for damage or missing parts. See that all the locking pins are present and locked in place.

If the slider is air-powered, check for air leaks.

Check that the fifth wheel is not so far forward that the tractor frame will hit the landing gear, or the cab will hit the trailer, during turns.

Landing Gear

The landing gear should be fully raised. Check for missing parts and damage. The crank handle should be in place and secured. If the landing gear is power-operated, check for air or hydraulic leaks.

Detachable Electrical Connections

Detachable electrical connections ("pigtail") carry current from the tractor to the trailer, where it powers the trailer lights. Like any other circuit, this connection can short out or break. Make sure the electrical cord is firmly plugged in and secure. Check the electrical lines for damage.

Converter Dolly

You may couple a trailer to a straight truck or another trailer using a converter dolly. This is a fifth wheel assembly on an axle. In inspecting it, you must check it as you would a fifth wheel and an axle. You must inspect the dolly's tires, wheels, suspension, lights, brakes, and mud flaps, if you have them.

The converter dolly might have its own air tank. If so, it requires the same maintenance and inspection as regular air tanks. That is, it must be checked for moisture and drained.

A converter dolly also has its own landing gear, called a dolly leg or support leg. It supports the tongue or drawbar when the dolly isn't connected to anything. Inspect it as you would the landing gear. Check the tongue as you would a tow bar.

If your converter dolly has a spare tire carrier, make sure the tire is secure.

The converter dolly must also have a set of safety chains. Check them for damage.

When you check the connection at the converter dolly, make sure the pintle-eye of the dolly is in place in the pintle hook of the trailer. The pintle hook must be latched and the safety chain secured to the trailer. The lights and brakes on the converter dolly must work.

Brakes

Perform these checks in addition to the other brake checks you do as part of your inspection.

Test Tractor Protection Valve

Charge the trailer air brake system. (That is, build up normal air pressure and push in the "Trailer Air Supply" knob). Shut the engine off. After releasing the parking brakes, step on and off the brake pedal several times to reduce the air pressure in the tanks. The tractor protection valve control should pop out (or go from the "Normal" to "Emergency" position) when the air pressure falls into the pressure range specified by the manufacturer (usually 20 to 45 psi).

 TIP   *Remember, the tractor protection valve is an eight-sided, red, push-pull knob.*

If the tractor protection valve doesn't work right, an air hose or trailer brake leak could drain all the air from the tractor. This would cause the emergency brakes to come on, and you could lose control of the vehicle.

Test Trailer Emergency Brakes

Charge the trailer air brake system and check that the trailer rolls freely. Then stop and pull out the tractor protection valve control (or trailer emergency valve) or place it in the "Emergency" position. Pull gently on the trailer with the tractor to check that the trailer emergency brakes are on.

Test Trailer Service Brakes

Check for normal air pressure and release the parking brakes. Move the vehicle forward slowly, and apply the trailer brakes with the hand control (trolley valve or Johnson bar), if so equipped. You should feel the brakes apply. This tells you the trailer brakes are connected and working. (The trailer brakes should be tested with the hand valve but controlled in normal operation with the foot pedal, which applies air to the service brakes at all wheels.)

It's a good idea to check the brakes after several moments of charging the air system. Look under the trailer at the brakes. You should be able to see if the brake lining is against the drum. If not, allow the compressor to continue charging the system for a little longer before you try to back under the trailer. Even a trailer brake system that has lost all of its air supply from sitting will be fully charged by the time you have completed all your safety checks and made all the connections, if the air supply hoses were connected early in the coupling procedure.

To test the brake system for crossed air lines, turn the tractor engine off. Apply and release the trailer brakes with the hand valve. Listen for brake movement and air release.

Brake Check for Doubles and Triples

Check that air flows to all trailers. Use the tractor parking brake or chock the wheels to hold the vehicle. Wait for air pressure to reach normal, then push in the red "Trailer Air Supply" knob. This will supply air to the emergency (supply) lines. Use the trailer hand brake to provide air to the service line.

Go to the rear of the vehicle combination. Open the emergency line shut-off valve at the rear of the last trailer. You should hear air escaping, showing the entire system is charged. Close the emergency line valve.

Open the service line valve to check that service pressure goes through all the trailers. For this test you'll need the trailer hand brake or the service brake pedal applied. Then close the valve. If you do not hear air escaping from both lines, check that the shut-off valves on the other trailer or trailers and dollies are in the "Open" position. You must have air all the way to the back for all the brakes to work.

SEVEN-STEP VEHICLE COMBINATION INSPECTION

There are seven main steps in this inspection routine. They are:

- approach the vehicle
- raise the hood or tilt the cab and check the engine compartment
- start the engine and inspect inside the cab
- check the lights
- walk all around the vehicle, inspecting as you go
- check the signal lights
- check the brakes

Step One: Vehicle Overview

As you walk toward the vehicle, notice its general condition. Look for damage. Note if the vehicle is leaning to one side. This could mean a flat tire. Cargo may be overloaded or may have shifted. Or, it could be a suspension problem. Look under the vehicle for fresh oil, coolant, grease, or fuel leaks. Check the area around the vehicle for people, other vehicles, objects, low hanging wires, or limbs. These could become hazards once the vehicle begins to move.

Look over the most recent vehicle inspection report. Drivers typically have to make a vehicle inspection report in writing each day. The vehicle owner then should see that any items in the report that affect safety are repaired. If the mechanic repaired the defects, or determined no repairs were needed, there will be a certification for that on the form.

You should look at the last report to find out what was the matter, if anything. Inspect the vehicle to find out whether problems were fixed, or whether repairs were not needed.

Step Two: Engine Compartment Check

Check that the parking brakes are on and/or wheels are chocked. You may have to raise the hood or open the engine compartment door. If you have a tilt cab, secure loose items in the cab before you tilt it. Then they won't fall and break something.

At this site, check the engine oil level. The oil should be above the "Low" or "Add" mark on the dipstick. See that the level of coolant is above the "Low" level line on the reservoir. Check the condition of coolant-carrying hoses.

Check the radiator shutters and winterfront, if your vehicle has one. Remove ice from the radiator shutters. Make sure the winterfront is not closed too tightly. If the shutters freeze shut or the winterfront is closed too much, the engine may overheat and stop. Inspect the fan. Make sure the blades are sound and not likely to catch on hanging wires or hoses.

If your vehicle has power steering, check the fluid level and the condition of the hoses. Check the dipstick on the oil tank. The fluid level should be above the "Low" or "Add" mark.

Here's where you check the windshield washer fluid level.

If your battery is located in the engine compartment, perform the battery check described earlier here.

If you have an automatic transmission, check the fluid level. You may have to do this with the engine running. The owner's manual will tell you if that is the case.

Check drive belts for tightness and excessive wear. Press on the center of the belt. The owner's manual will tell you how much slack there should be. Also, if you can easily slide a belt over a pulley, it is definitely too loose or too worn. Look for leaks in the engine compartment. These could be fuel, coolant, oil, power steering fluid, hydraulic fluid, or battery fluid.

Look for cracked or worn insulation on electrical wiring.

Check the compressor for the air brakes. Check the oil supply. If it's belt-driven, check the belt. It should be in good condition. Check belt tightness. Lower and secure the hood, cab, or engine compartment door.

Next, check any handholds, steps, or deck plates. Remove all water, ice and snow, or grease from handholds, steps, and deck plates that you must use to enter the cab or to move about the vehicle. This will reduce the danger of slipping.

Step Three: Inside the Cab

Get in the cab. Inspect inside the cab. Start the engine. Make sure the parking brake is on. Shift into neutral, or "Park" if your transmission is automatic. Start the engine and listen for unusual noises.

Check the gauge readings. The oil pressure should come up to normal within seconds after the engine is started. The ammeter and/or voltmeter should give normal readings. Coolant temperature should rise gradually until it reaches the normal operating range. The temperature of the engine oil should also rise slowly to the normal operating range.

Oil, coolant, and charging circuit warning lights will come on at first. That tells you the lights are working. They should go out right away unless there's a problem.

Make sure your controls work. Check all of the following for looseness, sticking, damage, or improper setting:

- steering wheel
- accelerator
- foot brake
- trailer brake
- parking brake
- retarder controls (if your vehicle has them)
- transmission controls
- interaxle differential lock (if your vehicle has one)
- horn or horns
- windshield wiper and washers
- headlights
- dimmer switch
- turn signal
- four-way flashers
- clearance, identification, and marker-light switch(es).

If your vehicle has a clutch, test it now. Depress the clutch until you feel a slight resistance. One to two inches of travel before you feel resistance is normal. More or less than that signals a problem.

Check your mirrors and windshield for the defects described earlier.

Check that you have the required emergency equipment and that it is in good operating condition.

Check for optional items, such as a tire changing kit, and items required by state and local laws, such as mud flaps.

Step Four: Check Lights

Next, check to see that the lights are working. Make sure the parking brake is set, turn off the engine, and take the key with you. Turn on the headlights (on low beams) and the four-way flashers, and get out. Go to the front of the vehicle. Check that the low beams are on and both of the four-way flashers are working. Push the dimmer switch and check that the high beams work. Turn off the headlights and four-way hazard warning flashers. Turn on the parking, clearance, side-marker, and identification lights. Turn on the right-turn signal, and start the "walk-around" part of the inspection.

Step Five: Walk-around Inspection

Walk all around the vehicle, inspecting as you go. Start at the cab on the driver's side. Cover the front of the vehicle. Work down the passenger side to the rear. Cover the rear of the vehicle. Work up the driver's side back to the starting position. Do it this way every time.

Clean all the lights, reflectors, and glass as you go along.

Left Front Side

The driver's door glass should be clean. Door latches or locks should work properly.

Perform a tire, wheel, and rim check at the left front wheel. Perform a suspension system and braking system check. Check the steering parts. Check the air hoses for leaks and breaks. Check the brake assembly. Look for parts that need replacement or repair: a cracked drum or a loose or worn brake lining. Check the slack adjuster. Have it adjusted, if necessary. Drain air tanks if necessary.

Front

Check the condition of the front axle. Perform a check of the steering system parts. You should actually take hold of steering system parts and shake them to say for sure that they are not loose.

Check the windshield. Pull on the windshield wiper arms to check for proper spring tension. Check the wiper blades for the defects described earlier.

Check the lights and reflectors (parking, clearance, and identification, and turn signal lights) at this site. Make sure you have all the required ones as described earlier. They should be clean and working.

Right Front Side

The right front has parts similar to the left front. Perform the same checks. The passenger door glass should be clean. Door latches or locks should work properly. Perform a tire, wheel, and rim check at the right front wheel. Perform a suspension system and braking system check. Check the steering parts. Check the air hoses for leaks and breaks. Check the brake assembly. Look for parts that need replacement or repair: a cracked drum or a loose or worn brake lining. Check the slack adjuster. Have it adjusted if necessary. Drain air tanks as needed. For COE trucks, check the primary and safety cab locks here.

Check the coupling to the trailer: the fifth wheel and locking jaw, the air and electrical connections, the landing gear, and other parts. Check for damaged and broken parts. Make sure the connection is secure. The lights and brakes should work.

Right Side

The fuel tank is located here. Check to see that it is mounted securely, not damaged or leaking. The fuel crossover line must be secure, and not hanging so low that it would be exposed to road hazards. The filler caps must be on firmly. Check the fuel supply. Perform these checks for each fuel tank your vehicle has.

Check all parts you can see at this site. See that the rear of the engine is not leaking. Check for leaks from the transmission. Check the exhaust system parts for defects as described earlier.

Inspect the frame and cross members for bends and cracks.

Electrical wiring should be secured against snagging, rubbing, and wearing.

If your vehicle has a spare tire carrier or rack, check it for damage. It should be securely mounted in the rack. Make sure you have the correct size spare tire and wheel. The spare tire must be in good condition and properly inflated.

Cargo Securement

Check the cargo securement at the right front of the vehicle. It must be properly blocked, braced, tied, chained, and so forth. If you need a header board, make sure it is strong enough and secure. If you have side boards, check them. The stakes must be strong enough for the load, free of damage, and properly set in place.

Make sure you haven't created handling problems for yourself by overloading the front axles.

Any canvas or tarp must be properly secured to prevent tearing and billowing.

If you are hauling an oversized load, have the necessary signs safely and properly mounted.

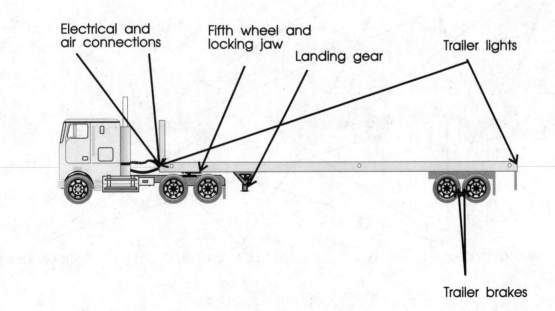

Figure 11-8. Checking the coupling should be a standard part of your trailer inspection.

Make sure you have the permits you need to haul this load.

Check that the curbside cargo compartment doors are securely closed, latched, and locked. See that the required security seals are in place.

Right Rear

At this location, check the wheels, rims, tires, brake system, and suspension parts for the defects described in the first part of this chapter. Check that all parts are sound, properly mounted, and secure, and not leaking. Check the axle. Powered axles should not leak.

If your vehicle has sliding tandem axles, check the lock mechanism. If the axles are air-powered, check for leaks. Check the lights and reflectors required at this location. They should be clean and working.

Rear

Check the lights and reflectors required at this location. They should be clean and working.

Make sure you have a license plate. It should be clean and securely mounted.

If mud flaps are required, check them now. They should be free of damage and fastened according to whatever regulations apply. If there are no legal guidelines, make sure that the flaps are at least well-fastened, not dragging on the ground or rubbing the tires.

If you're pulling doubles or triples, check the coupling between trailers. Check the converter gear. Look for loose or damaged parts. The lights and brakes should work. Make sure the safety chain is on and the pintle hook is latched.

Check the air shut-off at the rear of the trailer. At the rear of the last trailer in the combination, the shut-off valve should be closed. If it's not, air will escape at the end of the trailer air lines. The shut-off valves at the rear of the middle trailer or trailers should be open. If they're not, air won't get to the last trailer.

Cargo Securement

Check the cargo securement at the right rear of the vehicle. It must be properly blocked, braced, tied,

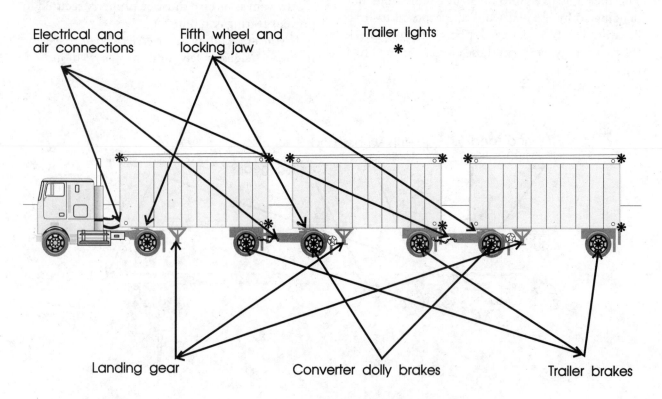

Figure 11-9. Check the coupling between trailers if you're pulling doubles or triples.

chained, and so forth. Check that the tailboard is strong enough and secure. If you have end gates, check them. The stakes must be strong enough, free of damage, and properly set in the stake pockets.

Any canvas or tarp must be properly secured to prevent tearing and billowing. Make sure the tarp doesn't cover up your rear lights.

Check to make sure you haven't distributed the load so as to overload the rear axles.

If you are hauling an oversized load, have the necessary signs safely and properly mounted. Have any lights or flags required for projecting loads. Make sure you have the permits you need to haul this load.

Check that the rear doors are securely closed, latched, and locked.

Left Rear and Front Side

At this location, check the wheels, rims, tires, brake system, and suspension parts just as you did for the right rear and front side. Check that all parts are sound, properly mounted and secure, and not leaking. Check the axle. Powered axles should not leak.

If your vehicle has sliding tandem axles, check the lock mechanism. If the axles are air-powered, check for leaks. Check the lights and reflectors required at this location. They should be clean and working.

Your vehicle's battery may be mounted here, rather than in the engine compartment. If so, perform the battery check now.

Check the coupling to the trailer: the fifth wheel and locking jaw, the air and electrical connections, the landing gear, and other parts. Check for damaged and broken parts. Make sure the connection is secure. The lights and brakes should work.

Step Six: Check Signal Lights

Get in the cab and turn off all the lights. Then, turn on the stop lights (apply the trailer hand brake. Or, have a helper step on the brake pedal while you watch to see if the light goes on).

Now turn on the left-turn signal lights. Then get out and check the left side lights, rear and front.

Step Seven: Check Brakes

Now you'll perform some tests with the engine running. Get back in the vehicle. Turn off any lights you don't need for driving. Check to see that you have all the required papers, trip manifests, permits, and so forth. Secure all loose articles in the cab. If they are free to roll around, they could hinder your operation of the controls, or hit you in a crash.

Start the engine. Perform air brake tests, numbers one through four. (Review Chapter 10 if you don't recall the tests). Remove the chocks. Perform brake tests five and six as described in Chapter 10. Test the trailer brakes.

PULLING TRAILERS SAFELY

Vehicle combinations are usually heavier, longer, and require more driving skill than single CMVs. If you're pulling trailers, you need more knowledge and skill than drivers of single vehicles.

FMCSR Part 393 says that you must know about the handling and stability of tractor-trailer combinations. You must understand such effects as

- off-tracking
- response to steering
- response to braking
- oscillatory sway
- rollover in steady turns
- yaw instability in steady turns

Off-tracking

When a vehicle goes around a corner, the rear wheels follow a different path than the front wheels. This is called off-tracking or "cheating." Figure 11-11 shows how off-tracking causes the path followed by a tractor-semitrailer to be wider than the vehicle combination itself.

Longer vehicles will offtrack more. The rear wheels of the power unit (truck or tractor) will offtrack some. The rear wheels of the trailer will offtrack even more. If there is more than one trailer, the rear wheels of the last trailer will offtrack the most.

Step 1: Vehicle overview for general condition

Step 2: Engine compartment checks
Fluid levels and leaks
Hoses and belts
Battery
Windshield wipers/washers
Wiring

Step 3: Inside the cab checks
Parking brake on
Gauge readings
Warning lights
Controls
Emergency equipment
Optional equipment

Step 4: Check lights
Headlights (low and high beams)
Four-way flashers
Parking lights
Clearance lights
Side marker lights
Identification lights
Right-turn signal

Step 5: Walk-around inspection
A. Left front side: wheel and tire, suspension, brakes, axle, steering, side marker lamp and reflector, door glass, latches and locks, mirrors
B. Front of cab: axle, steering system, windshield, lights and reflectors
C. Right front side: door glass, latches and locks, wheels and tires, suspension, brakes, axle, steering, condition of bed or body, side marker lamp and reflector
D. Right side: fuel tank, exhaust system, frame and cross members, wiring, spare tire, lights and reflectors, mirrors
E. Cargo securement, curbside doors
F. Right rear: wheels and tires, suspension, brakes, axle, lights and reflectors
G. Rear: lights and reflectors, license plate, mudflaps, coupling between trailers, landing gear, doors
H. Cargo securement
I. Left rear and left front: wheels and tires, suspension, brakes, steering, axle, lights and reflectors, battery, coupling to trailer

Step 6: Check signal lights

Step 7: Brake check

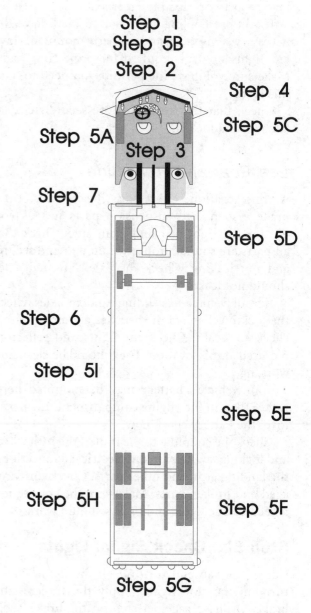

Figure 11-10. Combination Vehicle Inspection Aid.

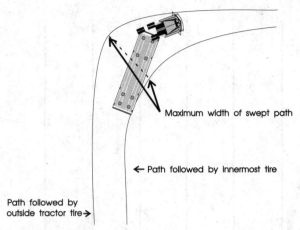

Figure 11-11. Off-tracking.

In a right turn, for example, the rear trailer wheels will tend to pass closer to the curb than the tractor wheels. You must steer the front end wide enough around a corner so the rear end does not run over the curb, pedestrians, other vehicles, and so forth. At the same time, keep the rear of your vehicle close to the curb. This will stop other drivers from passing you on the right.

If you cannot complete your turn without entering another traffic lane, turn wide as you complete the turn (Figure 11-12). This is better than swinging wide to the left before starting the turn. It will keep other drivers from passing you on the right. If drivers pass on the right, you might crash into them when you turn.

Response to Steering

Trucks with trailers have a dangerous "crack-the-whip" effect. When you make a quick lane change, the crack-the-whip effect can turn the trailer over. There are many accidents where only the trailer has overturned. "Rearward amplification" causes the crack-the-whip effect. This is to say that a small steering correction at the front of the vehicle will be a large movement at the end of the last trailer. It can be large enough and forceful enough to turn over the vehicle.

Figure 11-13 shows eight types of vehicle combinations and the rearward amplification each has in a quick lane change. Combinations with the least crack-the-whip effect are shown at the top and those with the most at the bottom. Rearward amplification of 2.0 in the chart means that the rear trailer is twice as likely to turn over as the tractor. You can see that triples have a rearward amplification of 3.5. This means you can roll the last trailer of triples 3.5 times as easily as a five-axle tractor-semitrailer.

More than half of truck driver deaths in crashes are from truck rollovers.

Remember, even small steering movements will have big results at the end of your vehicle combination. You must steer gently. Avoid sudden changes in direction. If you make a sudden move-

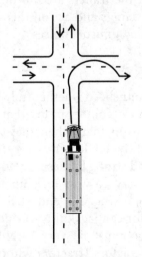

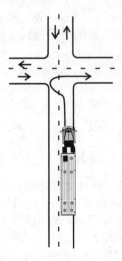

Figure 11-12. Turning right while pulling a trailer.

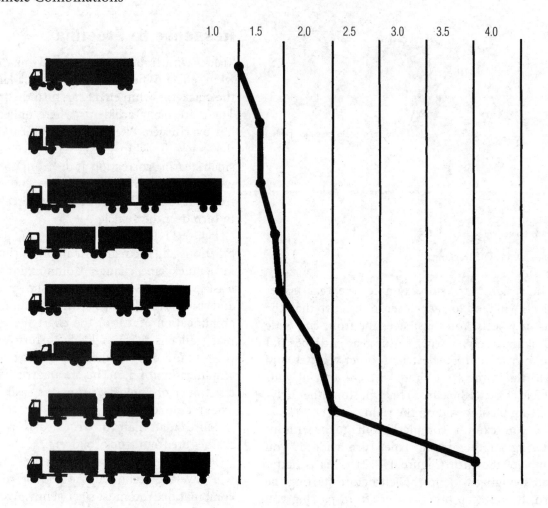

Figure 11-13. Rearward amplification of different vehicle combinations.

ment with your steering wheel, you could tip over a trailer.

Follow far enough behind other vehicles. Follow at least one second for each 10 feet of your vehicle length, plus another second for speeds over 40 mph. Look far enough down the road to avoid being surprised and having to make a sudden lane change.

At night, drive slowly enough to see obstacles with your headlights before it's too late to change lanes or stop gently. Slow to a safe speed before going into a turn.

Sensory Feedback

"Sensory feedback" means getting information from your senses. Skilled tractor-trailer drivers can see, feel, and hear what's going on behind them. Of course, use your sense of sight. Check your mirrors often. When they're properly adjusted, you can see if your trailers are following in a straight path.

Learn to use your sense of "feel" to make gentle steering corrections and gradual turns. Your body will feel if the trailers are out of control. Be alert to those feelings and act on them. Keep your ears tuned to any unusual noises.

Braking

Control your speed whether fully loaded or empty. Large vehicle combinations that are empty take longer to stop than when they are fully loaded. When lightly loaded, the very stiff suspension springs and strong brakes give poor traction. That makes it very easy to lock up the wheels. Your trailer can swing out and strike other vehicles. Your tractor can jackknife very quickly.

You also must be very careful about driving "bobtail" tractors (tractors without semitrailers).

Tests have shown that bobtails can be very hard to stop smoothly. It takes them longer to stop than a tractor-semitrailer loaded to maximum gross weight.

In any vehicle combination, allow lots of following distance. Look far ahead, so you can brake early. Don't be caught by surprise and have to make a "panic" stop. You're more likely to lock up the wheels when you stop short.

When the wheels of a trailer lock up, the trailer will tend to swing around. This is more likely to happen when the trailer is empty or lightly loaded. This type of jackknife is often called a "trailer jackknife." This is similar to a rear-wheel skid on a straight vehicle.

You stop a trailer skid much the same way you stop a rear-wheel skid. First, realize you've lost traction. The earliest and best way to recognize that the trailer has started to skid is by seeing it in your mirrors. Any time you apply the brakes hard, check the mirrors to make sure the trailer is staying where it should be. Once the trailer swings out of your lane, it's very difficult to prevent a jackknife.

If you have started to skid, stop using the brake. Release the brakes to get traction back. Do not use the trailer hand brake (if you have one) to "straighten out the rig." This is the wrong thing to do. It's the brakes on the trailer wheels that caused the skid in the first place. Once the trailer wheels grip the road again, the trailer will start to follow the tractor and straighten out.

Oscillatory Sway

The top of the trailer can tilt, or oscillate. This could be caused by a banked road or by making a turn. You saw how a high center of gravity can worsen this effect. The more trailers you have, the harder it is to keep them all under control. On a very uneven road, one trailer could be leaning left, the other right.

Rollover in Steady Turns

When more cargo is piled up in a truck, the "center of gravity" moves higher up from the road. You know this means the truck becomes easier to turn over. Fully loaded tractor-trailers are 10 times more likely to roll over in a crash than empty vehicles.

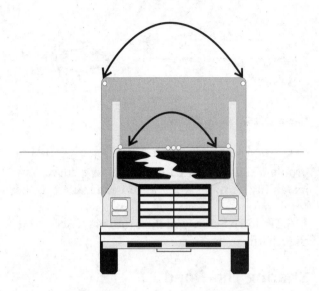

Figure 11-14. Oscillatory sway.

There are two things you can do to help prevent rolling over:

- keep the cargo (and the center of gravity) as close to the ground as possible
- go slowly around turns

Keeping cargo, and thus the center of gravity, low is important in straight trucks, as you have seen. It's even more important in vehicle combinations. Remember, the center of gravity is where the weight acts as a force. It stands to reason that the heavier the weight, the greater that force is going to be.

Also, keep the load centered. If the load is to one side so it makes a trailer lean, a rollover is more likely. Make sure your cargo is centered and spread out as much as possible.

Yaw Stability in Steady Turns

Yawing describes the side-to-side movement at the back of the trailer. This can be a real problem in a steady turn. To understand why, you have to know something about another natural force: centrifugal force. This is the tendency of objects to move in one direction. Here's how it works when you take a curve.

You have been driving forward in a straight line. You enter an off-ramp and steer the vehicle around the curve. Because of centrifugal force, the vehicle tends to continue in a straight line. If you could not control the vehicle, it would drive right off the curve.

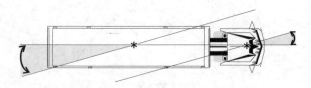

Figure 11-15. Yaw instability.

Straight vehicles don't have quite as big a problem fighting centrifugal force in a curve. Yaw instability becomes worse as you add trailers to the combination.

Slow down before you enter the curve. Accelerate slightly through the curve.

Sharing the Road

Vehicle combinations cause extra problems for other motorists. Vehicle combinations are slower on upgrades. They take longer to pass, and it takes longer to pass them.

When you're in the right-most or merging lane with your triple combination, you make it hard for other motorists to enter or exit. You could easily block the entrance and exit ramp.

The splash and spray from your vehicle combination can make it difficult for other motorists to see. Also be aware that as your vehicle cuts through the wind, it creates drafts. For small vehicles coming up alongside yours, it can be like getting hit by a crosswind. This is called "aerodynamic buffeting."

Last, vehicle combinations take longer to change lanes. The steer-countersteer involved in changing lanes can cause the vehicle combination to become unstable, as you have seen. Doubles and triples must be driven very smoothly to avoid a rollover or jackknife. To avoid accidents, you must look far ahead so you can slow down or change lanes gradually when necessary.

Good space management is vital. Doubles and triples take up more space than other commercial vehicles. They are longer and need more space because they can't be turned or stopped suddenly.

Allow more following distance. Make sure you have large enough gaps before entering or crossing traffic. Be certain you are clear at the sides before changing lanes.

Be more careful in adverse conditions. In bad weather, slippery conditions, and mountain driving, you must be especially careful if you drive

doubles and triples. You will have greater length and more dead axles to pull with your drive axles than other drivers. There is more chance for skids and loss of traction.

A three-axle tractor pulling a two-axle semi-trailer is the most popular, heavy, over-the-road vehicle used in the trucking industry. Nearly all the nation's highway freight is hauled in the "18-wheeler." So, many CDL applicants (including you?) will need a Group A type license. You know this means you must pass the basic knowledge and skills tests. You must take your skills test in an 18-wheeler. Then you must answer special knowledge questions about pulling a trailer.

Your 18-wheeler will probably have air brakes. In that case, you can't afford an air brake restriction on your CDL. Be sure to master Chapter 10 so you can pass the air brake tests.

The other popular vehicle combinations are doubles and triples. Pulling doubles and triples only makes the CMV driver's job more complicated. There are simply more vehicle parts to know about, to inspect, and to control. Drivers of double and triple combinations have an enormous responsibility. Drivers who get the doubles/triples endorsement have shown they've probably earned it because of their knowledge and skill.

PASS POST-TRIP

Instructions: For each true/false test item, read the statement. Decide whether the statement is true or false. If it is true, circle the letter A. If it is false, circle the letter B. For each multiple-choice test item, choose the answer choice—A, B, C, or D—that correctly completes the statement or answers the question. There is only one correct answer. Use the answer sheet provided on page 265.

1. Before coupling, trailer height should be
   _____.
   A. about three inches higher than the tractor fifth wheel
   B. just higher than the center of the tractor fifth wheel
   C. just low enough to be raised slightly as the tractor is backed under it
   D. just below the top of the tractor frame rails

2. When the coupling is complete, there should be _____ between the upper and lower fifth wheel.
   A. about two inches
   B. no more than an inch
   C. no space
   D. just enough space to see through

3. Air lines should be connected to the trailer during _____.
   A. coupling
   B. uncoupling
   C. coupling and uncoupling
   D. at no time during coupling or uncoupling

4. When uncoupling, you should pull the tractor partly clear of the trailer and stop because _____.
   A. the air lines are still connected at this time
   B. the electrical cable is still connected
   C. the support of the tractor frame will keep the trailer from falling if the landing gear collapses or sinks
   D. you need to know if the trailer will need to have chocks

5. In a triple combination, the heaviest trailer should be in the middle.
   A. True
   B. False

6. The pre-trip inspection of a double or triple combination should include draining the converter dolly air reservoir.
   A. True
   B. False

7. The emergency air line to the trailer _____.
   A. supplies air to the trailer air tanks and controls the emergency brakes on vehicle combinations
   B. is usually color-coded red
   C. will cause serious problems if it fails or comes loose on the highway
   D. all of the above

8. A fully loaded tractor-trailer is 10 times more likely to roll over in a crash than an empty one.
   A. True
   B. False

9. When a tractor-trailer makes a right turn, _____.
   A. the trailer wheels will pass closer to the curb than the tractor wheels
   B. the tractor wheels will pass closer to the curb than the trailer wheels
   C. the trailer wheels will track the tractor wheels exactly
   D. it's accepted that the trailer wheels will run up over the curb.

10. On a curved off-ramp, centrifugal force _____.
    A. gives you more traction
    B. helps you stay in the curve
    C. tends to force you out of the curve
    D. decreases the effects of rearward amplification

CHAPTER 11 ANSWER SHEET

Test Item		Answer Choices			Self - Scoring	
					Correct	Incorrect
1.	A	B	C	D	☐	☐
2.	A	B	C	D	☐	☐
3.	A	B	C	D	☐	☐
4.	A	B	C	D	☐	☐
5.	A	B	C	D	☐	☐
6.	A	B	C	D	☐	☐
7.	A	B	C	D	☐	☐
8.	A	B	C	D	☐	☐
9.	A	B	C	D	☐	☐
10.	A	B	C	D	☐	☐

Cut Here

Chapter 12

Tank Vehicle Endorsement

In this chapter, you will learn about:

- the causes of cargo surge
- the effects of cargo surge on vehicle handling
- how to prevent cargo surge
- the braking response of tank vehicles
- baffled and compartmented tank vehicles
- the different types of tank vehicles
- liquid cargo density and cargo surge
- tank emergency systems
- the retest and marking requirements for DOT specification vehicles

To complete this chapter, you will need:

- a dictionary
- a pencil or pen
- blank paper, a notebook, or an electronic notepad
- colored pencils, pens, markers, or highlighters
- a CDL preparation manual from your state Department of Motor Vehicles, if one is offered
- the Federal Motor Carrier Safety Regulations pocketbook (or access to U.S. Department of Transportation regulations, Parts 383, 393, and 396 of Subchapter B, Chapter 3, Title 49, Code of Federal Regulations)
- the owner's manual for your vehicle, if available

PASS PRE-TRIP

Instructions: Read the statements. Decide whether each statement is true or false. If it is true, circle the letter A. If it is false, circle the letter B.

1. The difference between cargo tanks and portable tanks is that portable tanks may be loaded or unloaded while off the vehicle.
   A. True
   B. False

2. Liquid surge can push a stopped truck into an intersection.
   A. True
   B. False

3. Forward-and-backward surge is a big problem with smooth bore tanks.
   A. True
   B. False

A tank vehicle is a vehicle used to transport cargo in bulk. You'll see tanks carrying lime, and you'll see tanks carrying milk.

As far as commercial driver licensing is concerned, though, "tanks" mean vehicles carrying liquids and gases in bulk. This definition is given in FMCSR Part 383.

Part 383 further defines a tank vehicle as a permanently attached tank, or a portable tank able to hold 1,000 gallons or more. That is, you can have a straight truck-type tanker. You can have a semitrailer tanker. You can have a portable tank on a flatbed trailer. Permanent tanks are loaded and unloaded while still on the vehicle. Portable tanks may be loaded or unloaded while off the vehicle.

As you know by now, you need a Tank Vehicle Endorsement on your CDL to transport liquids in bulk. This chapter will help you get that endorsement.

EFFECTS OF LIQUID LOADS ON VEHICLE HANDLING

It takes special knowledge and skill to haul bulk liquids. This cargo presents two handling problems:

- high center of gravity
- liquid surge

High Center of Gravity

You recall that "high center of gravity" means that much of the load's weight is carried high up off the road. This makes the vehicle top-heavy and rollover more likely. Rollover is particularly likely with liquid tankers. Tests have shown that tankers can turn over at the speed limits posted for curves. Take highway curves and ramps well below the posted speeds.

Liquid Surge

Liquid surge results from movement of the liquid in partially filled tanks. This movement can have bad effects on handling. For example, when the vehicle is coming to a stop, the liquid will surge back and forth. When the wave hits the end of the tank, it tends to push the truck in the direction in which the wave is moving. If the truck is on a slippery surface like ice, the wave can shove a stopped truck out into an intersection.

Free-flowing lightweight liquids like water and milk tend to slosh more than thicker liquids. Thicker liquids like corn syrup tend to ride better than free-flowing liquids. The surging back and forth of any liquid in a tanker can cause serious driving problems.

The less you move the liquid around, the less effect liquid surge will have. So you must make all your movements—turning, shifting, stopping—gradually and slowly.

BRAKING RESPONSE OF TANKS

The brakes on most vehicles work best when the vehicle is fully loaded. The weight of the load gives you more traction. When the vehicle is empty, or bobtailing, it takes longer to stop. You could need as much as twice the stopping distance. Also, the brakes are more likely to lock up when you make an emergency stop in an empty vehicle.

The worst case would be a vehicle with a partial, liquid load. First, your vehicle does not have the traction it would have if it were full. Second, the surge factor is worse with a partial load than a full load.

The surge factor is at its worst in a tanker that is 80 percent full. This is because there is enough liquid to be quite heavy. At the same time, there is enough space for it to surge. If the tanker is 90 percent full, there is more liquid and more weight, but not enough space for it to move around. If the tanker is 40 percent full, there is plenty of space. However, there is not enough liquid and weight to throw the tanker too far out of control.

So, with a partial load, you lose the advantage of traction you would have with a full load. Then, the surging of the liquid can be so forceful, it can overcome the stopping power of your brakes. Your vehicle could travel quite a distance before it comes to a complete stop.

Avoid braking in turns. You must drive slowly and carefully at all times. That way, if you must use the brakes, you can do so gradually. This reduces the effect of liquid surge. You can keep the vehicle under control as you stop.

BAFFLES AND COMPARTMENTS

Some tankers are one long cylinder inside. These are called "smooth bore" or tanks. The smooth bore tanker is designed to carry one particular

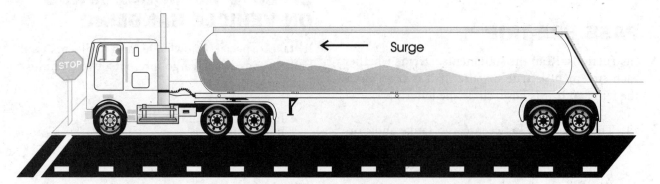

Figure 12-1. Liquid surge in a tank vehicle.

cargo at a time. Because it is a long hollow tube with no sharp corners, it is easy to load, unload, and clean.

A problem is there's nothing inside to slow down the flow of the liquid. Therefore, forward-and-backward surge is very strong. Smooth bore tanks usually those that transport food products (milk, for example). That means you must be aware of the surge factor. You must be very careful when driving smooth bore tanks, especially when starting and stopping.

Some liquids are heavier than others. That is, an amount of one liquid will weigh more than the same amount of another. This is due to density. You'll find liquid densities discussed in greater detail later in this chapter. For now, know that

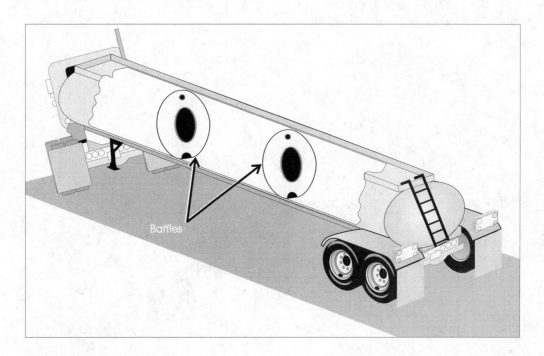

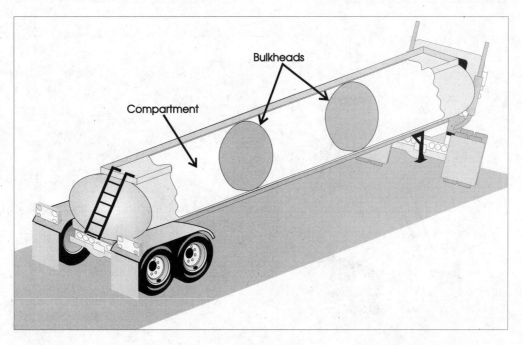

Figure 12-2. The baffles in the trailer in the top picture control the movement of the liquid cargo. The compartments in the trailer in the bottom picture completely separate different types of liquids.

when you have a heavy liquid in your tanker, the tanker should be either more or less than 80 percent full. As you just read, at 80 percent, 20 percent of the tank is left empty for movement within the tank. If you are more than 80 percent full, there is less empty space for movement. If you have less liquid, the surge factor is also lessened because there's simply less liquid to move.

Some liquid tanks are divided into several smaller areas by baffles or bulkheads. You must pay attention to weight distribution when loading and unloading the smaller areas. Don't put too much weight on the front or rear of the vehicle.

Bulkheads

Sometimes tankers are divided into compartments by bulkheads. These allow more than one type of product to be carried at the same time. For exam-

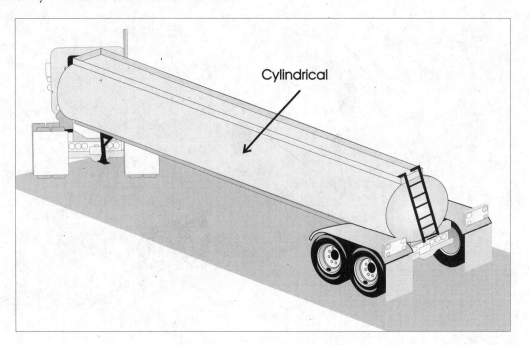

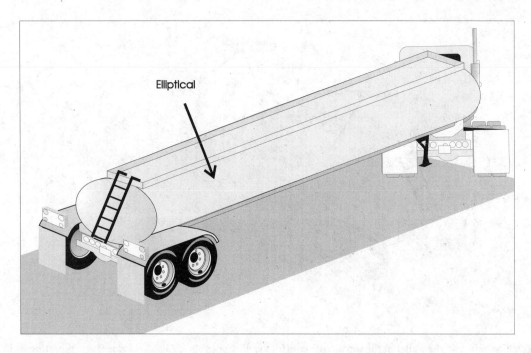

Figure 12-3. Cylindrical and elliptical are the two basic shapes you will see most often in liquid tankers.

ple, a compartmented tanker would allow you to haul both diesel fuel and gasoline in one trip.

Some tankers have double-walled bulkheads. The law requires that diesel fuel and gasoline be divided by a double bulkhead.

Bulkheads also help to control the surge factor somewhat.

Baffles

Baffled liquid tanks also have bulkheads in them. Baffles are dividers that simply break up the tank area into smaller areas. There is a hole at the top of the baffle that allows air movement through the trailer. The hole on the bottom allows the liquid to flow through for unloading. The middle hole is big enough that a person can crawl inside for cleaning. A tanker may have several baffles.

The baffles help to control the forward-and-backward liquid surge. However, side-to-side surge can still occur. This can cause a rollover. Be extremely cautious (slow and careful) in taking curves or making sharp turns with a partially or fully loaded liquid tanker.

TYPES OF TANKS

In general, liquid tankers are either elliptical or cylindrical in shape. The elliptical is egg-shaped at the end and has the lower center of gravity. It's more stable on the road. The elliptical shape cuts down the surge factor.

The cylindrical is round. It's considered the strongest and most versatile. There are also variations of these shapes since the liquids they haul have differing needs for drainage and weight distribution. For example, the double conical is a kind of cylindrical tanker.

Within these two classes, there are many different types of tankers. This is because a tanker has design and construction features that suit it to haul a particular liquid. Some tanks are lined (cladded). This protects the tank from corrosive cargo. Some tanks are heated or cooled. Others are insulated. That helps to keep the cargo within a temperature range without actually heating or cooling it. Still other tanks hold the cargo under pressure.

LIQUID DENSITY AND CARGO SURGE

Some liquids weigh more than others. That is, an amount of one liquid can weigh more than the same amount of another. The heavier liquid has a higher density.

The higher the density of the liquid, the less the cargo surge will be. A high density liquid will be less likely to move in the first place. This is because of inertia.

 TIP *Remember inertia? If not, review Chapter 8.*

It would take a lot of force to overcome the inertia of a high density liquid.

A small amount of a high density liquid can be very heavy. You can't load much of it into the tanker without being overweight. So you will have a lower center of gravity.

On the other hand, it is fairly easy to get a light liquid sloshing around.

The density of plain water is 8.3 pounds per gallon.

DRIVING TANKS

A tanker on the road is most stable when it is empty. A tanker that is 80 percent full is the least stable.

Except for a high center of gravity, a tanker is somewhat stable when it is full. However, you will rarely fill your tank completely full. Liquids expand as they warm up. You must leave room for the liquid to expand. This is called outage. Different liquids expand by different amounts. So, each liquid requires a different amount of outage. You must know the outage requirement when hauling liquids in bulk. You'll get this information from your dispatcher. You may also have an outage chart in the driver's manual your employer gives you.

Your employer may also want you to know about innage. This is the depth of liquid in the tank. The measurement is taken from the surface of the liquid to the bottom of the tank.

Also, a full tank of dense liquid (such as some acids) may exceed legal weight limits. For that reason you may often only partially fill tanks with heavy liquids. The amount of liquid to load into a tank depends on:

- the amount the liquid will expand in transit, and
- the weight of the liquid, and
- legal weight limits

How will you know the level of liquid in your tank? Your vehicle is required to have a liquid level sensing device. A probe inside the tank is wired to a controller. You'll usually find the controller on the curb side of the vehicle near other operator controls.

You have read several general cautions now about driving carefully when transporting liquids. Here are some specific guidelines.

Never Tailgate

This is always a dangerous practice, but it is much more dangerous with a liquid load. You've seen what can happen in a panic stop. Cargo surge adds to the stopping distance. A collision is almost always bound to result.

Shifting

When upshifting, always release the clutch after the surge has hit the rear of the tank.

Ramps

Always slow down and downshift before entering a freeway entrance or exit ramp. This risk of a roll-over is very great on a ramp. A slow speed in the curve will reduce this risk.

Downgrades

Always downshift at the top of a long hill or mountain. Downshift before you start down the grade. Use the snub braking technique to control your speed.

You know by now that you should not fan the brakes on a downgrade. This is especially important with liquid loads. Pumping or fanning the brakes causes the vehicle to rock. This increases the liquid surge.

Evasive Action

You know you should never make sharp or sudden changes in direction at high speed. This is even more important when hauling liquids because of the surge factor and high center of gravity.

If you drop off the edge of the pavement, never return abruptly to the roadway. Get the vehicle under control and reduce your speed. If possible, avoid using the brakes until your speed has dropped to about 20 mph. Then brake very gently to avoid skidding on a loose surface. If the shoulder is clear, stay on it until your vehicle has come to a stop. Then you can safely bring the vehicle back onto the roadway.

If you are doing a good job of speed and space management, you should not have to take evasive action in the first place.

Adverse Conditions

On slippery roads, nearly all vehicles lose traction. This is even worse when you have a liquid load. The liquid surge itself is enough to cause your vehicle to lose traction. You may have to stop driving and wait out the weather, whereas you might have gone on with a different type of load.

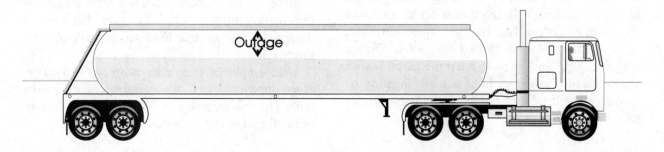

Figure 12-4. Outage.

Uncoupling

Let the load settle before you uncouple your trailer. If you don't, liquid surge could cause the landing gear to collapse.

EMERGENCY SYSTEMS

Cargo tanks are designed to reduce the chance of accidental leaks. If the tank becomes damaged, design features work to keep the cargo from leaking out. For example, fittings and attachments that don't themselves hold any cargo are attached to the tank in a special way. They're designed so that if they break off, cargo won't leak out. The rear bumper protects the tank and piping from rear-end collisions. Filling, manhole, and inspection openings are protected from being damaged in a vehicle rollover. Safety devices prevent the manhole and fill covers from opening fully when the tank is pressurized.

Most cargo tanks are also equipped with emergency systems that help keep the cargo from leaking out if there's a major accident. Some systems operate automatically. Others must be operated by the driver.

Piping that isn't protected from accidental damage must have a stop-valve and a shear section.

Stop Valves

A stop valve is a valve that stops the flow of lading (the cargo). You'll find stop valves on the tank loading and unloading outlets. These are held in the closed position by their own energy supply. When they're closed, the cargo is kept inside the tank.

The valves can be internal or external. Internal valves are self-closing. Their power supply is inside the tank. External valves self-close in emergencies, such as a fire or broken hose. (During normal operations, they can be operated manually.) The power for external self-closing stop valves is located outside the tank.

Each valve has a remote control that is located more than 10 feet from the valve. A cable links the control to the valve. This is a part of the emergency system the driver can operate manually. You operate the cable by a lever or other control. You'll usually find this lever in an operator's cabinet on the curb side of the vehicle. You may also find an emergency release trip at the road side toward the front of the vehicle.

Cargo tanks that carry flammable, pyrophoric, oxidizing, or poisonous liquids must have manual or mechanical controls for the stop valves. The stop valves must also have controls that will respond to heat, in case of fire. Thermal valves will close before the temperature gets above 250 degrees Fahrenheit.

Shear Sections

A shear section is a "sacrificial" device. That means it will fail under pressure.

You have already seen one example of a "sacrificial" device. A fuse is such a device. It fails under a load, and thereby protects another device.

A shear section will break away under pressure. In a rollover, the weight of the vehicle and cargo will press on a section of the vehicle not meant to bear so much weight. The shear section will break off under this pressure. This will preserve the important part of a pipe and its attachment to the tank, preventing a leak. In other words, the shear section will break before the pipe will.

You'll find the shear section outside the stop valve but inside the accident damage protection device.

Pressure Relief Systems

Cargo tanks also have pressure relief systems. They keep the tank from bursting or collapsing from too much or too little pressure. This helps keep the lading from leaking or escaping if the vehicle rolls over or is hit.

There's a primary and a secondary pressure relief system. The primary system has at least one reclosing valve. The secondary system has another valve that will back-up or assist the primary valve. These systems will be marked with the pressure at which they're set to discharge and with the flow rate. (You'll also see the manufacturer's name and the model number.)

Which of these devices your vehicle will have and where they are located depends on the tank's specification. Check the owner's manual for your own vehicle if you're not certain of what you have or where it is.

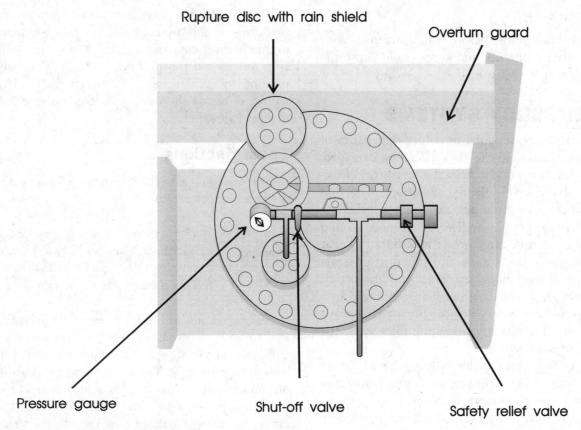

Figure 12-5. Tank vehicle safety features.

Portable Tank Emergency Systems

Portable tanks don't have all the emergency systems that cargo tanks do. Here are the features they do have.

The outlets of portable tanks that carry liquefied compressed gas have excess-flow valves. That is, the outlet valves are rated for flow by the manufacturer. Once the flow rate exceeds this limit, the excess-flow valve will close automatically. Portable tanks will have safety relief valves that work automatically. Filling and discharge lines will have manually operated shut-off valves.

RETEST AND MARKING REQUIREMENT

Hazardous materials may only be hauled in "authorized" vehicles. The DOT has set specifications for the design and construction of vehicles that will haul hazmat. These specifications state what material can be used to build the tank. A corrosive, heavy, or radioactive material calls for a different design and different construction materials than a flammable one does. These specifications also provide guidelines for designing and building tanks that will carry hazardous materials under pressure or at a certain temperature.

A tank that meets DOT specifications is "authorized" to carry hazmat. Regulations in title 49 CFR Parts 178 and 180 state that DOT specification tanks must be retested. These parts set the retest schedule. The regulations further state that DOT specification tanks must be marked with retest information. As the driver, you won't usually be involved in testing tanks. You must know how to read the markings, though, so you can comply with the cargo tank regulations when you operate one.

Retest

Title 49 CFR Part 180 deals with the manufacture of cargo tanks. It also covers the operation, maintenance, repair, and requalification of all specification cargo tanks. It includes improved standards for inspecting and testing tanks.

PASS BILLBOARD

Cargo and portable tanks may be built to DOT specifications. If so, they receive a DOT specification code number. Coded tankers are those that have been approved to haul hazardous liquids. As you know by now, you can't haul a placarded load of hazardous liquids without a Hazardous Materials Endorsement.

Cargo tanks are numbered in the 300s and 400s. These are:

- MC 300
- MC 301
- MC 303
- MC 304
- MC 305
- MC 306
- MC 307
- MC 310
- MC 311
- MC 312
- MC 330
- MC 331
- MC 338
- MC 406
- MC 407
- MC 412

Portable tanks are numbered:

- 51
- 56
- 57
- 60
- IM 101
- IM 102

Cargo Tank Specifications

MC 300 and 301 Tanks. These tanks are no longer built, but some may still be in service. They are made of mild steel or mild steel, high-tensile steel, or stainless steel combined. They are used mainly for transporting flammable or poisonous liquids.

MC 303, 304, and 305 Tanks. Tanks meeting this specification are no longer built. However, some are still in service. MC 303 tanks are made of welded ferrous alloy (high-tensile steel). They can also be made of stainless steel. They are used mainly to haul flammable or poisonous liquids.

MC 304 tanks are made of mild steel, welded ferrous alloy or stainless steel and aluminum. They also carry flammable or poisonous liquids. These tanks were built to keep liquids under a pressure of 25 psi.

MC 305 tanks are built of aluminum alloys. They carry flammable or poisonous liquids at atmospheric pressure.

MC 306 Tanks. MC 306 tanks are made of aluminum or stainless steel. They carry flammable liquids and combustible fuels like No. 2 diesel. They also carry flammable solvents and non-regulated chemicals. They carry all these liquids at low or atmospheric pressure.

MC 307 Tanks. MC 307 tanks are "general chemical trailers." They are made of stainless steel and aluminum. They hold flammable liquids with vapor pressure of at least 25 psi. This is a moderate pressure, but it's higher than what MC 306 tanks can handle. They also carry poisonous and flammable liquids that are also corrosive or oxidizing.

MC 310 and 311 Tanks. These tanks are no longer built, but may still be in service. MC 310 and 311 tanks are made of steel and stainless steel. They haul mainly corrosive liquids.

MC 312 Tanks. MC 312 tanks are made of steel and stainless steel. They transport corrosive liquids, oxidizers, organic peroxides, and poisonous liquids, as well as some solids.

MC 312 tanks are used to carry hydrogen peroxide. Mild steel, copper, and brass won't stand up to hydrogen peroxide, so these tanks must be pure aluminum, or aluminum/magnesium alloys. Some are stainless steel.

This is a high pressure tanker.

MC 330 Tanks. These tanks are no longer built, but may still be in service. They are made of steel, and are mainly for hauling compressed gases.

MC 331 Tanks. These are seamless steel. They are used mainly to haul compressed gases.

MC 338 Tanks. MC 338 tanks are insulated, and built with a design pressure of 25.3 to 500 pounds per square inch gauge (psig). They are used mainly to haul cryogenic liquids.

 The pressure of liquids and gases is measured with a manometer. The measurement is stated as "pounds per square inch gauge" (psig).

TOFC Tanks. A TOFC tank is used for railroad flat car service (piggyback). "TOFC" stands for Tank On Flat Car. This is an MC 307 or 312 tank made of stainless steel with a design pressure of 35 psi.

Non-spec Tanks. "Non-spec" tanks are not built to any one of the US DOT MC specifications. They transport non-regulated loads.

MC 406, 407, and 412 Tanks. These tanks are all designed to haul cargo under pressure.

Portable Tanks

Specification 51. These are steel portable tanks. They can be seamless or welded, or both. They have a design pressure of between 100 psig and 500 psig, and carry mainly corrosive liquids.

Specification 56. These are metal portable tanks with a maximum GWR of 7,000 pounds.

Specification 57. These metal portable tanks can hold between 110 and 660 gallons.

Specification 60. These are welded steel cylindrical portable tanks. Sometimes they are lined.

IM 101 and 102. These are steel portable tanks. Sometimes they are lined. They are designed to carry liquids for intermodal transport.

Tanks are inspected and tested two ways. One is a visual inspection of both the outside and inside of the tank. The other tests the tank for leaks. Some tanks are pressure-tested. This can be a pneumatic or hydrostatic test. Yet another test measures the thickness of the metal that the tank is made of.

Figure 12-6 shows the inspection and retest schedule for most tanks.

MC 330 and MC 331 cargo tanks in sodium metal service don't have to be pressure-tested. Certain uninsulated lined or clad cargo tanks also don't have to be pressure-tested. These are tanks designed to haul cargo at a pressure of less than 15 psig. These must, however, get an external visual inspection and a lining inspection at least once each year.

Marking

The month and year of the last test or inspection, and what type of test or inspection was done, is marked on the tank itself. Or, it may be stamped on the certification plate. You may see these abbreviations:

- V for external visual inspection and test
- I for internal visual inspection
- P for pressure retest
- L for lining test
- K for leakage test
- T for thickness test

For example, the marking "10-95, P, V, L" would tell you that in October 1995, the cargo tank received and passed a pressure retest, external visual inspection and test, and a lining inspection.

Portable Tanks

Portable tanks are also inspected, retested, and marked. The schedule is given in Title 49 CFR Part 180. These tanks are at least pressure-tested and visually inspected. Some are also tested to see how well they hold up to vibration and being dropped.

Test or Inspection	Cargo Tank Configuration, and Service	Period	Test or Inspection	Cargo Tank Configuration, and Service	Period
Inspections			**Tests**		
External Visual	All cargo designed tanks to be loaded by vacuum with full opening rear head	six months	Leakage	MC 330 and MC 331 cargo tanks in chlorine service	two years
	All other cargo tanks	one year		All other cargo tanks except MC 338	one year
Internal Visual	All insulated cargo tanks except MC 330, MC 331, MC 338	one year	Pressure (hydrostatic or pneumatic)	All cargo tanks that are insulated with no manhole or insulated and lined, except MC 338	one year
	All cargo tanks transporting lading corrosive to the tank	one year		All cargo tanks designed to be loaded by vacuum with full opening rear head	two years
	All other cargo tanks except MC 338	five years		MC 330 and MC 331 cargo tanks in chlorine service	two years
				All other cargo tanks	five years
Lining/Cladding	All lined or clad cargo tanks transporting lading corrosive to the tank	one year	Thickness over entire tank	All unlined cargo tanks in corrosive service except MC 338	two years

Figure 12-6. Inspection and retest schedule.

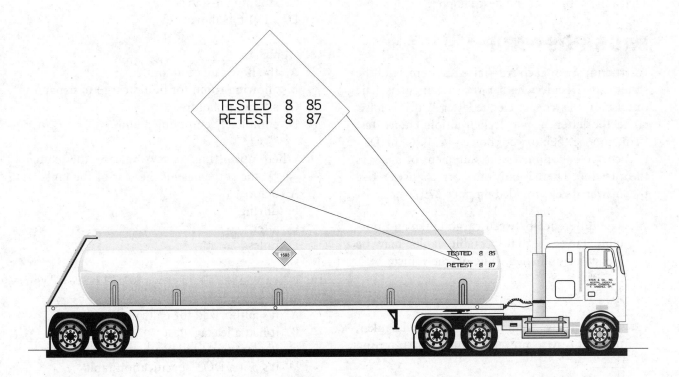

Figure 12-7. A tanker with testing and retest markings.

Briefly, the testing and inspection schedule is as follows:

- Specification 51 once every five years
- Specification 56 and 57 once every two and one-half years. Specification 60 tanks are tested at the end of the first four-year period after the original test. Then they are tested at least once every two years up to a total of 12 years of service. After that, Specification 60 portable tanks are retested once a year.

The date of the most recent retest is marked on the tank, on or near the metal certification plate.

Hauling bulk liquids can be dangerous work. That's not because of the type of liquid. After all, what could be more harmless than milk or water? It's the handling problems the liquid cargo creates that make hauling bulk liquids such a challenge. Even the most simple maneuvers must be made with extreme caution when you have a liquid load.

Of course, some liquid loads are hazardous. Bulk liquid tanks are often filled with dangerous substances that can explode, burst into flame, even kill if not handled properly. To transport these, you not only need a Tank Vehicle Endorsement but also a Hazardous Materials Endorsement. That's the subject of the next chapter.

PASS POST-TRIP

Instructions: For each true/false test item, read the statement. Decide whether the statement is true or false. If it is true, circle the letter A. If it is false, circle the letter B. For each multiple-choice test item, choose the answer choice—A, B, C, or D—that correctly completes the statement or answers the question. There is only one correct answer. Use the answer sheet provided on page 279.

1. The difference between cargo tanks and portable tanks is that portable tanks may be loaded or unloaded while off the vehicle.
   A. True
   B. False

2. When taking a curve with a loaded tanker, the best way to prevent rolling is by traveling _____.
   A. at a speed above the posted limit
   B. at the posted speed limit
   C. at a speed below the posted limit
   D. with the brakes on

3. Liquid surge can push a stopped truck into an intersection.
   A. True
   B. False

4. The brakes on a bulk liquid tanker work best when the vehicle is _____.
   A. fully loaded
   B. partly loaded
   C. empty
   D. bobtailing

5. Forward-and-backward surge is a big problem with smooth bore tankers.
   A. True
   B. False

6. Baffles help control _____ in a tanker.
   A. the high center of gravity
   B. side-to-side surge
   C. forward-and-backward surge
   D. weight distribution

7. The density of liquid cargo affects _____.
   A. weight distribution
   B. cargo surge
   C. stopping distance
   D. all of the above

8. Outage is _____.
   A. the flow rate of an outlet
   B. allowing room for liquid cargo to expand
   C. a vacuum failure
   D. completely emptying a tank

9. When upshifting, always release the clutch _____ the surge has hit the rear of the tank.
   A. before
   B. during
   C. after
   D. unless

10. When you see a tanker marked "12-89, K," you know _____.
    A. it's authorized for chlorine service
    B. it had a leakage test
    C. it's carrying food products
    D. it's not a DOT specification tank

CHAPTER 12 ANSWER SHEET

Test Item	Answer Choices				Self - Scoring Correct	Incorrect
1.	A	B	C	D	☐	☐
2.	A	B	C	D	☐	☐
3.	A	B	C	D	☐	☐
4.	A	B	C	D	☐	☐
5.	A	B	C	D	☐	☐
6.	A	B	C	D	☐	☐
7.	A	B	C	D	☐	☐
8.	A	B	C	D	☐	☐
9.	A	B	C	D	☐	☐
10.	A	B	C	D	☐	☐

Cut Here

Chapter 13

Hazardous Materials Endorsement

In this chapter you will learn about:

- hazardous materials regulations
- handling hazardous materials
- hazardous materials emergencies
- hazardous materials emergency response procedures

To complete this chapter you will need:

- a dictionary
- a pencil or pen
- blank paper, a notebook, or an electronic notepad
- colored pencils, pens, markers, or highlighters
- a CDL preparation manual from your state Department of Motor Vehicles, if one is offered
- the Federal Motor Carrier Safety Regulations pocketbook with Hazardous Materials Compendium (or access to U.S. Department of Transportation regulations, Parts 171, 172, 173, 177, 178, 180, 383, 392, and 397 of Title 49, Code of Federal Regulations)
- the owner's manual for your vehicle, if available

PASS PRE-TRIP

Instructions: Read the statements. Decide whether each statement is true or false. If it is true, circle the letter A. If it is false, circle the letter B.

1. A written plan is something you must apply for before making a trip with explosives.
   A. True
   B. False

2. Hazardous materials markings can be a label, tag, or sign.
   A. True
   B. False

3. In addition to a special license and training, you may need a special permit from the federal, state, or local government to haul hazardous materials.
   A. True
   B. False

Hazardous materials present a risk to health, safety, and property. This is especially true when they are being transported. If you plan to drive a vehicle with a placarded load of hazardous materials, you must have a Hazardous Materials Endorsement on your CDL.

There's only one exception to this. You don't need a hazardous materials endorsement to operate a vehicle with a hazmat load that is not placarded. When would a hazmat load not need placards? Title 49 CFR Section 177.823 lists three conditions:

- the vehicle is escorted by a representative of a state or local government
- the carrier has permission from the Department of Transportation, or
- the vehicle must be moved to protect life or property

At all other times, a vehicle with hazmat must have placards.

HAZARDOUS MATERIALS REGULATIONS

Hazardous materials handled improperly in transport can injure or kill people. Regulations exist to prevent this. The point of the regulations is to:

- contain the material
- communicate the risk
- assure safe drivers and equipment

Containment

To protect drivers and others, the rules tell shippers how to package safely. Similar rules tell drivers how to load, transport, and unload bulk tanks. These are containment rules.

Communication

Shippers must warn drivers and others about a material's hazardous qualities. They put warning labels on packages and describe materials in a way that clearly warns of the risk. There are rules for drivers too. They must warn others if there is an accident or a leak. Placards are another way to communicate the risk.

Safe Drivers and Equipment

As the driver, you must complete a Transportation Security Administration security screening. If you have a CDL and want to upgrade to one with a hazmat endorsement, you must complete this screening. In addition, you must complete this screening to renew a hazmat endorsement you may already have. Hazmat endorsements must be renewed every five years so that this screening may take place.

Your state will administer the screening. Your state must provide the TSA with a driver's fingerprints, criminal history, alien status, and information from CDLIS and international databases going back seven years.

The TSA requires that you provide proof of citizenship or immigration status. Proof should include one document from List A and one document from List B, both of which follow:

List A: ONE Document Is Acceptable from the Following Forms of Identification

- Unexpired U.S. Passport (book or card)—demonstrates U.S. Citizenship
- Unexpired Enhanced Tribal Card (ETC)—demonstrates U.S. Citizenship
- Unexpired Merchant Mariner Document (MMD)—designates U.S. Citizenship
- Unexpired Free and Secure Trade (FAST) Card—designates U.S. Citizenship
- Unexpired NEXUS Card—designates U.S. Citizenship
- Unexpired Secure Electronic Network for Travelers Rapid Inspection (SENTRI) Card—designates U.S. Citizenship
- Unexpired U.S. Enhanced Driver's License (EDL)—designates U.S. Citizenship
- Permanent Resident Card (I-551), often referred to as a "Green Card"—demonstrates LPR status
- Unexpired Foreign Passport AND immigrant visa with I-551 annotation of "Upon Endorsement Serves as Temporary Evidencing Permanent Resident of 1 Year"—demonstrates LPR status
- Unexpired Re-entry Permit (I-327)—demonstrates LPR status

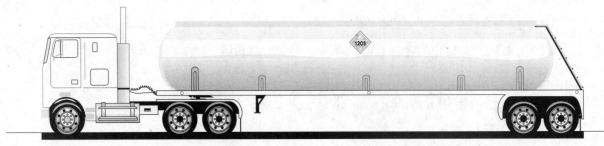

Figure 13-1. Placards on vehicles communicate the risk of hazardous materials transport.

*At least two documents** from List B are required if you do not have a document from List A. The documents must include:
1) a valid photo ID, and
2) a document that meets citizenship requirements. For example, the combination of a driver's license and a U.S. Social Security card contains a photo ID, but does not demonstrate citizenship status and is not a valid combination. However, the combination of a valid driver's license and a U.S. birth certificate would meet the document requirements for enrollment.

List B: You MUST Have at Least ONE Document from Each Category Below for a Total of TWO Documents

The names on all documents MUST MATCH with your application name, unless you provide a valid proof of name change*

(*Photo ID + Proof of U.S. Citizenship + Name Change Document {if needed} = Valid Combination*)

Valid Photo ID/Valid Proof of Citizenship
- *Unexpired* driver's license issued by a state or an outlying possession of the United States
- *Unexpired* photo ID card issued by a state or an outlying possession of the United States. This must include a state or state agency seal or logo (such as a state university ID)
- U.S. military ID card or U.S. retired military ID
- U.S. military dependent's card
- Native American tribal document with photo
- *Unexpired* Merchant Mariner License (MML) bearing an official raised seal, or a certified copy
- *Unexpired* Department of Homeland Security (DHS)/Transportation Security Administration (TSA) Transportation Worker Identification Credential (TWIC)
- *Unexpired* Merchant Mariner Credential (MMC)

Valid Proof of Citizenship
- Original or certified copy of birth by a state, county, municipal possession of the United States, bearing an official seal
- U.S. Certificate of Citizenship
- U.S. Certificate of Naturalization
- U.S. Citizen Identification Card
- Consular Report of Birth Abroad

- Certification of Report of Birth Abroad
- Expired U.S. passport within 12 months of expiration**

*Valid Proof of Name Change (if needed)
Original or Certified Copy of a Court Ordered Name Change Document (to include marriage certificates and divorce decrees) if the names on your photo ID and citizenship status document do not exactly match.
**An expired U.S. passport may not be presented by itself. It must be presented with at least one other List B document (and a name change document if needed). You may provide a second List B document, or one of the following: Voter's card, U.S. Social Security card, U.S. military discharge papers (DD-214), or a Department of Transportation (DOT) medical card.

You must pass a written test about transporting hazardous materials. To pass the test, you must know how to:

- recognize shipments of hazardous materials
- safely load shipments
- correctly placard your vehicle
- safely transport shipments

Learn the rules and follow them. Following the rules reduces the risk of injury from hazardous materials. Taking shortcuts by breaking rules is unsafe. Rule breakers can be fined and put in jail.

Inspect your vehicle before and during each trip. Law enforcement officers may stop and inspect your vehicle. They can also check your shipping papers. They will look for a hazardous materials endorsement on your driver's license.

You can see that nearly everyone involved in the transport of hazardous materials has responsibilities.

Shipper

The shipper sends products from one place to another by truck, railroad, ship, or airplane. Shippers use the hazardous materials regulations to decide the product's:

- proper shipping name
- hazard class
- identification number

- correct packaging
- correct label and markings
- correct placard

Shippers also package the materials and label and mark the package. They prepare the shipping paper and supply placards. Shippers certify on the shipping paper that they have prepared the shipment according to the rules. Exceptions to this are if you are pulling cargo tanks supplied by you or your employer, or if these materials and supplies are being transported by the carrier who owns them.

Carrier

Carriers take the shipment from the shipper to its destination. They check that the shipper correctly named, labeled, and marked the shipment. They refuse improper shipments. Carriers report accidents and incidents involving hazardous materials to the proper government agency.

Driver

You, the driver, make sure all who have been involved with the hazardous material—the packagers, loaders, shippers, and carriers—have done their job. You check that the shipper has identified, marked, and labeled the product. You refuse leaking shipments. You placard your vehicle when loading, if needed.

You make your best effort to transport the shipment safely and without delay. You follow all special rules about transporting hazardous material. Also, you keep hazardous materials shipping papers in the proper place.

Your responsibilities as a driver transporting hazardous materials are to:

- recognize a hazardous materials shipment
- identify the hazards the shipment presents
- communicate the danger to others
- know what to do in an emergency

How do you fulfill these responsibilities? To answer that, you must know the hazards materials regulations. As you know, all CDL drivers must know these regulations, even if they don't plan to haul hazmat. That way, they can recognize a hazardous materials load. They can determine if they are licensed to transport that load or not.

The hazardous materials regulations you must know are:

- Title 49 CFR Part 397 (found in the FMCSR)
- Title 49 CFR Parts 171, 172, 173, 177, and 178 (the Hazardous Materials Regulations, or HMR)

FMCSR PART 397

Some of the hazardous materials regulations are found in FMCSR Part 397. You read about them in Chapter 5. Here's a very brief review. These regulations cover:

- state and local laws
- attendance of motor vehicles
- parking
- routes
- fires
- smoking
- fueling
- tires
- instructions and documents

State and Local Laws

Remember "higher standard of care?" When there are two or more sets of regulations, you should follow the strictest set. Local hazmat regulations may be more strict than the federal regulations. If so, you would follow the local regulations.

Attendance of Motor Vehicles

With few exceptions, a vehicle hauling Division 1.1, 1.2, or 1.3 explosives must never be abandoned. There must always be a qualified person with the vehicle. This is called "attendance." You "attend" the vehicle when you are in the vehicle, awake, and not in the sleeper. If you are within 100 feet of the vehicle and have a clear view of it, you are also attending the vehicle.

A "qualified" person is someone the carrier has put in charge of attending the vehicle. To be qualified, you must be informed about the hazmat in the vehicle. You must know what to do in an emergency. You must be authorized and able to move the vehicle.

A vehicle with explosives may sometimes be left unattended, if:

- the vehicle is parked in a safe haven, or on a construction or survey site
- the lawful bailee of the explosives knows they are explosive and what to do in an emergency

The "lawful bailee" is the person who is officially in charge of or responsible for the explosives.

- the vehicle is in clear sight of the bailee, or is in a safe haven

There is one reason why you may leave a vehicle with other kinds of hazmat unattended. That's if you must leave the vehicle to perform some maintenance, put out warning devices, or some other necessary part of your driving job.

Parking

Never park with Division 1.1, 1.2, or 1.3 explosives within five feet of the traveled part of the road. Unless your work requires it, do not park within 300 feet of

- a bridge, tunnel, or building,
- a place where people gather, or
- an open fire

If you must park to do your job, do so only briefly. Don't park on private property unless the owner is aware of the danger. Someone must always watch the parked vehicle. You may let someone else watch it for you only if your vehicle is

- on the shipper's property, or
- on the carrier's property, or
- on the consignee's property

You can, of course, leave your vehicle unattended in a safe haven.

You may park a placarded vehicle (not carrying explosives) within five feet of the traveled part of the road only if your work requires it. Do so only briefly. Someone must always watch the vehicle when parked on a public roadway or shoulder. Do not uncouple a trailer and leave it with hazardous

material on a public street. Do not park within 300 feet of an open fire.

Routing

Vehicles hauling hazardous cargo don't always follow the easiest and fastest highways. Instead, hazmat routing must avoid roads that go through heavily populated areas. The only exception to this is when there is no other possible way to get where you're going.

For Division 1.1, 1.2, or 1.3 explosives, the carrier or driver must have a written route plan. Usually the carrier writes up the plan. You may have to do it yourself if you start your trip from somewhere other than your home terminal.

This written plan must include curfews and permits from those cities that require them. Any such permit must be applied for prior to the trip.

The written plan may list certain hours of the day when a vehicle hauling hazmat may travel through a city. Some cities have ordinances as to what time of day hazardous materials are allowed to pass through them. Other cities require permits that must be applied for in advance. More and more cities and states are passing laws that deal with hazardous materials.

Fires

You must not park a vehicle with hazmat within 300 feet of open fire. Don't even drive past one unless you are sure you can do so without stopping.

Smoking

Vehicles with hazardous cargo must be kept away from flames. This includes flames used to light cigarettes. So, smoking is not allowed within 25 feet of a vehicle carrying Class 1 materials, Class 5 materials, or flammable materials classified as Division 2.1, Class 3, Divisions 4.1 and 4.2. It's also not permitted around an empty tank motor vehicle that has been used to transport Class 3, flammable materials, or Division 2.1 flammable gases, which when it was loaded, was required to be marked or placarded.

Fueling

Special care must be taken when fueling a vehicle loaded with hazardous cargo. Turn the engine off first. Then be sure that someone is at the nozzle throughout the fueling process, controlling the flow.

Tires

A tire fire could be especially dangerous when you're hauling hazmat. Therefore, frequent tire inspections are required en route. If you have dual tires, check them at the start of each trip and when you park. Check the tires at the start of each trip and when you park. Check for proper tire inflation.

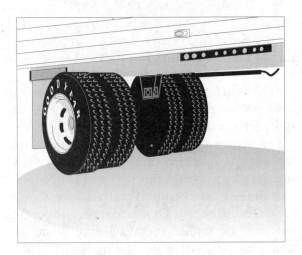

Figure 13-2. When hauling hazmat, check dual tires at the start of each trip and every time you park.

If a tire is low, flat, or leaking, you must get it repaired before going any further. If your check shows you have a hot tire, you must remove it. Place it at a safe distance from your vehicle. An overheated tire can burst into flame. Then inspect your vehicle until you find why the tire overheated in the first place. Get the problem fixed before you continue to drive.

Instructions and Documents

If you're hauling Division 1.1, 1.2, or 1.3 explosives, your carrier must give you certain instructions and documents. You must have a copy of FMCSR Part 397. You must get written instructions on what to do in an emergency or if you are delayed. These instructions must include the names and telephone numbers of people to contact about the emergency or delay. You must also get written information about the nature of the explosives you're transporting. The instructions must also tell you what precautions to take in case of fire, accidents, or leaks.

You must sign a receipt for these documents. The carrier keeps the receipt on file for a year.

When you get these documents, you must keep them with you. You must know everything they contain and obey the instructions.

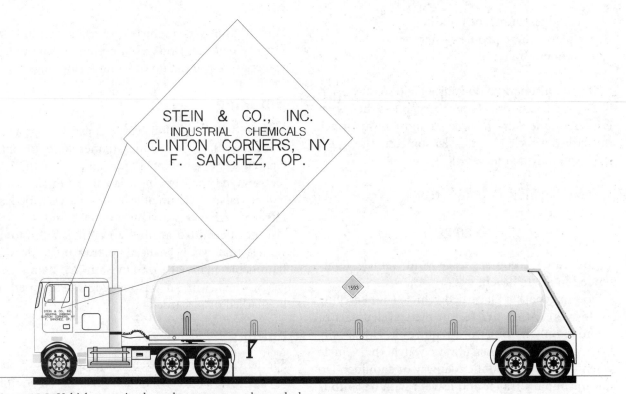

Figure 13-3. Vehicles carrying hazardous waste must be marked.

TITLE 49 CFR PART 171

Title 49 CFR Part 171 states the DOT requirements for the transport of hazardous materials and waste by rail car, aircraft, and vessel. The definition of "transportation" is broad. It includes pretransportation activities like filling hazardous materials packaging, loading hazardous materials, and reviewing shipping papers. It also covers interstate, intrastate, and foreign motor vehicle transport. We'll mention the rail, air, and vessel regulations only when they also have to do with motor vehicle transport. This part also covers the transport of flammable cryogenic liquids in portable tanks and cargo tanks. Last, this part states requirements for manufacturing, marking, maintaining, repairing, and testing hazmat packaging.

General

Hazmat may not be transported in commerce unless it's been properly classed, described, packaged, marked, labeled, and prepared for shipment the way the regulations require.

Hazardous waste may not be hauled in a vehicle that isn't marked as follows:

- the legal name or trade name of the private carrier operating the vehicle
- the motor carrier identification number (a number with the letters USDOT) issued by the FMCSA
- the name of the person operating the vehicle, if that's not the carrier, accompanied by the operating carrier

When hazardous waste is delivered, the entire shipment must be delivered. If the shipment is to be transferred to another carrier, the entire shipment must be transferred. That is to say, you can't deliver some of it and dump the rest. You could be fined, even jailed, for discharging hazardous waste anywhere but the place named on the manifest.

Definition of Terms

This part contains general information and definitions used in other hazmat regulations. Here's a brief rundown of some that relate to transporting hazardous materials by motor vehicle.

"Approved" means approval issued or recognized by the Department of Transportation.

"Atmospheric gases" means argon, krypton, neon, nitrogen, oxygen, and xenon.

"Btu" means British thermal unit. It's a measurement of heat energy.

"Bulk packaging" includes a transport vehicle or freight container in which hazardous materials are loaded with no other container. Such a vehicle can hold more than 119 gallons of liquid, 882 pounds of solid, or 1,000 pounds (water capacity) of gas.

"C" means Celsius or Centigrade. It's one scale used for measuring temperature.

"Cargo tank" means any tank permanently attached to or forming a part of any motor vehicle. A straight truck with a tank cargo compartment would fit this definition. It also means any bulk liquid or compressed gas packaging not permanently attached to any motor vehicle. An example would be a tank semitrailer. Because of its size, construction, or the way it's attached to a motor vehicle, it can be loaded or unloaded without being removed from the motor vehicle. Cylinders are not cargo tanks. (Cargo tanks are discussed in more detail in Chapter 12.)

"Carrier" means a person engaged in the transportation of passengers or property by land or water, as a common, contract, or private carrier. It also covers civil aircraft.

"Combustible liquid" is defined in 173.120 (and later in this chapter).

"Compressed gas" is defined in Part 173.115 (and later in this chapter).

"Consumer commodity" means a material that is packaged and distributed for retail sale for use by individuals. The material is meant for personal care or household use. This term also includes drugs and medicines.

"Corrosive material" is defined in Part 173.136. It's also described in more detail later in this chapter.

"Cryogenic liquid" is super-cold liquid. It's further defined in Part 173.115.

"Cylinder" means a pressure vessel designed for pressures higher than 40 psia (you'll find "psia" defined later in this section). A portable tank or cargo tank is not considered a cylinder.

"Designated facility" means a hazardous waste treatment, storage, or disposal facility that has been named on the manifest by the generator (creator of the waste).

"DOD" means the U.S. Department of Defense.

"EPA" means U.S. Environmental Protection Agency.

"Etiologic agent" is defined in Part 173.134 (and later in this chapter).

"Explosives" are defined in Part 173.50.

"°F" means degrees Fahrenheit. It's another scale for measuring temperature.

"Filling density" refers to how full tankers and cylinders should be loaded with bulk liquids and compressed gases. These are defined in Parts 173.304, 173.314, 173.315, 173.316, 173.318, and 173.319 (or review Chapter 12).

"Flammable gas," "flammable liquid," and "flammable solid" are defined in Part 173 (and later in this chapter).

"Flash point" means the lowest temperature at which a substance gives off flammable vapors. These vapors will ignite if they come in contact with a spark or flame. The flash point of liquids is given in Part 173.120.

"Freight container" means a reusable container with a volume of 64 cubic feet or more. It's designed and built so it can be lifted with its contents intact. It's mainly used to hold packages (in unit form) during transport.

"Fuel tank" means a tank other than a cargo tank used to transport flammable or combustible liquid. It's also a tank that holds compressed gas used to fuel the transport vehicle it's attached to, or equipment on that vehicle. This is different from, say, a tank semitrailer that transports fuel as cargo.

"Gas poisonous by inhalation" is defined in Part 173.115.

"Gross weight" means the weight of packaging plus the weight of its contents.

"Hazardous material" means a substance or material, including a hazardous substance, that the Secretary of Transportation judges to pose an unreasonable risk to health, safety, and property when transported in commerce.

"Hazardous substance" means a material listed in the Appendix to Part 172.101. This appendix lists how much of the material, in one package, is considered hazardous. If the hazmat is in a mixture or solution, the solution is measured to see how much of it is hazmat. This measurement is compared against a table of RQ (reportable quantities). In short, some solutions and mixtures have enough hazmat to be hazardous. Others don't.

Solutions/Mixture Concentration By RQ pounds (kilograms)	Table Weight Percent	PPM (parts per mil)
5000 (2270)	10	100,000
1000 (454)	2	20,000
100 (45.4)	0.2	2,000
10 (4.54)	0.02	200
1 (0.454)	0.002	20

"Hazardous waste" means any material that is subject to the Hazardous Waste Manifest Requirement of the U.S. Environmental Protection Agency, specified in 40 CFR Part 262.

"ICAO" means International Civil Aviation Organization.

"Intermodal container" means a freight container designed and built to be used in two or more modes of transport.

"Intermodal portable tank" or "IM portable tanks" means a specific class of portable tanks designed mainly for international intermodal use.

"Irritating material" is further defined in Part 173.132(a) (and later in this chapter).

"Limited quantity" means the maximum amount of a hazardous material for which there is a specific labeling and packaging exception. In other words, some amounts of hazmat don't have to be labeled or packaged according to the usual regulations. That's a limited quantity of that hazmat. There are no exceptions for poison-by-inhalation materials.

"Marking" means putting the descriptive name, instructions, cautions, weight, or specification marks required by these regulations on outer containers of hazmat.

"Mixture" means a material composed of more than one chemical compound or element.

"Mode" means any of the following transportation methods: rail, highway, air, or water.

"Motor vehicle" includes a vehicle, machine, tractor, trailer, or semitrailer, or any combinations of these. This term assumes the vehicle is driven or drawn by mechanical power, is on the highway, and is transporting passengers or property. It doesn't refer to vehicles used on rail. It also doesn't include a trolley bus running on a fixed overhead electric wire and carrying passengers.

"Name of contents" means the proper shipping name specified in Part 172.101.

"N.O.S." means not otherwise specified.

"Non-bulk packaging" means a package smaller than the capacities stated for "bulk packaging."

"Operator" means a person who controls the use of an aircraft, vessel, or vehicle.

"Organic peroxide" is further defined in Part 173.128.

"ORM" means Other Regulated Materials. It is defined in Part 173.144.

"Outage" or "ullage" means the amount by which a packaging falls short of being full of liquid. It's usually stated in percent by volume. "Outage" was discussed in Chapter 12. Refer to that chapter for more details about outage.

"Outer packaging" means the outermost enclosure used to transport a hazardous material. This is other than a freight container.

"Overpack" means an enclosure that is used by a single consignor to provide protection or convenience in handling of a package or to combine two or more packages. "Overpack" does not include a freight container. Overpack has a special meaning when used to refer to packaging specified by Part 178.

"Oxidizer" or "oxidizing material" is defined in 173.127 (and later in this chapter).

"Packaging" means the assembly of one or more containers and any other part needed to comply with the minimum packaging requirements of this subchapter. It includes containers (other than freight containers or overpacks), portable tanks, cargo tanks, tank cars, and multi-unit tank car tanks. Radioactive materials are covered in Part 173.403.

"Poisonous material" other than a gas is defined in Part 173.132.

"Portable tank" means any packaging (except a cylinder having a 1,000-pound or less water capacity) designed to be loaded into or on or temporarily coupled to a transport vehicle or ship. These tanks are equipped with skids, mounting, or other parts that make mechanical handling possible. Not included in this definition is any cargo tank, tank car, multi-unit tank car tank, or trailers that carry 3AX, 3AAX, or 3T cylinders.

"Preferred route" or "Preferred highway" is a highway for shipment of "highway route controlled quantities" of radioactive materials. These routes are named by state routing agencies. When there is no other option (as described in Part 177.103), any interstate system highway can be named a preferred route.

"Proper shipping name" means the name of the hazardous material shown in Roman print (not italics) in Part 172.101.

"P.s.i." or "psi" means pounds per square inch.

"P.s.i.a." or "psia" means pounds per square inch absolute.

What's the difference between psi and psia? Psi is a measurement taken using atmospheric pressure as a yardstick. Psia uses vacuum (or zero pressure) as a reference point.

"P.s.i.g." or "psig" means pounds per square inch gauge. That's a measurement of fluid or gas pressure taken with a gauge called a manometer.

"Pyrophoric liquid" is defined in Parts 173.124(b) (and later in this chapter).

Hazardous Substances Other than Radionuclides	
Paraformaldehyde	1000 (454)
Paraldehyde	1000 (454)
Parathion	10 (4.54)
PCBs	1 (0.454)
PCNB	100 (45.4)
Pentachlorobenzene	10 (4.54)
Pentachloroethane	10 (4.54)
Pentachloronitrobenzene	100 (45.4)
Pentachlorophenol	10 (4.54)
1,3-Pentadiene	100 (45.4)
Perchloroethylene	100 (45.4)
Perchloromethyl mercaptan @	100 (45.4)
Phenacetin	100 (45.4)
Phenanthrene	5000 (2270)
Phenol	1000 (45.4)
Phenol, 2-chloro-	100 (45.4)

Figure 13-4. Part of the list of Hazardous Substances Other than Radionuclides.

"Radioactive materials" are defined in Part 173.403.

"Reportable Quantity" (RQ) means the quantity specified in Column 2 of the table in Part 172.101, for any material identified in Column 1.

"Shipping paper" means a shipping order, bill of lading, manifest, or other shipping document serving the same purpose. These documents must have the information required by Subpart C of Part 172.

"Solution" means any uniform liquid mixture of two or more chemical compounds or elements that won't separate while being transported under normal conditions.

"Spontaneously combustible materials (solid)" means a solid substance that may heat up or ignite by itself while being transported under normal conditions. This includes sludges and pastes. It also includes solids that will heat up and ignite when exposed to air.

"SCF" (standard cubic foot) means one cubic foot of gas measured at 60°F and 14.7 psia.

"Technical name" means a recognized chemical name currently used in scientific and technical handbooks, journals, and texts. Some generic descriptions are used as technical names. These are "organic phosphate," "organic phosphorus compound mixture," "methyl parathion," and "parathion."

"TOFC" means trailer-on-flat-car.

"Transport vehicle" means a cargo-carrying vehicle such as an automobile, van, tractor, truck, semitrailer, tank car, or rail car used to transport cargo by any mode. Each cargo-carrying body is a separate transport vehicle.

"United States" means the fifty states, the District of Columbia, the Commonwealth of Puerto Rico, the Virgin Islands, American Samoa, or Guam.

"Viscous liquid" means a liquid material that has a measured viscosity in excess of 2500 centistokes at 25 degrees Centigrade (77 degrees Fahrenheit). Viscosity is determined using procedures in ASTM Method D 445-72 "Kinematic Viscosity of Transparent and Opaque Liquids (and the Calculation of Dynamic Viscosity)" or ASTM Method D 1200-70 "Viscosity of Paints, Varnishes, and Lacquers by Ford Viscosity Cup."

 TIP *"Viscosity" is a measurement of a liquid's tendency to resist flowing. A liquid with a high viscosity doesn't flow as easily as one with a low viscosity.*

"Volatility" refers to how fast materials evaporate and become vapor.

That's a partial list of terms defined in Part 171.8. You should be aware of the official definition of what is hazardous, as well as the meaning of other terms used in the business of transporting hazardous materials. Should you run into other terms for which you need definitions, look in Part 171.8.

This part contains other regulations you must know if you plan to haul hazmat. Some of them are mainly of interest to shippers, carriers, and employers. We'll cover the ones the driver should know. If you are a carrier or employer as well as a driver, you should inform yourself of all the hazardous materials regulations.

Keep this in mind as well. State and local governments may have their own hazmat regulations. You must know the regulations of every local area in which you plan to transport hazardous materials.

North American Shipments

Part 171.12 deals with North American shipments. Shipments to or from Mexico must comply with the regulations in this subchapter.

Canada has its own hazmat regulations. These are the Transportation of Dangerous Goods or TDG regulations. They're similar to DOT hazmat regulations. Canadian shipments that meet TDG requirements usually also meet DOT hazmat regulations. Sometimes, special action is required.

Read Title 49 CFR Parts 171.12, 171.22, and 171.23 in depth if you handle Canadian hazardous materials shipments.

Reporting Hazardous Materials Incidents

Carriers are responsible for reporting hazardous materials incidents. Since the driver is the one on the scene, your employer will likely need your help with this. You may be given special instructions or duties regarding hazardous materials incidents such as leaks and spills.

TITLE 49 CFR PART 172

This part includes the Hazardous Materials Table. Here you will also find information on how to understand and communicate the danger. This is done through shipping papers and manifests, package marking, labels, and placards. This part of the HMR also tells about special provisions, emergency response, and training requirements.

The Hazardous Materials Table lists hazardous materials, along with their description, proper shipping name, and class. It lists the label, packing group, marking, and placards for each material.

The hazmat table also shows if a material is to be transported by truck, rail, air, or vessel. All this information helps to communicate the hazard presented by the material—and prevent trouble. You'll use this table to:

- decide if the material is regulated for transport by truck
- determine if the proper shipping name has been used on shipping papers and package markings
- identify the material's hazard class
- find the proper identification number
- find the packing group
- determine if the label is correct
- determine if the correct package was used

You'll find part of the table in Figure 13-5.

Column 1 of the table tells you if a hazardous material is regulated for truck transport. This is done by codes.

The letter "A" means the transport of the hazmat is regulated when done by air. "W" marks hazmat that will be restricted when transported by vessel. The letter "D" identifies proper shipping names appropriate only for domestic shipments, while "I" is for international commerce. The letter "G" identifies proper shipping names for which one or more technical names of the hazardous material must be entered. They will then be shown in parentheses, in association with the basic description.

Some materials are regulated because they are "RQ"—reportable quantity. If a certain amount of the material is spilled or released into the environment, the EPA wants to know about it. The EPA determines the quantity. When material is shipped in this quantity, it is regulated, even though it may not be when shipped in a lesser quantity.

The plus sign (+) in column 1 shows that the proper shipping name and hazard class for the entry has been fixed and can't be changed. In other words, this is the name to use whether or not the material meets the definition of that class.

The different hazard classes were first discussed in Chapter 5. You've also just seen them mentioned in the definitions section of Part 171. Each class is defined in greater detail in different parts of the HMR. You'll find most of them in Part 173.

In column 2 you'll find the proper shipping name. Entries are in alphabetical order. Names shown in regular type are proper shipping names. Use these names to determine whether the shipper has used the proper names on shipping papers and packages. Names shown in slanted type (italics) are not proper shipping names.

When hazmat is shipped in a mixture or solution, the proper shipping name itself may identify the material as a solution or mixture. Or you may see the word "solution" or "mixture." Either the exact concentration or the range of concentration will be stated.

You may see "N.O.I." or "N.O.I.B.N." "N.O.I." stands for "not otherwise indexed." "N.O.I.B.N." and "N.O.S." stand for "not otherwise indexed by name." An example of when this type of entry might be used would be a mixture of flammable liquids. No specific product shipping name can

§172.101 Hazardous Material Table									
Symbols	Hazardous materials descriptions and proper shipping names	Hazard class or Division	Identification Numbers	Packing Group	Label(s) Required (if not excepted)	Special provisions	(8) Packaging Authorizations (§ 173.***)		
							Exceptions	Non-bulk packaging	Bulk packaging
(1)	(2)	(3)	(4)	(5)	(6)	(7)	(8A)	(8B)	(8C)
—	Poisonous, solids, self heating, n.o.s ...	6.1	UN3124	1	POISON, SPONTANEOUSLY COMBUSIBLE	A5__	None	211	241

Note: Columns 9 and 10 do not apply to transportation by highway.

Figure 13-5. Part of the Hazardous Materials Table.

be established, so this mixture might be listed as "Flammable Liquid N.O.S."

Column 3 gives the hazard class or division. The hazard class gives you an idea about the type of danger presented by the material. It's easy to guess the danger presented by something like "detonators for ammunition." However, if you are hauling "cymenes" and don't know what that is, the table will help. You'll learn that this is a Class 3 material, and so is a flammable liquid.

If you find "Forbidden" in column 3 for something you are to transport, contact your employer or dispatcher immediately. A "Forbidden" item may not be transported.

Some materials fit into more than one class. Then, the number of the class that presents the highest danger should be used. Part 173.2 will help you to decide how to classify such a material. Part 173.2a lists the classes from the highest danger to the least danger.

Some materials will always have the hazards of, and therefore the classification of, a single class. These are:

- Class 1 (explosive) material that meets any other hazard class or division as defined in this part shall be assigned a division in Class 1
- Division 5.2 (organic peroxide) material that meets the definition of any other hazard class or division shall be classed as Division 5.2
- Division 6.2 (infectious substance) material that also meets the definition of another hazard class or division, other than Class 7, or that also is a limited quantity Class 7 material, shall be classed as Division 6.2
- material that meets the definition of a wetted explosive in Part 173.124(a)(1) of this subchapter (Division 4.1). Wetted explosives are either specifically listed in the Hazardous Materials Table or are approved by the Associate Administrator.
- a limited quantity of a Class 7 (radioactive) material that meets the definition for more than one hazard class or division shall be classed in accordance with Part 173.423. It's classed for the additional hazard, packaged to conform with the requirements specified in Part 173.421, and offered for transportation in accordance with the requirements that apply to the hazard for which it is

classed. A limited quantity Class 7 (radioactive) material that is classed other than Class 7 is excepted from some of the requirements if the entry "Limited quantity radioactive material" appears on the shipping paper in association with the basic description.

Materials that meet the definition of other classes but that are poisonous when inhaled require special marking, labeling, placarding, and shipping paper descriptions.

You may find a letter after some of the entries in column 3. For example, those labeled "detonators for ammunition" have the letter B after the hazard class number. These letters refer to compatibility groups and segregation requirements. These relate to loading requirements. You'll learn more about segregation later in this chapter.

Hazmat will be listed by a technical name or a generic name. Some items on the table are hazardous wastes. Either the name will identify the material as hazardous waste, or the word "waste" will be used with the name.

Column 4 lists the proper identification number that must be on the packaging or shipping paper, bulk and non-bulk packages, portable tanks, and cargo tanks. Materials for international shipment will have a UN number. Materials with an NA number can't be shipped outside the United States, except to Canada.

Column 5 names the section of Part 173 that describes the packing requirements for the entry. The packing group is chosen by the degree of danger presented by the material. Compare the packing group entry on the table with the entry on the shipping papers. Note that Class 2 (Gases), Class 7 (Radioactive material), Combustible liquids, and ORM-D materials do not have packing groups.

Use column 6 to check that the packages have the proper labels. You may see more than one label listed. The first label listed indicates the main hazard. Other labels indicate secondary hazards. All the labels listed must be displayed.

Column 7 lists any special provisions, such as packaging requirements, that may apply to the entry. The entries in column 7 are codes. You'll find what the codes stand for in Part 172.102. In some states, codes containing the letter "H" refers to a special provision that applies only to transportation by highway. A code containing the letter "N" refers to a special provision that applies only to non-bulk packaging requirements, while "B" codes are for bulk-packaging requirements. A code containing the letter "T" refers to a special provision that applies only to transportation in IM portable tanks. Codes with the letter "W" refer to a special provision that applies only to transportation by water; "R" codes are for transport by rail, and "A" codes are for air transport.

Column 8 will help you to be sure that the shipper has put the hazmat in the proper package for truck transport. There are three segments to column 8. Put "173" in front of any three-digit number that you find in any of these segments. That gives you the number of the federal regulation part that provides additional information. In segment A of column 8, the entry refers to packaging exceptions. In segment B of column 8, the entry refers to authorized non-bulk packaging. Segment C lists those sections that authorize bulk packing.

The final columns, columns 9 and 10, refer to shipments regulated when transported other than by truck. Column 9 lists quantity limits for transport aboard aircraft or railcar. Column 10 refers to requirements for water transport.

Another list that's used is the List of Hazardous Substances and Reportable Quantities. It's appendix A to the hazmat table. As you've learned, the DOT and EPA want to know about spills or leaks of some products in certain quantities. This list names those products and their "reportable quantities," or RQ. You may see RQs on the shipping paper.

Some items have more than one name. The other name is listed in the "Synonyms" column on the list.

Shipping Papers and Manifests

Being able to recognize a hazardous materials shipment is the first step in handling it correctly and safely. All licensed CMV drivers must be able to recognize a hazardous materials shipment. One way to do this is by the shipping paper and manifest listings. Filling out the shipping paper is mainly the responsibility of the shipper. As the driver, you must know how to read it to know what you're carrying. Part 172 states how hazardous materials are to be described on the shipping paper. This covers all hazardous materials (other than those for air or water transport, and ORM-D unless it is for transport by air or water).

Shipping Papers

Not all shipping papers are alike. When it comes to hazardous materials, though, they all must have certain information:

- the proper shipping name from column 2 on the Hazardous Materials Table
- the hazard class or division from column 3 on the Hazardous Materials Table
- the identification number from column 4 on the Hazardous Materials Table
- the packing group from column 5 on the Hazardous Materials Table
- total quantity by weight, volume, or whatever is appropriate for the material
- an emergency response telephone number for each hazardous material, or one number if it will cover all the hazmat listed

Some shipping papers may include both hazardous and non-hazardous shipments. In such cases, the hazardous material must be described first. Or, it can be highlighted in a different color from the other entries. Another option is to place an "X" in the "HM" (hazardous materials) column before the shipping name. You may see the letters RQ instead of X if the item is not hazardous but is a reportable quantity. If the hazardous material is hazardous waste, the shipper must put the word "waste" on the paper, before the shipping name.

Figure 13-6 is a sample shipping paper. Note that there are page numbers that tell whether there is more than one page. The total number of pages appears on the first page.

In column 1 of each entry, you'll see the quantity.

Column 2 has the description of the hazardous product. This includes the proper shipping name, hazard class, and identification number, in that order. The proper shipping name from column 2 of the Hazardous Materials Table should always be used. Then comes the hazard class from column 3 of the table. That's followed by the ID number from the Hazardous Materials Table, column 4.

Sometimes this name doesn't give much of a clue as to the nature of the materials or wastes. So, descriptive terms are added. This could be "ignitability," "corrosivity," or "reactivity." These terms are called ICR terms (perhaps you can guess why). Other descriptive terms may limit the transport of certain materials to a particular DOT specification tank.

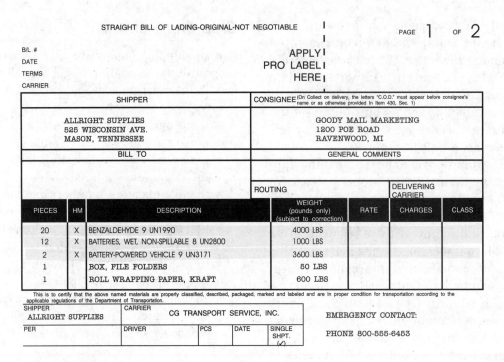

STRAIGHT BILL OF LADING-ORIGINAL-NOT NEGOTIABLE

PAGE 1 OF 2

B/L #
DATE
TERMS
CARRIER

APPLY
PRO LABEL
HERE

SHIPPER	CONSIGNEE (On Collect on delivery, the letters "C.O.D." must appear before consignee's name or as otherwise provided In Item 430, Sec. 1)
ALLRIGHT SUPPLIES 525 WISCONSIN AVE. MASON, TENNESSEE	GOODY MAIL MARKETING 1200 POE ROAD RAVENWOOD, MI
BILL TO	GENERAL COMMENTS

ROUTING | DELIVERING CARRIER

PIECES	HM	DESCRIPTION	WEIGHT (pounds only) (subject to correction)	RATE	CHARGES	CLASS
20	X	BENZALDEHYDE 9 UN1990	4000 LBS			
12	X	BATTERIES, WET, NON-SPILLABLE 8 UN2800	1000 LBS			
2	X	BATTERY-POWERED VEHICLE 9 UN3171	3600 LBS			
1		BOX, FILE FOLDERS	50 LBS			
1		ROLL WRAPPING PAPER, KRAFT	600 LBS			

This is to certify that the above named materials are properly classified, described, packaged, marked and labeled and are in proper condition for transportation according to the applicable regulations of the Department of Transportation.

SHIPPER ALLRIGHT SUPPLIES	CARRIER	CG TRANSPORT SERVICE, INC.			EMERGENCY CONTACT:
PER	DRIVER	PCS	DATE	SINGLE SHPT. (✓)	PHONE 800-555-6453

Figure 13-6. Sample shipping paper.

Cargo and portable tanks may be built to DOT specifications. If so, they receive a DOT specification code number. Coded tankers are those that have been approved to haul hazardous liquids. Refer to Chapter 12 for more details on DOT specifications for tanks.

You may also see a note that alerts you to special dangers. Examples are "dangerous when wet" and "poison inhalation hazard." "Inhalation hazard" tells you that you can be poisoned by simply breathing in the hazardous material or its vapors. Knowing that, you would be able to protect yourself if it leaked or spilled.

Shipping papers must have entries for unusual circumstances. A shipment made under a DOT exemption must be marked "DOT-E" (before October 1, 2007) or "DOT-SP" (after October 1, 2007) followed by the exemption number that was assigned. The basic description of limited quantity of a hazmat must be followed by the words "Limited Quantity" (or the abbreviation "LTD. QTY.").

Information for radioactive material must include the name of the radionuclide, its physical and chemical form, and its activity in curies, milli-curies, or microcuries (these are all measurements of radioactivity). Also, the words "Highway Route Controlled Quantity" must appear, along with the category label and the transport index for each Yellow II or Yellow III package.

Some packages contain only the remains of a hazardous material. In that case, no quantity is listed. Still, these remains can be hazardous themselves. The description on the shipping paper will then read "Residue: Last contained..." with the name of the hazardous material that was in the package and the letters "RQ."

The last column for the entry gives the weight. On the bottom right of this shipping paper, you'll find the emergency contact phone number. You can see that it is clearly marked as an emergency contact number, as required by law.

Last, you'll find the shipper's certification on the shipping paper. This is the shipper's statement that the shipment was prepared according to regulations.

Manifest

Manifests are used for transporting hazardous wastes. The shipper prepares the manifest, dates, and signs it. If the shipper is not the one who created the waste, whoever did must sign and date the manifest. The first carrier involved in the transport of the waste also signs and dates it. A copy of the manifest must be signed by each carrier who transports the waste. The facility receiving the waste dates and signs the copy.

A copy of the manifest with all the dates and signatures goes to each person representing each carrier transporting the waste. One of those could be you, the driver. A copy also goes to the person who represents the facility that receives the waste. Last, the manifest copy is returned to the shipper or generator (creator) of the waste, and all the carriers. Your employer will keep the manifest on file for three years. This sets up a "paper trail" of all the steps and stops in the transport of the waste.

"Waste" must be entered before the proper shipping name on the manifest. The manifest will also include this information:

- document number
- generator's U.S. EPA 12-digit identification number
- total number of pages used to complete the manifest, and any continuation pages
- name, mailing address, and emergency contact phone number of the generator
- name and identification number of the first transporter who will transport the waste, plus names and identification numbers of any additional transporters
- special handling instructions
- shipper's name, mailing address, phone number, and Environmental Protection Agency (EPA) identification number
- name and EPA identification number of each carrier
- name, address, and EPA identification number of the designated facility (plus an alternate facility, if given)
- description of the material
- quantity (by units of weight or volume)
- type and number of containers loaded on the vehicle

Marking

Every hazmat package, freight container, and transport vehicle must be marked as described in this part of Title 49 CFR. Most of the marking requirements are the shipper's responsibility. They are different for bulk and non-bulk packages. Since the markings can be important to you, you should know something about them.

Remember the definition of bulk and non-bulk packaging? "Bulk packaging" includes a transport vehicle or freight container in which hazardous materials are loaded with no other container, such as an inner lining. Such a vehicle can hold more than 119 gallons of liquid, 882 pounds of solid, or 1,000 pounds (water capacity) of gas. "Non-bulk packaging" means a package smaller than the capacities stated for "bulk packaging."

Marking must be long-wearing, in English, and printed on or attached to the surface of the package. Markings can also be put on a label, tag, or sign. They must stand out against the background color. They must not be hidden by other labels or fixtures. They must be placed away from other markings such as advertisements.

"ORM" stands for "Other Regulated Materials," as you now know.

Packages of radioactive materials may be marked "Type A" or "Type B." These stand for different levels of hazard.

Packages of liquid hazardous materials are marked with arrows to show you how to keep the package upright. That way it's less likely to leak. Arrows may be used along with or instead of words.

Tanks

Hazardous materials may not be carried in portable or cargo tanks or tank cars unless they are marked. They must be marked at least on each side with the UN identification number specified for the cargo to be hauled. Portable tanks holding over 1,000 gallons and cargo tanks must also be marked on each end. If the tractor hides the front of the cargo tank, the tractor should be marked. The name of the owner or shipper must be displayed on a portable tank carrying hazmat.

The shipper is supposed to give the carrier the necessary placards with these identification numbers if the tank isn't already marked. Orange panels with four-inch high black numbers can be used instead of diamond-shaped placards.

The name of the material may have to be marked on the tank. If so, it must be in letters at least two inches high. Both sides of the tank must be marked. This is required when the tank will be carrying certain Class 2 materials such as flammable and non-flammable gas and cryogenic liquid. "QT" and "NQT" markings are required for MC 330 and MC 331 cargo tanks. The letters stand for "Quenched and Tempered" or "Not Quenched and Tempered." This describes the type of metal used to build the tank.

Cargo tanks carrying more than one hazmat must be marked on two sides with the ID number of each one. The markings should be in the same order as the compartments in which the material is being carried. If an "Elevated Temperature Material" is the cargo, the tank must be marked on each side and both ends with the word "Hot."

Labels

You know that the packages holding hazardous materials must be labeled. This is required by Title 49 CFR Part 172 with few exceptions. Special labels are used for this. Shippers are responsible for labeling packages. You, the driver, must be able to recognize the labels and check the shipping papers to make sure that the shipment is prepared properly.

Labels are not used on compressed gas cylinders that are not overpacked. Instead, the label is placed on a hang tag hung around the neck of the cylinder. Or, a decal may be used. The cylinders themselves may be stamped as holding compressed gas. Compressed gas cylinders that are permanently mounted on a vehicle don't need labels. ORM packages don't need labels. Neither do packages of combustible liquids. Packages with low activity radioactive materials don't need labels if they're hauled in a special vehicle used only by the consignor. Cargo tanks and single unit tank cars don't need labels.

Cargo tanks don't need labels if they are placarded.

What if a package contains more than one type of hazmat? In some cases, the package must have more than one label.

Even empty packages may have to have labels. When a package is empty, any hazardous material shipping name and identification number markings, any hazard warning labels or placards, and any other markings indicating that the material is hazardous (e.g., RQ, Inhalation Hazard) should be removed, obliterated, or securely covered in transportation. If they aren't, then labels are still needed unless the packaging is not visible in transportation and the packaging is loaded by the shipper and unloaded by the shipper or consignee. Empty packages aren't merely empty. They are:

- unused
- cleaned of residue and purged of vapors to remove any potential hazard
- refilled with a material that is not hazardous to such an extent that any residue remaining in the packaging no longer poses a hazard, or
- contain only the residue of an ORM-D material, a Division 2.2 non-flammable gas, other than ammonia, anhydrous, and with no subsidiary hazard, at an absolute pressure less than 280 kPa (40.6 psia); at 20°C (68°F); and any material contained in the packaging does not meet the definitions for a hazardous substance, a hazardous waste, or a marine pollutant.

A non-bulk packaging containing only the residue of a hazardous material covered by Table 2 does not have to be included in determining the applicability of the placarding requirements, and is not subject to the shipping paper requirements of this subchapter when collected and transported by a contract or private carrier for reconditioning, remanufacture, or reuse. (You'll find Table 2 later in this chapter). The description on the shipping paper for a packaging containing the residue of a hazardous material may include the words "Residue: Last Contained . . ." along with the basic description of the hazardous material last contained in the packaging.

A package that contains the remains of an elevated temperature material may remain marked in the same manner as when it contained a greater quantity of the material even though it no longer meets the definition for an elevated temperature material.

DOT 106 and 110 tanks must be labeled on each end.

There are three different "Radioactive" labels for different levels of radioactivity. White I is the lowest level. Next comes Yellow II. The highest level is Yellow III.

Hazmat package labels are diamond-shaped and four inches square. They have a solid black line border. The labels show the class of hazardous materials to which the package belongs. Any numbers that might be on the labels are part of a worldwide system used to identify hazardous materials. Each number stands for a different type of hazmat. Figure 13-7 shows the hazmat labels and placards.

U.S. Department of Transportation

Pipeline and Hazardous Materials Safety Administration

DOT CHART 15
Hazardous Materials Markings, Labeling and Placarding Guide

Refer to 49 CFR, Part 172:

Marking - Subpart D

Labeling - Subpart E

Placarding - Subpart F

NOTE: This document is for general guidance only and should not be used to determine compliance with 49 CFR, Parts 100-185.

HAZARDOUS MATERIALS MARKINGS

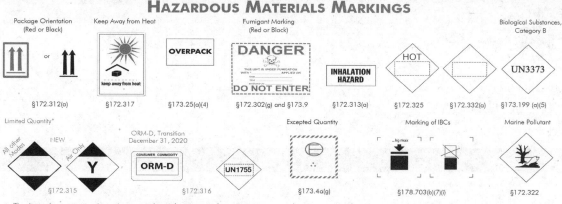

Package Orientation (Red or Black) — §172.312(a)

Keep Away from Heat — §172.317

OVERPACK — §173.25(a)(4)

Fumigant Marking (Red or Black) — §172.302(g) and §173.9

INHALATION HAZARD — §172.313(a)

HOT — §172.325

§172.332(a)

Biological Substances, Category B — UN3373 — §173.199 (a)(5)

Limited Quantity* — §172.315

ORM-D, Transition December 31, 2020 — ORM-D / UN1755 — §172.316

Excepted Quantity — §173.4a(g)

Marking of IBCs — §178.703(b)(7)(i)

Marine Pollutant — §172.322

* The limited quantity marking designates hazardous materials packages meeting the requirements for transportation as a limited quantity by air (Y mark) and packages meeting the requirements for transport as a limited quantity by surface modes (no Y). A Y-marked package meeting the requirements for transport by air may be transported by all modes. In some instances packages bearing the surface mark (no Y) may also be acceptable for transport by air provided the packages meet all relevant requirements for air transport (for example UN0012, UN0014, or UN0055).

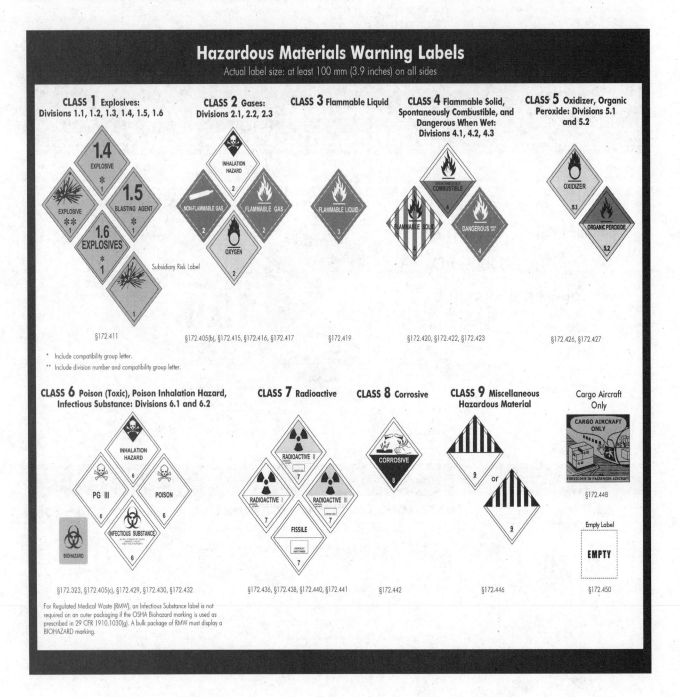

Hazardous Materials Warning Labels

Actual label size: at least 100 mm (3.9 inches) on all sides

CLASS 1 Explosives:
Divisions 1.1, 1.2, 1.3, 1.4, 1.5, 1.6

CLASS 2 Gases:
Divisions 2.1, 2.2, 2.3

CLASS 3 Flammable Liquid

CLASS 4 Flammable Solid,
Spontaneously Combustible, and
Dangerous When Wet:
Divisions 4.1, 4.2, 4.3

CLASS 5 Oxidizer, Organic
Peroxide: Divisions 5.1
and 5.2

§172.411

§172.405(b), §172.415, §172.416, §172.417

§172.419

§172.420, §172.422, §172.423

§172.426, §172.427

* Include compatibility group letter.

** Include division number and compatibility group letter.

CLASS 6 Poison (Toxic), Poison Inhalation Hazard,
Infectious Substance: Divisions 6.1 and 6.2

CLASS 7 Radioactive

CLASS 8 Corrosive

CLASS 9 Miscellaneous
Hazardous Material

Cargo Aircraft
Only

Empty Label

§172.323, §172.405(c), §172.429, §172.430, §172.432

§172.436, §172.438, §172.440, §172.441

§172.442

§172.446

§172.448

§172.450

For Regulated Medical Waste (RMW), an Infectious Substance label is not
required on an outer packaging if the OSHA Biohazard marking is used as
prescribed in 29 CFR 1910.1030(g). A bulk package of RMW must display a
BIOHAZARD marking.

Hazardous Materials Warning Placards
Actual placard size: at least 250 mm (9.84 inches) on all sides

CLASS 1 Explosives

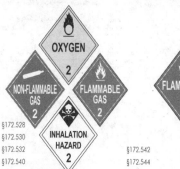

§172.522
§172.523
§172.524
§172.525

* For Divisions 1.1, 1.2, or 1.3, enter division number and compatibility group letter, when required; placard any quantity. For Divisions 1.4, 1.5, and 1.6, enter compatibility group letter, when required; placard 454 kg (1,001 lbs) or more.

CLASS 2 Gases

§172.528
§172.530
§172.532
§172.540

For NON-FLAMMABLE GAS, OXYGEN (compressed gas or refrigerated liquid), and FLAMMABLE GAS, placard 454 kg (1,001 lbs) or more gross weight. For POISON GAS (Division 2.3), placard any quantity.

CLASS 3 Flammable Liquid and Combustible Liquid

§172.542
§172.544

For FLAMMABLE, placard 454 kg (1,001 lbs) or more. GASOLINE may be used in place of FLAMMABLE placard displayed on a cargo tank or portable tank transporting gasoline by highway. Placard combustible liquid transported in bulk. See §172.504(f)(2) for use of FLAMMABLE placard in place of COMBUSTIBLE. FUEL OIL may be used in place of COMBUSTIBLE on a cargo or portable tank transporting fuel oil not classed as a flammable liquid by highway.

CLASS 4 Flammable Solid, Spontaneously Combustible, and Dangerous When Wet

§172.546, §172.547, §172.548

For FLAMMABLE SOLID and SPONTANEOUSLY COMBUSTIBLE, placard 454 kg (1,001 lbs) or more. For DANGEROUS WHEN WET (Division 4.3), placard any quantity.

CLASS 5 Oxidizer & Organic Peroxide

Organic Peroxide, Transition-2011 (rail, vessel, and aircraft) 2014 (highway)

§172.550, §172.552

For OXIDIZER and ORGANIC PEROXIDE (other than TYPE B, temperature controlled), placard 454 kg (1,001 lbs) or more. For ORGANIC PEROXIDE (Division 5.2), Type B, temperature controlled, placard any quantity.

CLASS 6 Poison (Toxic) and Poison Inhalation Hazard

§172.504(f)(10), §172.554, §172.555

For POISON (PGI or PGII, other than inhalation hazard) and POISON (PGIII), placard 454 kg (1,001 lbs) or more. For POISON-INHALATION HAZARD (Division 6.1), inhalation hazard only, placard any quantity.

CLASS 7 Radioactive

§172.556

Placard any quantity - packages bearing RADIOACTIVE YELLOW-III labels only. Certain low specific activity radioactive materials in "exclusive use" will not bear the label, but the radioactive placard is required for exclusive use shipments of low specific activity material and surface contaminated objects transported in accordance with §172.504(e) Table 1 and §173.427(a)(6).

CLASS 8 Corrosive

§172.558

For CORROSIVE, placard 454 kg (1,001 lbs) or more.

CLASS 9 Miscellaneous

§172.560

Not required for domestic transportation. A bulk packaging containing a Class 9 material must be marked with the appropriate ID number displayed on a Class 9 placard, an orange panel, or a white square-on-point display.

Dangerous

§172.521

A freight container, unit load device, transport vehicle, or rail car which contains non-bulk packages with two or more categories of hazardous materials that require different placards specified in Table 2 §172.504(e) may be placarded with DANGEROUS placards instead of the specific placards required for each of the materials in Table 2. However, when 1,000 kg (2,205 lbs) or more of one category of material is loaded at one loading facility, the placard specified in Table 2 must be applied.

Limited Quantity Marking

§172.315(a)(2)
(Vessel transport only).

Safety begins with communication!

General Guidelines on Use of Warning Labels and Placards

LABELS

See 49 CFR, Part 172, Subpart E, for complete labeling regulations.

- The Hazardous Materials Table [§172.101, Col. 6] identifies the proper label(s) for the hazardous material listed.

- Any person who offers a hazardous material for transportation MUST label the package, if required [§172.400(a)].

- Labels may be affixed to packages when not required by regulations, provided each label represents a hazard of the material contained in the package [§172.401].

- For labeling mixed or consolidated packages, see §172.404.

- The appropriate hazard class or division number must be displayed in the lower corner of a primary and subsidiary hazard label [§172.402(b)].

- For classes 1,2,3,4,5,6, and 8, text indicating a hazard (e.g., "CORROSIVE") is NOT required on a primary or subsidiary label. The label must otherwise conform to Subpart E of Part 172 [§172.405].

- Labels must be printed on or affixed to the surface of the package near the proper shipping name marking [§172.406(a)].

- When primary and subsidiary labels are required, they must be displayed next to each other [§172.406(c)].

- For a package containing a Division 6.1, PG III material, the POISON label specified in §172.430 may be modified to display the text PG III instead of POISON or TOXIC. Also see §172.405(c).

- The ORGANIC PEROXIDE label [§172.427] indicates that organic peroxides are highly flammable. Use of the ORGANIC PEROXIDE label eliminates the need for a flammable liquid subsidiary label. The color of the border must be black and the color of the flame may be black or white.

PLACARDS

See 49 CFR, Part 172, Subpart F, for complete placarding regulations.

- Each person who offers for transportation or transports any hazardous material subject to the Hazardous Materials Regulations must comply with all applicable requirements of Subpart F [§172.500].

- Placards may be displayed for a hazardous material, even when not required, if the placarding otherwise conforms to the requirements of Subpart F of Part 172 [§172.502(c)].

- For other than Class 7 or the DANGEROUS placard, text indicating a hazard (e.g., "FLAMMABLE") is not required. Text may be omitted from the OXYGEN placard only if the specific ID number is displayed on the placard [§172.519(b)(3)].

- For a placard corresponding to the primary or subsidiary hazard class of a material, the hazard class or division number must be displayed in the lower corner of the placard [§172.519(b)(4)].

- Except as otherwise provided, any bulk packaging, freight container, unit load device, transport vehicle or rail car containing any quantity of material listed in Table 1 must be placarded [§172.504].

- When the aggregate gross weight of all hazardous materials in non-bulk packages covered in Table 2 is less than 454 kg (1,001 lbs), no placard is required on a transport vehicle or freight container when transported by highway or rail [§172.504(c)].

- Notes: See §172.504(f)(10) for placarding Division 6.1, PG III materials.

- Placarded loads require registration with USDOT. See §107.601 for registration regulations.

- The new ORGANIC PEROXIDE placard became mandatory 1 January 2011 for transportation by rail, vessel, or aircraft and becomes mandatory 1 January 2014 for transportation by highway. The placard will enable transport workers to readily distinguish peroxides from oxidizers [§172.552].

PLACARDING TABLES
[§172.504(e)]
TABLE 1

Category of material (Hazard Class or division number and additional description, as appropriate)	Placard name
1.1	EXPLOSIVES 1.1
1.2	EXPLOSIVES 1.2
1.3	EXPLOSIVES 1.3
2.3	POISON GAS
4.3	DANGEROUS WHEN WET
5.2 (Organic peroxide, Type B, liquid or solid, temperature controlled)	ORGANIC PEROXIDE
6.1 (Materials poisonous by inhalation (see §171.8))	POISON INHALATION HAZARD
7 (Radioactive Yellow III label only)	RADIOACTIVE[1]

[1]RADIOACTIVE placard also required for exclusive use shipments of low specific activity material and surface contaminated objects transported in accordance with §173.427(b)(4) and (5) or (c) of the subchapter.

TABLE 2

Category of material (Hazard Class or division number and additional description, as appropriate)	Placard name
1.4	EXPLOSIVES 1.4
1.5	EXPLOSIVES 1.5
1.6	EXPLOSIVES 1.6
2.1	FLAMMABLE GAS
2.2	NON-FLAMMABLE GAS
3	FLAMMABLE
Combustible Liquid	COMBUSTIBLE
4.1	FLAMMABLE SOLID
4.2	SPONTANEOUSLY COMBUSTIBLE
5.1	OXIDIZER
5.2 (Other than organic peroxide, Type B, liquid or solid, temperature controlled)	ORGANIC PEROXIDE
6.1 (Other than materials poisonous by inhalation)	POISON
6.2	(None)
8	CORROSIVE
9	Class 9 (See §172.504(f)(9))
ORM-D	(None)

IDENTIFICATION NUMBER DISPLAYS

 and or

§172.332

Appropriate placard must be used with orange panel.

IDENTIFICATION NUMBER MARKINGS ON ORANGE PANELS OR APPROPRIATE PLACARDS MUST BE DISPLAYED ON: (1) Tank Cars, Cargo Tanks, Portable Tanks, and other Bulk Packagings; (2) Transport vehicles or freight containers containing 4,000 kg (8,820 lbs) in non-bulk packages of only a single hazardous material having the same proper shipping name and identification number loaded at one facility and transport vehicle contains no other material, hazardous or otherwise; and (3) transport vehicles or freight containers containing 1,000 kg (2,205 lbs) of non-bulk packages of materials poisonous by inhalation in Hazard Zone A or B. See §§172.301(a)(3), 172.313(c), 172.326, 172.328,172.330, and 172.331.

Square white background required for placard for highway route controlled quantity radioactive material and for rail shipment of certain explosives and poisons, and for flammable gas in a DOT 113 tank car (§172.507 and §172.510).

§172.527

This Chart is available online at the following link:
http://phmsa.dot.gov/hazmat

U.S. Department of Transportation

Pipeline and Hazardous Materials Safety Administration

USDOT/PHMSA/OHMIT/PHH-50
1200 New Jersey Avenue, SE
Washington, D.C. 20590
Phone: (202) 366-4900
Email: training@dot.gov

PHH50-0143-0214

Figure 13-7. General Guidelines on Use of Warning Labels and Placards.

Placards

Placards are required by the regulations in Part 172. A placard is similar to a label. However, a placard is attached to the outside of the vehicle to show clearly that it contains a load of hazardous materials. The placard for a certain material will be of the same color and wording as its label. Look at Figure 13-8.

You will check to make sure that you have the right placard or placards for the load.

Placards go on each end and each side of the vehicles.

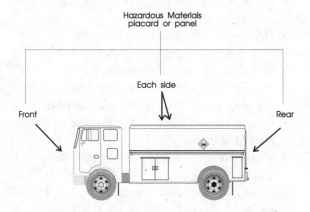

Figure 13-8. How to properly affix placards to a vehicle.

When more than one type of hazmat is being hauled, the vehicle must have more than one type of placard. A single placard reading "Dangerous" is enough when two or more non-bulk packages of material Placard Table 2 are hauled together under these conditions:

- the total weight of any single hazardous material loaded at a single facility doesn't exceed 2,205 pounds
- the cargo isn't being hauled in portable tanks, cargo tanks, or tank cars

When more than one division placard is required for Class 1 materials, you only need the placard for the lowest division number.

Tank vehicles used to haul any hazardous materials must have placards at all times, even when empty. After the tank has been completely cleaned, or reloaded with material that doesn't need placards, the placards can come off. Empty non-bulk packages don't need placards.

If your shipping paper shows your cargo presents a "poison-inhalation hazard," your vehicle must have "Poison" placards on each side and each end, as well as the class placard.

The law requires the shipper to provide the carrier with the right placards for the load. The only exception is if the vehicle is already correctly placarded.

As you have seen, most vehicles must be placarded on all four sides. Here are some that don't:

- tanks with valves, fittings, regulators, or gauges on one end that would hide the placard don't have to placarded on that end
- portable tanks able to hold less than 1,000 gallons need to be placarded on only opposite sides

You've also seen that cargo tanks and portable tanks that were placarded for hazardous materials must still be placarded when emptied. Placards may be removed only when the tank is completely cleaned, or when it's reloaded with a material not needing placards.

There are additional placarding exceptions:

When more than one division placard is required for Class 1 materials on a transport vehicle, rail car, freight container, or unit load device, only the placard representing the lowest division number must be displayed.

A FLAMMABLE placard may be used in place of a COMBUSTIBLE placard on a cargo tank or portable tank or a compartmented tank car that contains both flammable and combustible liquids.

A NON-FLAMMABLE GAS placard is not required on a transport vehicle that contains non-flammable gas if the transport vehicle also contains flammable gas or oxygen and it is placarded with the required FLAMMABLE GAS or OXYGEN placards.

OXIDIZER placards are not required for Division 5.1 materials on freight containers, unit load devices, or transport vehicles that also contain Division 1.1 or 1.2 materials and that are placarded with EXPLOSIVES 1.1 or 1.2 placards.

For transportation by transport vehicle or rail car only, an OXIDIZER placard is not required for Division 5.1 materials on a transport vehicle, rail car, or freight container that also contains Division 1.1 or 1.2 explosives and is placarded with EXPLOSIVES 1.1 or 1.2 placards.

The EXPLOSIVE 1.4 placard is not required for those Division 1.4 Compatibility Group S (1.4S) materials that are not required to be labeled 1.4S.

For domestic transportation of oxygen or compressed oxygen, or refrigerated liquid, the OXYGEN placard may be used in place of a NON-FLAMMABLE GAS placard.

For domestic transportation, a POISON INHALATION HAZARD placard is not required on a transport vehicle or freight container that is already placarded with the POISON GAS placard.

For domestic transportation, a Class 9 placard is not required. A bulk packaging containing a Class 9 material must be marked with the appropriate identification number displayed on a Class 9 placard, an orange panel, or a white-square-on-point display configuration as required by subpart D of this part.

For Division 6.1, Packing Group III materials, a POISON placard may be modified to display the text "PG III" below the mid-line of the placard.

For domestic transportation, a POISON placard is not required on a transport vehicle or freight container required to display a POISON INHALATION HAZARD or POISON GAS placard.

Otherwise, placards must be placed so they can be seen from all directions the vehicle faces. When you're pulling a trailer, one side faces the rear of the tractor or trailer in front of it. That side doesn't need a placard, since no one can see it anyway when the vehicles are coupled.

Placards must be securely attached. They must be clear of ladders, pipes, doors, and tarps that could hide them. They must be placed high enough on the vehicle so the wheels don't spray dirt or water on them. They must be at least three inches away from any other marking on the vehicle. They must be maintained in good condition so they can always be read clearly. Make sure the placard holder doesn't hide any part of the placard except maybe the borders.

Like labels, some placards may include the UN identification number. This should be in the lower corner of the diamond. Placarding requirements are summed up on pages 301–302. Note that the materials in Table 1 (Figure 13-9) must always be placarded as shown in the table. Materials in Table 2 (Figure 13-10) don't have to be placarded when the gross weight of all hazardous materials is less than 1,000 pounds.

TITLE 49 CFR PART 173

This part includes definitions of hazardous materials for the purpose of transporting them. It also gives the requirements for preparing hazardous materials shipments. Inspecting, testing, and retesting requirements are given for those involved in these procedures. Most of the requirements in this part are of interest to shippers. We'll discuss the ones that are important to you, the driver.

Hazardous materials classes are listed in order from most to least dangerous:

- Class 7 (radioactive materials, other than limited quantities)
- Division 2.3 (poisonous gases)
- Division 2.1 (flammable gases)
- Division 2.2 (non-flammable gases)
- Division 6.1 (poisonous liquids), Packing Group I, poisonous-by-inhalation only
- a material that meets the definition of a pyrophoric material
- a material that meets the definition of a self-reactive material
- Class 3 (flammable liquids), Class 8 (corrosive materials), Division 4.1 (flammable solids), Division 4.2 (spontaneously combustible materials), Division 4.3 (dangerous when wet materials), Division 5.1 (oxidizers), or Division 6.1 (poisonous liquids or solids other than Packing Group I, poisonous-by-inhalation)
- Combustible liquids
- Class 9 (miscellaneous hazardous materials)

Fumigated Trucks

Trucks and trailers sometimes haul cargo that has been fumigant or treated with poisonous liquid, solid, or gas. Such vehicles must be placarded with a special "Fumigant" placard. These vehicles may not be transported until the vehicle has been aired out and no longer presents a danger.

Standard Requirements for Packages

In the definition section, you learned that some hazardous materials react dangerously with water, air, heat, or flame—even each other. These materi-

TABLE 1		
Category of material (Hazard class or division number and additional description, as appropriate)	Placard name	Placard design section reference (§)
1.1	EXPLOSIVES 1.1	172.522
1.2	EXPLOSIVES 1.2	172.522
1.3	EXPLOSIVES 1.3	172.522
2.3	POISON GAS	172.540
4.3	DANGEROUS WHEN WET	172.548
5.2 (Organic peroxide, Type B, liquid or solid, temperature controlled)	ORGANIC PEROXIDE	172.552
6.1 (Material poisonous by inhalation (see §171.8 of this subchapter))	POISON INHALATION HAZARD	172.555
7 (Radioactive Yellow III label only)	RADIOACTIVE[1]	172.556

[1]RADIOACTIVE placard also required for exclusive use shipments of low specific activity material and surface contaminated objects transported in accordance with §173.427(b)(4) and (5) or (c) of the subchapter.

RADIOACTIVE placards are also required for: All shipments of unpackaged LSA-I material or SCO-I; all shipments required by §§173.427, 173.441, and 173.457 of this subchapter to be operated under exclusive use; and all closed vehicles used in accordance with §173.443(d).

Figure 13-9. Table 1.

als come in packages that prevent such reactions. These are marked with DOT package specification marks.

You'll use column 8 of the hazmat table (located on page 291) to make sure that materials are packaged in the proper packaging for truck transport.

Qualification, Maintenance, and Use Of Tanks

Cargo and portable tanks used to haul hazardous materials must meet DOT specifications. Tanks must also meet inspection, retest, and marking requirements. The month and year of the most recent test are stamped into the metal of each tank. Or, it may be stamped on or near the metal certification plate.

Retest and marking of tanks is discussed in more detail in Chapter 12. That chapter also includes the retest schedule for cargo and portable tanks.

Portable tanks that have become worn out or damaged must be retested. So must tanks that have not been used to transport hazardous materials for

a year. Portable tanks must be loaded onto trucks so that no part of the tank sticks out or overhangs.

Hazardous materials that will react with each other must not be loaded together in the same cargo tank.

Qualification, Maintenance, and Use Of Cylinders

Cylinders are built to specification just like tanks. DOT numbers show what the cylinder may be used for. Refer to Figure 13-11 for more details.

Cylinders are also tested. Most of them are retested every five years. Specification 3HT cylinders are retested every three years. Retests are not required for Specification 3C, 4C, and 4L cylinders. The cylinder will be marked with the month and year of the test.

Definitions and Preparation

The remainder of Part 173 has details about the hazardous materials themselves. These include definitions and packaging descriptions that are mostly of interest to shippers. What most driv-

TABLE 2		
Category of material (Hazard class or division number and additional description, as appropriate)	Placard name	Placard design section reference (§)
1.4	EXPLOSIVES 1.4	172.523
1.5	EXPLOSIVES 1.5	172.524
1.6	EXPLOSIVES 1.6	172.525
2.1	FLAMMABLE GAS	172.532
2.2	NON-FLAMMABLE GAS	172.528
3	FLAMMABLE	172.542
Combustible liquid	COMBUSTIBLE	172.544
4.1	FLAMMABLE SOLID	172.546
4.2	SPONTANEOUSLY COMBUSTIBLE	172.547
5.1	OXIDIZER	172.550
5.2 (Other than organic peroxide, Type B, liquid or solid, temperature controlled)	ORGANIC PEROXIDE	172.552
6.1 (Other than material poisonous by inhalation)	POISON	172.554
6.2	(None)	
8	CORROSIVE	172.558
9	Class 9 (see §172.504(f)(9))	172.560
ORM-D	(None)	

(f) *Additional placarding exceptions.*

(1) When more than one division placard is required for Class 1 materials on a transport vehicle, rail car, freight container, or unit load device, only the placard representing the lowest division number must be displayed.

(2) A FLAMMABLE placard may be used in place of a COMBUSTIBLE placard on—

(i) A cargo tank or portable tank.

(ii) A compartmented tank car which contains both flammable and combustible liquids.

(3) A NON-FLAMMABLE GAS placard is not required on a transport vehicle which contains non-flammable gas if the transport vehicle also contains flammable gas or oxygen and it is placarded with FLAMMABLE GAS or OXYGEN placards, as required.

(4) OXIDIZER placards are not required for Division 5.1 materials on freight containers, unit load devices, transport vehicles, or rail cars which also contain Division 1.1 or 1.2 materials and which are placarded with EXPLOSIVES 1.1 or 1.2 placards, as required.

Figure 13-10. Table 2.

(5) For transportation by transport vehicle or rail car only, an OXIDIZER placard is not required for Division 5.1 materials on a transport vehicle, rail car, or freight container which also contains Division 1.5 explosives and is placarded with EXPLOSIVES 1.5 placards, as required.

(6) The EXPLOSIVE 1.4 placard is not required for those Division 1.4 Compatibility Group S (1.4S) materials that are not required to be labeled 1.4S.

(7) For domestic transportation of oxygen, compressed or oxygen, refrigerated liquid, the OXYGEN placard in §172.530 of this subpart may be used in place of a NON-FLAMMABLE GAS placard.

(8) For domestic transportation, a POISON INHALATION HAZARD placard is not required on a transport vehicle or freight container that is already placarded with the POISON GAS placard.

(9) For Class 9, a CLASS 9 placard is not required for domestic transportation, including that portion of international transportation, defined in §171.8 of this subchapter, which occurs within the United States. However, a bulk packaging must be marked with the appropriate identification number on a CLASS 9 placard, an orange panel, or a white square-on-point display configuration as required by subpart D of this part.

(10) For Division 6.1, PG III materials, a POISON placard may be modified to display the text "PG III" below the mid line of the placard.

(11) For domestic transportation, a POISON placard is not required on a transport vehicle or freight container required to display a POISON INHALATION HAZARD or POISON GAS placard.

(g) For shipments of Class 1 (explosive materials) by aircraft or vessel, the applicable compatibility group letter must be displayed on the placards, or labels when applicable, required by this section. When more than one compatibility group placard is required for Class 1 materials, only one placard is required to be displayed, as provided in paragraphs (g)(1) through (g)(4) of this section. For the purposes of paragraphs (g)(1) through (g)(4), there is a distinction between the phrases *explosive articles* and *explosive substances*. *Explosive article* means an article containing an explosive substance; examples include a detonator, flare, primer, or fuse. *Explosive substance* means a substance contained in a packaging that is not contained in an article; examples include black powder and smokeless powder.

(1) Explosive articles of compatibility groups C, D, or E may be placarded displaying compatibility group E.

(2) Explosive articles of compatibility groups C, D, or E, when transported with those in compatibility group N, may be placarded displaying compatibility group D.

(3) Explosive substances of compatibility groups C and D may be placarded displaying compatibility group D.

(4) Explosive articles of compatibility groups C, D, E, or G, except for fireworks, may be placarded displaying compatibility group E.

Figure 13-10. (cont'd.)

Cylinders made in compliance with-	Used exclusively for-
DOT-3A, DOT-3AA, DOT-3A480X, DOT-4AA480	Anhydrous ammonia of at least 99.95% purity.
DOT-3A, DOT-3AA, DOT-3A480X, DOT-3B, DOT-4B, DOT-4BA, DOT-4BW	Butadiene, inhibited, which is commercially free from corroding components.
DOT-3A, DOT-3A480X, DOT-3AA, DOT-3B, DOT-4AA480, DOT-4B, DOT-4BA, DOT-4BW	Cyclopropane which is commercially free from corroding components.
DOT 3A, DOT 3AA, DOT 3A480X, DOT 4B, DOT 4BA, DOT 4BW, DOT 4E	Chlorinated hydrocarbons and mixtures thereof that are commercially free from corroding components.
DOT-3A, DOT-3AA, DOT-3A480X, DOT-3B, DOT-4BA, DOT-4BW, DOT-4E	Fluorinated hydrocarbons and mixtures thereof which are commercially free from corroding components.
DOT-3A, DOT-3AA, DOT-3A480X, DOT-3B, DOT-4B, DOT-4BA, DOT-4BW, DOT-4E	Liquified hydrocarbon gas which is commercially free from corroding components.
DOT-3A, DOT-3AA, DOT-3A480X, DOT-3B, DOT-4B, DOT-4BA, DOT-4BW, DOT-4E	Liquified petroleum gas that meets the detail requirements limits in Table 1 of ASTM 1835, Standard Specification for Liquefied Petroleum (LP) Gases (incorporated by reference; see Sec. 171.7 of this subchapter) or an equivalent standard containing the same limits.
DOT-3A, DOT-3AA, DOT-3B, DOT-4B, DOT-4BA, DOT-4BW, DOT-4E	Methylacetylene-propadiene, stabilized, which is commercially free from corroding components.
DOT-3A, DOT-3AA, DOT-3B, DOT-4B, DOT-4BA, DOT-4BW	Anhydrous mono, di, trimethylamines which are commercially free from corroding components.
DOT4B240, DOT-4BW240	Ethylene imine, inhibited.

Figure 13-11. Chart of cylinder use (from Title 49 CFR 180.209).

ers need to know about the hazardous materials themselves was covered in Chapter 6, and in the "definitions" section of this chapter. What follows are some other important points of information not covered in Chapter 6.

Explosives. Explosives are chemical compounds, mixtures, or devices that explode. Division 1.1 explosives are the most dangerous. Division 1.2 explosives are less dangerous. They don't explode. Instead, they burn rapidly. Of the three, Division 1.6 explosives present the least danger. They are the least dangerous of all. They're not likely to explode by accident.

Leaking or damaged packages of explosives are forbidden. So are loaded firearms. Other explosives forbidden for transport are:

- An explosive mixture or device containing a chlorate and also containing: an ammonium salt, including a substituted ammonium or quaternary ammonium salt; or an acidic substance, including a salt of a weak base and a strong acid
- propellants that are unstable, condemned, or deteriorated
- nitroglycerin, diethylene glycol dinitrate, or any other liquid explosives not specifically authorized by this subchapter
- fireworks that combine an explosive and a detonator, or fireworks containing yellow or white phosphorus
- a toy torpedo, the maximum outside dimension of which exceeds 23 mm (0.906 inch), or a toy torpedo containing a mixture of potassium chlorate, black antimony (antimony sulfide), and sulfur, if the weight of the explosive material in the device exceeds 0.26 g (0.01 ounce)
- an explosive article with its means of initiation or ignition installed, unless otherwise approved

Don't transport these hazardous materials. Explosives can be very dangerous if not handled properly. The packages may be marked with special instructions and warnings. Examples are "This Side Up" and "Do Not Store Or Load With Any High Explosive." These warnings help all those involved in the shipment to handle the explosives safely.

Hazard zones. Certain materials always present an inhalation danger. The degree of danger is described by its hazard zone rating. Hazard Zone A is the most dangerous, while Hazard Zone D is the least.

Flammable, combustible, and pyrophoric liquids. A flammable liquid is one with a flash point that is not more than 140 degrees Fahrenheit, or any material in a liquid phase with a flash point at or above 100 degrees Fahrenheit that is intentionally heated and offered for transportation or transported at or above its flash point in a bulk packaging. A combustible liquid has a flash point at or above 140 degrees and below 200 degrees Fahrenheit. Liquids with higher flash points can still burn. They are not termed flammable or combustible in this setting. Still, you would not use "non-flammable" or "non-combustible" markings on your vehicle when hauling these materials.

A pyrophoric liquid is one that will ignite by itself within five minutes of coming into contact with the air.

The flash points determine the packing group (column 5 on the hazmat table).

Flammable solids, oxidizers, and organic peroxides. A flammable solid is a solid material, other than an explosive, that can cause a fire. The fire can start from friction. Or, the material might retain heat from manufacturing or processing. This heat can start a fire. These materials may also burn when combined with water.

Oxidizers throw off oxygen. This in turn causes other materials to burn.

Organic peroxides are not all that dangerous by themselves. But they do release a lot of oxygen. Around a spark or flame, they become very dangerous. Having organic peroxides near a fire is like adding gasoline to the flames. Sparks from your vehicle's electric system or friction from an overheated tire could start a fire. Organic peroxides could turn a small fire into a major disaster. That makes them dangerous to transport.

Corrosive materials. A corrosive is a solid or liquid that harms human skin on contact. Corrosives can dissolve metal and other materials, too. In fact, corrosion rates are determined by the material's effect on steel. Still, it's the damage to human skin that is used in the definition.

Gases. Compressed gases are held under pressure. When in the container, compressed gases have an absolute pressure of more than 40 psi (that is, 40 psia) at 68 degrees Fahrenheit.

Compressed gas is said to be flammable if:

- a mixture of 13 percent or less will form a flammable mixture when combined with air
- it passes an American Society for Testing and Materials explosives test for flammability

Compressed gas in solution. A compressed gas in solution is a nonliquefied compressed gas that is dissolved in a solvent.

Cryogenic liquid is a coolant gas. It will boil at minus 130 degrees Fahrenheit when it's at 14.7 psi absolute.

This class of hazardous materials also includes refrigerant gases and dispersant gases. These are compressed gases. Some are flammable, and some are non-flammable.

Title 49 CFR Part 173 lists filling densities for transporting compressed gases in cargo and portable tank containers. Filling densities were covered in Chapter 12.

Poisons, irritating materials, etiologic agents, and radioactive materials. There are two divisions of poison: 6.1 and 6.2. Poison Division 6.1 is a material other than a gas. Of the two divisions, 6.1 is the most dangerous. Just a small amount in the air can kill. Poison Division 6.2 are biohazards.

Irritating materials are liquid or solid. When exposed to air or flames, they give off dangerous fumes.

Etiologic agents are microorganisms (germs). They cause disease.

Radioactive materials give off dangerous rays. Some radioactive materials give off strong rays. Others are weak. Different type packages will be used based on the type of radioactive material they are to contain.

Other regulated materials. "Other regulated materials" are those that don't meet any of the other definitions. Still, they are dangerous when transported in commerce. So, they must be regulated to protect the public. ORM-D is an ordinary consumer item. It's dangerous in transport because of its form, amount, or because of the way it's packaged.

Part 173 contains other regulations and exceptions for a variety of materials, such as nitric acid, wet batteries, chemical kits, matches, asbestos, cigarette lighters, fire extinguishers—the list goes on. As a driver hauling hazmat, you should be familiar with Part 173 as it relates to the cargo you're carrying.

TITLE 49 CFR PART 177

Part 177 lists more requirements for private, common, or contract carriers for the transport of hazardous materials by motor vehicle.

General Requirements

Carriers must comply with state and federal regulations. They must instruct their employees about these regulations and how to comply with them.

Packages, vehicles, documents, shippers, carriers, and drivers must all meet the requirements of Title 49 CFR Parts 170 to 189, and 390 to 397 (the FMCSR).

Shipping paper accessibility. Shipping papers must have the shipper's certification. They must be marked differently from shipping papers for regular cargo. Or, the hazmat shipping papers can be placed on the top of a stack of cargo documents.

Drivers must carry these shipping papers so they are easy to get. When you're at the vehicle's controls, the shipping paper must be within your easy reach when you're wearing a seat belt. They can also be kept in the holder mounted on the inside of the driver's door. They must be where they can be easily seen by someone looking into the cab.

When you're not at the vehicle's controls, the shipping paper must be in the holder or on the driver's seat. These requirements are so that authorities can get to the shipping papers in case of an accident, or for an inspection.

Loading and Unloading

General requirements. Like any other cargo, hazardous materials packages must be loaded securely. Never load hazardous materials on a pole trailer. Don't you smoke or let anyone nearby smoke while loading and unloading explosives, flammable liquids, solids or compressed gas, or oxidizers. Don't let any fire or flame get near these hazardous materials. That includes matches, lighters, cigars, and pipes, as well as cigarettes. Keep the hand brake on at all times when loading and unloading hazardous materials.

Don't use any tools that could puncture, dent, or damage hazmat packages.

You must be aware of the temperature of the hazmat containers and their contents and prevent any increase. Some materials must be maintained at a specific temperature called the control temperature. These are transported in vehicles specially designed for that purpose.

A cargo tank must be attended by a qualified person at all times when it is being loaded. The person responsible for loading the cargo tank is also responsible for seeing that it is attended.

A cargo tank must be attended by a qualified person at all times when it is being unloaded, too. This is not necessary when the cargo tank is on the consignee's property. This responsibility also ends if the power unit has been separated from the cargo tank and removed. When the job of transporting the materials is complete, the carrier is no longer responsible for them. This includes having to attend the unloading.

The meaning of "attendance" here is not quite the same as in FMCSR Part 397. Remember what that was? Here, you must have a clear view of the vehicle. You must be qualified. This term, too, has a specific meaning. It means that you must know about the danger and know emergency procedures. You must be authorized and able to move the vehicle. The difference with hazmat is you must be within 25 feet of the vehicle to attend it.

Some hazardous materials may not be loaded together with other hazardous materials. The prohibited combinations are listed in the Segregation and Separation Chart. Figure 13-12 shows some examples.

Keep cargo tank manholes and valves closed while driving.

You may not use a cargo heater when you transport certain hazardous materials. When you're hauling explosives, you have to go farther. You must drain or remove the heater fuel tank. Then

you must disconnect the heater's power source. This is so that there is no chance the heater will be used at all, even by mistake.

Explosives. Don't load explosives with the engine running. Don't use metal tools when you handle explosives. Don't roll the packages. Don't throw them or drop them. Keep packages of explosives away from sparks, flames, even the heat of exhaust gases. If you tarp a load of explosives, use rope or wire tiedowns.

Don't load Division 1.1 or 1.2 explosives into combinations of more than two cargo carrying vehicles. You may also not load Division 1.1 or 1.2 explosives when one of your trailers is a full trailer with a wheel base shorter than 184 inches. Also, if one of your vehicles is a cargo tank with placards, you may not load Division 1.1 or 1.2 explosives. You can't transport Division 1.1 or 1.2 explosives if your other cargo is any of these:

- explosive substances, N.O.S., Division 1.1A (explosive) material (Initiating explosive),
- packages of Class 7 (radioactive) materials bearing "Yellow III" labels,
- Division 2.3 (poisonous gas) or Division 6.1 (poisonous) materials, or
- hazardous materials in a portable tank or a DOT specification 106A or 110A tank

Check the inside of the vehicle. Don't load explosives if there are sharp bolts, screws, nails, or the like sticking out. They could puncture the packages of explosives. The floors must be tight, and lined to keep dust, powder, or vapor from leaking. The lining must not be made of iron or steel.

Don't load detonators with Division 1.1, 1.2, 1.3, 1.4, or 1.5 explosives unless the detonators are specially packaged for this.

Protect the cargo. Load explosives so they are completely contained in the vehicle, or covered by a tarp. They should not stick out at any point. If your vehicle has a tailgate, you must be able to close it securely. Make sure any other cargo that makes up the load won't damage the explosives.

You may transfer explosives from one container or vehicle to another while on the public highway only in an emergency. If such an emergency occurs, you must remember to put out your warning devices.

Flammable liquids. Stop the engine. Load and unload flammable liquids with the engine off. You

Segregation Table for Hazardous Materials																				
Class or division		Notes	1.1 1.2	1.3	1.4	1.5	1.6	2.1	2.2	2.3 gas zone A	2.3 gas Zone B	3	4.1	4.2	4.3	5.1	5.2	6.1 liquids PG I zone A	7	8 liquids only
Explosives	1.1 and 1.2	A	*	*	*	*	*	X	X	X	X	X	X	X	X	X	X	X	X	X
Explosives	1.3		*	*	*	*	*	X		X	X	X		X	X	X	X	X		X
Explosives	1.4		*	*	*	*	*	O		O	O	O		O					O		O
Very insensitive explosives	1.5	A	*	*	*	*	*	X	X	X	X	X	X	X	X	X	X	X	X	X
Extremely insensitive explosives	1.6		*	*	*	*	*													
Flammable gases	2.1		X	X	O	X				X	O							O	O	
Non-toxic, non-flammable gases	2.2		X			X														
Poisonous gas Zone A	2.3		X	X	O	X		X				X	X	X	X	X	X			X
Poisonous gas Zone B	2.3		X	X	O	X		O				O	O	O	O	O	O			O
Flammable liquids	3		X	X	O	X				X	O					O		X		
Flammable solids	4.1		X			X				X	O							X		O
Spontaneously combustible materials	4.2		X	X	O	X				X	O							X		X
Dangerous when wet materials	4.3		X	X		X				X	O							X		O
Oxidizers	5.1	A	X	X		X				X	O	O						X		O
Organic peroxides	5.2		X	X		X				X	O							X		O
Poisonous liquids PG I Zone A	6.1		X	X	O	X		O				X	X	X	X	X	X			X
Radioactive materials	7		X			X		O												
Corrosive liquids	8		X	X	O	X				X	O		O	X	O	O		X		

Figure 13-12. Segregation and Separation Chart.

Instructions for using the Segregation Table for Hazardous Materials are as follows:

(1) The absence of any hazard class or division or a blank space in the table indicates that no restrictions apply.

(2) The letter "X" in the table indicates that these materials may not be loaded, transported, or stored together in the same transport vehicle or storage facility during the course of transportation.

(3) The letter "O" in the table indicates that these materials may not be loaded, transported, or stored together in the same transport vehicle or storage facility during the course of transportation unless separated in a manner that, in the event of leakage from packages under conditions normally incident to transportation, commingling of hazardous materials would not occur. Notwithstanding the methods of separation employed, Class 8 (corrosive) liquids may not be loaded above or adjacent to Class 4 (flammable) or Class 5 (oxidizing) materials; except that shippers may load truckload shipments of such materials together when it is known that the mixture of contents would not cause a fire or a dangerous evolution of heat or gas.

(4) The "*" in the table indicates that segregation among different Class 1 (explosive) materials is governed by the compatibility table.

(5) The note "A" in the Notes column of the table means that, notwithstanding the requirements of the letter "X," ammonium nitrate (UN 1942) and ammonium nitrate fertilizer may be loaded or stored with Division 1.1 (explosive) or Division 1.5 materials.

(6) When the §172.101 table or §172.402 of this subchapter requires a package to bear a subsidiary hazard label, segregation appropriate to the subsidiary hazard must be applied when that segregation is more restrictive than that required by the primary hazard. However, hazardous materials of the same class may be stowed together without regard to segregation required for any secondary hazard if the materials are not capable of reacting dangerously with each other and causing combustion or dangerous evolution of heat, evolution of flammable, poisonous, or asphyxiant gases, or the formation of corrosive or unstable materials.

DO NOT LOAD . . .	IN THE SAME VEHICLE WITH . . .
Division 6.1 or 2.3 (POISON or poisonous inhalation hazard) labeled material	animal or human food unless the poison package is overpacked in an approved way. Foodstuff is anything you swallow. However, mouthwash, toothpaste, and skin creams are not foodstuff.
Division 2.3 (POISONOUS) Zone A or 6.1 PG1 Zone A	oxidizers, flammables, corrosives, organic peroxides.
Charged storage batteries	Division 1.1 explosives.
Class 1 detonating primers	any other explosives unless in authorized containers or packaging.
Division 6.1 cyanides or cyanide mixtures	acids, corrosive materials, or other acidic materials that could generate hydrogen cyanide. Cyanides are materials with the letters CYAN as part of their shipping name. For example: Acetone Cyanohydrin, Silver Cyanide, Trichloroisocyanuric acid.
Nitric acid (Class B)	Other corrosive liquids in carboys, unless separated from them in an approved way.

Figure 13-12. (cont'd.)

may only have the engine running if you need it to run a pump.

Guard against static. You must have a ground if the flammable liquids are in separate metal containers.

TIP *Review Chapter 6 for an explanation of grounding. Otherwise, static electricity might create enough of a spark to ignite the flammable liquid.*

Close manholes and valves. Do not drive a cargo tank carrying flammable liquid with the manhole cover or valves open. Make sure all liquid discharge systems are free of leaks.

Flammable solids and oxidizers. Protect the cargo. As with explosives, flammable solids and oxidizers must be completely contained and secured in the vehicle, or covered by a tarp. If your vehicle has tailgates, you must be able to close and secure them. You should have no part of the cargo sticking out.

Keep the cargo dry. Flammable solids and oxidizers may be especially dangerous when wet.

Guard against heat. This type of cargo can ignite or explode if it gets hot. Make sure you have room around the packages for air to circulate. This helps keep them from overheating. Also, it reduces the chance of friction between the packages creating heat and starting a fire.

Load bagged ground, crushed, granulated, or pulverized charcoal (known as charcoal screening) so that the bags are laid horizontally and piled so that there will be spaces for effective air circulation. Those spaces shall not be less than 3.9 inches wide. Maintain air space between the rows. Don't pile the bags; and air spaces shall be maintained between rows of bags. Bags should be no closer than 5.9 inches from the top of any motor vehicle with a closed body.

Don't load nitrates in an all-metal vehicle other than a closed one of aluminum or aluminum alloy.

Corrosive liquids. Nitric acid must be loaded on the floor of the vehicle. Don't load packages of nitric acid on top of anything else.

You may load such containers of corrosive liquids higher than one tier, but only if the containers are in a protective crate or box. Don't load them so high that the stack is unstable. Make sure the floor is even so these containers won't tip over

and break. You might need a false floor to provide a level surface.

All storage batteries containing any electrolyte must be loaded so that all such batteries will be protected against any other lading falling onto or against them. Protect and insulate battery terminals against short circuits.

Compressed Gases. Start with a flat surface. Compressed gas cylinders must have a flat, level surface. Your vehicle might have racks designed to carry compressed gas cylinders. In that case, you don't have to be concerned about the floor.

Secure cylinders in an upright position. Strap them in, or block and brace them in place. The same rules apply to portable tanks.

There are special rules for cylinders of hydrogen and cryogenic liquids. You may only transport these in special racks. You may only have a certain amount in your vehicle at one time. You must never enter a tunnel when you have these cylinders on your vehicle. Stop before crossing a railroad if your vehicle carries any amount of chlorine.

Only private and contract carriers are allowed to transport these materials. Then, the cargo must travel directly from origin to destination.

Load and unload these materials with the engine off. You may only have it on if you need it to run a pump. Then, you must turn it off before you disconnect the filling or discharge lines. If your vehicle has shut-off valves in the discharge end of the delivery hose, you may leave the engine running.

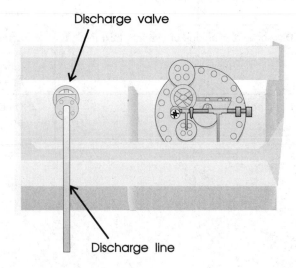

Figure 13-13. Close the cargo tank liquid discharge valve, except when loading and unloading.

You must have a gas mask when you're hauling chlorine in a cargo tank. Also, you must have an emergency kit to control leaks. Before you couple or uncouple your tank, you must detach any loading and unloading connections. Never leave the tank uncoupled from the tractor without chocking it first. Always close the liquid discharge valves on cargo tanks. The only time you should have them open is during loading and unloading. The unloading of these vehicles is done under close attendance by a qualified person.

Avoid delays. You may haul flammable cryogenic liquid in a package that holds more than 125 gallons (of water). When you do, you must be extra careful to avoid trip delays. Should you be delayed and notice a rise in pressure, get to a remote and safe location. Then vent the tank to relieve the excess pressure.

For each shipment of this type, you must make a written record of the cargo tank pressure and the outside temperature. Make this record at these times:

- at the start of each trip
- right before and after any manual venting
- at least once every five hours
- when you reach your destination

When you transport flammable cryogenic liquid in a cargo tank, the pressure must not exceed the marked rated hold time (MRHT). You must complete your trip from start to finish within the one-way travel time (OWTT). You'll find these figures marked on the tank. Your trip may take longer than the OWTT if you can stop and get the tank equilibrated. This is a pressure-balancing operation that may only be done where cryogenic liquids are loaded and unloaded. It's performed under tight controls.

Poisons. You must load arsenic only at places where the poisonous dust won't get into the air and hurt people. You must never do this on a public highway or in a public place.

You may not use a vehicle that had arsenic in it for anything else until it's been completely cleaned.

You may not haul a Division 2.3 (poisonous gas) or Division 6.1 (poisonous) material if there is any interconnection between packaging.

Never load materials labeled POISON or POISON INHALATION HAZARD with other cargo meant to be eaten by people or animals. The only exception is when the poison is overpacked in a dust-proof and liquid-tight package.

Radioactive material. The number of packages of radioactive materials you can have in your vehicle is controlled by a transport index number. You must not load more than 50.

Some radioactive packages are labeled "Yellow II" or "Yellow III." Load these according to the separation distances table shown below, Figure 13-14.

Neither you nor anyone else should be in a vehicle with radioactive materials unless they must.

Don't load fissile Class 7 materials with other fissile radioactive shipments. Load them at least 20 feet from other "radioactive" packages.

Separation and Segregation. As you saw earlier, you must not load certain combinations of hazardous materials. Refer to Figure 13-12 to see what some of these are.

Shipments in Transit, Accidents

Delays in delivery. When you're hauling hazardous materials, avoid delays in making delivery.

Total transport index	Minimum separation distances in feet to nearest undeveloped film for various times of transit					Minimum distance in feet to area of persons, or minimum distance in feet from dividing partition of cargo compartments
	Up to 2 hours	2-4 hours	4-8 hours	8-12 hours	Over 12 hours	
None	0	0	0	0	0	0
0.1 to 1.0	1	2	3	4	5	1
1.1 to 5.0	3	4	6	8	11	2
5.1 to 10.0	4	6	9	11	15	3
10.1 to 20.0	5	8	12	16	22	4
20.1 to 30.0	7	10	15	20	29	5
30.1 to 40.0	8	11	17	22	33	6
40.1 to 50.0	9	12	19	24	36	7

Note: The distance in the table must be measured from the nearest point on the packages of radioactive materials.

Figure 13-14. Separation distances table.

Disabled vehicle. In spite of your best efforts, your vehicle could break down en route. If this happens, make an extra special effort to protect your load of hazmat. Protect others from your load. Make every attempt to move the vehicle out of harm's way.

Broken packages. If you find broken or leaking packages en route, you may try to repair them if it will be safe to do that. You may send them on to the destination or back to the shipper in a salvage drum. If you do repair or repackage the hazmat, you must do such a good job that the hazmat is now safe to transport. The repaired package must not leak and spoil other cargo. Mark the repaired package clearly with the consignee's name.

If you can't repair the package, you must find a safe place to store the hazmat as quickly as possible. Don't dispose of it yourself.

Emergency stops. When hauling certain hazardous materials, you must not stop on the traveled part of the highway for any reason except for normal traffic stops. The hazardous materials in question are:

- flammable liquids and solids
- oxidizers
- corrosives
- compressed gases
- poisons

If you do stop, you must put out your warning devices.

If your vehicle is a cargo tank carrying flammable liquids or flammable compressed gases, you must do the same. This is true even if the vehicle is empty.

Put out your warning devices if you stop for any reason other than a normal traffic stop while hauling Division 1.1, 1.2, and 1.3 explosives.

Vehicle repair and maintenance. You may not use heat, flames, or devices that produce sparks to repair or maintain the cargo area or fuel system of your vehicle when you have a placarded load. No one is allowed to work on the vehicle inside a building unless the cargo area and fuel system are closed and not leaking. While the work is being done, there must be someone able to move the vehicle immediately if that becomes necessary. The vehicle must be removed from the building as soon as the work is done.

Special rules apply when your vehicle has Division 1.1, 1.2, or 1.3 explosives, or Class 3 (flammable liq-uids) or Division 2.1 (flammable gases). In that case, any source of flames, sparks, or glowing heat inside the building must be turned off.

Your cargo tank must be free of gas before it can be welded or repaired with a flame-producing device. This applies to cargo tanks that transport Class 3 or Division 6.1 hazmat.

TITLE 49 CFR PART 178

The regulations in Part 178 deal with shipping containers. They set specifications for manufacturing and testing containers used in hazardous materials transportation. In this part, a shipping container can be anything from one carboy to an entire cargo tank.

Most of the regulations in this part are of concern to manufacturers and shippers. We'll go into detail on those of particular interest to the driver. As always, though, you should be thoroughly familiar with all the regulations that relate to your driving job.

Part 178 requires manufacturers to mark authorized containers with a DOT specification. This includes not only portable and cargo tanks, but cylinders, drums, and so forth. These containers may also be marked with the UN symbols and packaging identification codes you've already read about.

The last section of Chapter 12 covered the design, construction, retesting, and marking of portable and cargo tanks in detail. We won't repeat that information here. If you plan to haul hazardous materials in these, you need a Tank Vehicle Endorsement. If you haven't already, read Chapter 12.

In this section, you'll find information about other hazmat containers that the driver needs to know.

Carboys, Jugs, and Drums

The DOT specification code for most carboys, jugs, and drums begins with the number 1, followed by a letter. These containers will be marked with the DOT specification code. You may find other markings, such as the manufacturer's name and the date of manufacture. You will find the capacity of the container marked on the outside as well. Markings may be printed or stamped on the container. You may also find them on a metal plate attached to the container.

These containers are tested when they are first made. The retesting requirements are different from the ones that apply to portable and cargo tanks. So you won't usually find a retest marking on them.

Inside Containers and Linings

The DOT specification code for inside containers and linings begins with the number 2, followed by a letter. Other markings include the manufacturer's name, date of manufacture, and capacity.

Cylinders

The DOT specification code for most cylinders is the number 3 or 4, plus one or two letters, and possibly some more numbers. Other markings include the service pressure and a serial number. You may also find the official mark of the official who inspected the cylinder, and the date of the test.

DOT specification requirements for compressed gas cylinders include pressure relief device systems. Most cylinders are retested for service pressure every five years. Some may be retested every 10 years instead. They're also inspected internally and externally. Then they're marked with the retest date and the retester's identification number. Figure 13-15 shows one way this marking may appear. You may also find this information near the serial number.

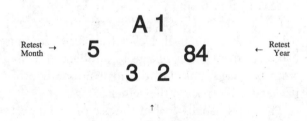

Retester's Identification Number
(A123 written clockwise from upper left)

Figure 13-15. Cylinder markings.

Portable Tanks

Portable tanks are DOT Specification 51, 56, 57, 60, IM 101, or IM 102.

Chapter 12 has more information on portable tank construction, retesting, and marking.

Cargo Tanks

Cargo tanks are DOT specification MC, followed by numbers.

Chapter 12 has more information on cargo tank construction, retesting, and marking.

General Packaging

General packaging is DOT Specification 7A. In addition to this code, you'll see the marking "USA DOT 7A Type A."

OTHER HAZARDOUS MATERIALS REGULATIONS

Title 49 CFR Part 179 describes the specifications for tanks that are to be mounted on or form part of a tank car and that transport hazardous materials in commerce. Some tank cars may have markings in addition to the ones you've already read about. The requirements for these markings are specified in this part. If you are pulling tank cars, you may wish to read this part to learn more about the construction, maintenance, repair, testing, and safety features of these vehicles.

Title 49 CFR Part 180 describes requirements pertaining to the maintenance, reconditioning, repair, inspection, and testing of cylinders, cargo tanks, and tank cars. These are requirements in addition to those contained in Parts 107, 171, 172, 173, and 178.

Title 49 CFR Part 392 deals with the driving of vehicles. Much of Part 392 was covered in Chapter 6. There is one section of this part that has to do with hauling hazardous materials, though. It tells you what to do at railroad grade crossings. These rules apply when your vehicle is placarded:

- Division 1.1
- Division 1.2 or 1.3
- Division 2.3 Poison gas
- Division 4.3
- Class 7
- Class 3 Flammable
- Division 5.1
- Division 2.2
- Division 2.3 Chlorine
- Division 6.1 Poison
- Division 2.2 Oxygen
- Division 2.1
- Class 3 Combustible liquid
- Division 4.1
- Division 5.1
- Division 5.2
- Class 8
- Division 1.4

This also applies any time you haul any amount of chlorine. Cargo tanks used for hazardous materials, whether full or empty, must also be brought to a stop before crossing railroad tracks.

You must not cross such a grade without stopping first. Stop within 50 feet of the track, but not any closer than 15 feet. Listen and look in both directions for an oncoming train. Make certain no train is approaching. When it is safe to proceed, you may cross the track. You must be in a gear that will let you get all the way across the track without changing gears. This is because you must not shift gears while crossing the track.

You need not stop at:

- a streetcar crossing or railroad tracks used exclusively for industrial switching purposes within a business district as defined in Part 390.5
- a railroad grade crossing when a police officer or crossing flagman directs traffic to proceed
- a railroad grade crossing controlled by a functioning highway traffic signal transmitting a green signal that, under local law, permits the CMV to proceed across the railroad tracks without slowing or stopping
- an abandoned railroad grade crossing that is marked with a sign indicating that the rail line is abandoned
- an industrial or spur line railroad grade crossing marked with an official sign reading "Exempt"

HAZARDOUS MATERIALS HANDLING

Many of the regulations you just read tell you how to handle hazardous materials safely. When you know and comply with these regulations, you can feel sure you are handling hazardous materials properly.

To review, here are some of the key issues in handling hazmat.

Forbidden Materials and Packages

Some materials are so hazardous, they must never be transported by common carrier by motor vehicle. Nitroglycerin is one example. If an item is "forbidden," it will be listed that way in column 3 (Hazard Class) of the Hazardous Materials Table.

Leaking or damaged packages of explosives are forbidden. So are loaded firearms. Don't transport these hazardous materials.

Loading and Unloading Materials

Do all you can to protect containers of hazardous materials. Don't use any tools that might damage containers or other packaging during loading. Don't use hooks.

Before loading or unloading, set the parking brake. Make sure the vehicle will not move.

Many products are more hazardous in the heat. Load hazardous materials away from heat sources.

Watch for signs of leaking or damaged containers. Do not transport leaking packages. Depending on the material, you, your truck, and others could be in danger.

Don't smoke! When loading hazardous materials, keep fire away. Don't let people smoke nearby. Never smoke around explosives, oxidizers, or flammables.

Secure packages against movement. Make sure containers don't move around in transit. Brace them so they will not fall, slide, or bounce around. Be very careful when loading containers that have valves or other fittings.

Don't open any package between the points of origin and destination. Never transfer hazardous products from one package to another. You may empty a cargo tank, but do not empty any other package while it is on the vehicle.

The rules usually forbid the use of cargo heaters, including automatic cargo heaters or air conditioner units. Unless you have read all the related rules, don't load the explosives, flammable liquids, and flammable gases in a cargo space that has a heater.

Use closed cargo space. You cannot have overhang or tailgate loads of explosives, flammable solids, and oxidizers. You must load these hazards into a closed cargo space unless all packages are:

- fire and water resistant, or
- covered with a fire and water resistant tarp.

Before loading or unloading any explosive, turn your engine off. Then check the cargo space. You must disable cargo heaters. Disconnect heater power sources and drain heater fuel tanks. There must be no sharp points that might damage cargo. Look for bolts, screws, nails, broken side panels, and broken floor boards.

Use a floor lining with Division 1.1, 1.2, or 1.3 explosives. The floors must be tight and the liner must not contain steel or iron.

Use extra care to protect explosives. Never use hooks or other metal tools. Never drop, throw, or roll the shipment. Protect explosive packages from other cargo that might cause damage.

Do not transfer a Division 1.1, 1.2, or 1.3 explosive from one vehicle to another on a public roadway except in an emergency. If safety requires an emergency transfer, set out red warning reflectors, flags, or electric lanterns. You must warn other highway users.

Never transport damaged packages of explosives. Do not take a package that shows any dampness or oily stain.

You might load breakable containers of corrosive liquids by hand. If so, load them one by one. Keep them right side up. Do not drop or roll the containers. Load them onto an even floor surface.

Load charged storage batteries so their liquid won't spill. Keep them right side up. Make sure other cargo won't fall against or short-circuit them.

If your vehicle doesn't have racks to hold cylinders, the cargo space floor must be flat. If not in a rack, the cylinders must be held upright or braced lying down flat, or in boxes that will keep them from turning over.

Cargo Segregation

Some types of hazardous materials should never be put with other types in the same load.

Do not transport Division 1.1 or 1.2 explosives in triples. Do not transport this class of explosives in vehicle combinations if:

- there is a placarded cargo tank in the combination, or
- the other vehicle in the combination contains initiating explosives, radioactive materials labeled "Yellow III," Division 2.3 (poison gas) or Division 6.1 (poison) or hazardous materials in a portable tank, DOT 106A or 110A tank.

Do not load nitric acid above any other product. Never load corrosive liquids next to or above:

- Class 4 (flammable)
- Class 5 (oxidizing)

Never transport Poison Division 6.1 or irritating materials in containers with interconnections. Never load a package labeled "Poison," "Poison Gas," or "Irritant" in the driver's cab or sleeper. Never load these materials with food meant for humans or animals.

Some packages of radioactive materials bear a number called the transport index. This is the amount of radiation that surrounds each package and passes through packaging nearby. The shipper prints the package's transport index on the label. The total transport index of all packages in a single vehicle must not exceed 50. There are also rules as to how long and how close you can load radioactive products to people, animals, or film.

The regulations name other materials you must keep apart. Refer to the Segregation and Separation Chart (see Figure 13-12) when in doubt about what's allowed.

Attendance and Safe Havens

You can leave your vehicle unattended in a safe haven. A safe haven is a government-approved place for parking unattended vehicles loaded with explosives.

Otherwise, someone must always be "attending" the vehicle. The person watching a placarded vehicle must:

- be in the vehicle, awake and not in the sleeper berth, or within 100 feet of the vehicle and within clear view of it,
- be aware of the hazards,
- know what to do in emergencies, and
- be able to move the vehicle if needed.

Parking

Never park with Division 1.1, 1.2, or 1.3 explosives within five feet of the traveled part of the road. Unless your work requires it, do not park within 300 feet of:

- a bridge, tunnel, or building,
- a place where people gather, or
- an open fire.

If you must park to do your job, do so only briefly.

Don't park on private property unless the owner is aware of the danger. Someone must always watch the parked vehicle. You may let someone else watch it for you only if your vehicle is

- on the shipper's property, or
- on the carrier's property, or
- on the consignee's property.

You may park a placarded vehicle (not carrying explosives) within five feet of the traveled part of the road only if your work requires it. Do so only briefly. Someone must always watch the vehicle when parked on a public roadway or shoulder. Do not park within 300 feet of an open fire.

Routes

Some states and counties require permits to transport hazardous material or waste. They may limit the routes you can use. Local rules about routes and permits change often. As the driver, it's your job to find out if you need permits or must use special routes. Make sure you have all needed papers before starting.

If you work for a carrier, ask your dispatcher about route limits or permits. If you are an independent and are planning a new route, check with state agencies where you plan to travel. Some localities prohibit transportation of hazardous materials through tunnels, over bridges, or other roadways. Check before you start.

When your vehicle is placarded, avoid heavily populated areas, crowds, tunnels, narrow streets, and alleys. Take other routes, even if inconvenient, unless there is no other way. Never drive a placarded vehicle near open fires unless you can safely pass without stopping.

If transporting Division 1.1, 1.2, or 1.3 explosives, you must have a written route plan and follow that plan. Carriers prepare the route plan in advance and give the driver a copy. You may plan the route yourself if you pick up the explosives at a location other than your employer's termi-

nal. Write out the plan in advance. Keep a copy of it with you while transporting the explosives. Deliver shipments of explosives only to authorized persons or leave them in locked rooms designed for explosives storage.

A carrier must choose the safest route to transport placarded radioactive material. After choosing the route, the carrier must tell the driver about the radioactive materials and show the route to be taken.

EMERGENCY EQUIPMENT

Protecting the Public

You know that any time you stop on the highway (other than a normal traffic stop), you must warn others by using your flashers and reflective devices. This is true no matter what type of load you have. When your load is hazardous materials, you must take special steps to warn other highway users and prevent accidents from happening.

Emergency stops. When hauling certain hazardous materials, you must not stop on the traveled part of the highway for any reason except for normal traffic stops. The hazardous materials in question are:

- flammable liquids and solids
- oxidizers
- corrosives
- compressed gases
- poisons

If you do stop, you must put out your warning devices.

If your vehicle is a cargo tank carrying flammable liquids or flammable compressed gases, you must do the same. This is true even if the vehicle is empty.

Put out your warning devices if you stop for any reason other than a normal traffic stop while hauling Division 1.1, 1.2, or 1.3 explosives.

You can only transfer explosives and flammable liquids from one container or vehicle to another in an emergency. Put out your warning devices and protect the scene if you must make this transfer.

Fighting Hazardous Material Fires

You might have to control minor truck fires on the road. The power unit of placarded vehicles must have a fire extinguisher rated 10 B:C or more. However, unless you have the training and equipment to do so safely, don't fight hazardous material fires. Dealing with hazardous material fires requires special training and protective gear.

When you discover a fire, send someone for help. You may use the fire extinguisher to keep minor truck fires from spreading to cargo before firefighters arrive. Feel trailer doors to see if they are hot before opening them. If so, you may have a cargo fire.

Don't open the doors. That could let air in and make the fire flare up. Without air, many fires only smolder. That does less damage than a fire in full flame.

If your cargo is already on fire, it is not safe to fight the fire. Keep the shipping papers with you to give to emergency personnel as soon as they arrive. Warn other people of the danger and keep them away.

Emergency Equipment for Loading and Unloading

Vehicles that use a pump to load and unload may have a manual by-pass. Manual by-passes control spills caused by excess pressure. Excess pressure can in turn cause hoses to rupture, disconnect, or separate from fittings.

Cargo tanks also have manual emergency valve shut-off controls. This is discussed in greater detail in Chapter 12.

Emergency Equipment for Tank Vehicles

When you transport chlorine in cargo tanks you must have an approved gas mask in the vehicle. You must also have an emergency kit for controlling leaks in dome cover plate fittings on the cargo tank.

EMERGENCY RESPONSE PROCEDURES

As you know, shippers must list an emergency response telephone number on the shipping paper. This number can be used by emergency responders to obtain information about any hazardous materials involved in a spill or fire.

Shippers also must provide emergency response information to the motor carrier for each hazardous material being shipped. The emergency response information must be able to be used away from the motor vehicle and must provide information on how to safely handle incidents involving the material. It must include information on the shipping name of the hazardous materials, risks to health, fire, explosion, and initial methods of handling spills, fires, and leaks of the materials.

Such information can be on the shipping paper or some other document that includes the basic description and technical name of the hazardous material. Or, it may be in a guide book such as the *Emergency Response Guidebook* (ERG). Motor carriers may keep an ERG on each vehicle carrying hazardous materials.

You, the driver, must provide the emergency response information to any federal, state, or local authority responding to a hazardous materials incident or investigating one.

There are some general rules to follow whenever you're involved in a hazardous materials accident. However, different hazardous cargos call for different responses.

Specific Emergency Procedures

Many states have specific emergency response guidelines. You should become familiar with what your state requires.

Explosives. If your vehicle breaks down or is in an accident while carrying explosives, warn others of the danger. Keep bystanders away. Do not allow smoking or open fire near the vehicle.

Remove all explosives before pulling apart vehicles involved in a collision. Place the explosives at least 200 feet from the vehicles and occupied buildings. If there is a fire, warn everyone of the danger of explosion. Stay a safe distance away.

Flammable liquids. If you are transporting a flammable liquid and have an accident or your vehicle breaks down, prevent bystanders from gathering. Warn people of the danger. Keep them from smoking.

Never transport a leaking cargo tank farther than needed to reach a safe place. If safe to do so, get off the roadway. Don't transfer flammable liquid from one vehicle to another on a public roadway except in an emergency.

Flammable solids and oxidizers. If a flammable solid or oxidizing material spills, warn others of the fire hazard. Do not open smoldering packages of flammable solids. Remove them from the vehicle if you can safely do so. Gather and remove any broken packages if safe to do so. Also remove unbroken packages if it will decrease the fire hazard.

Corrosives. If corrosives spill or leak in transit, be careful to avoid further damage or injury when handling the containers.

If further transportation of a leaking tank would be unsafe, get off the road. If safe to do so, try to contain any liquid leaking from the vehicle. Keep spectators away from the liquid and its fumes. Do everything possible to prevent injury to other highway users.

Compressed gases. If compressed gas is leaking from your vehicle, warn others of the danger. Only permit those involved in removing the hazard or wreckage to get close. You must notify the shipper of the compressed gas of any accident.

Unless you are fueling machinery used in road construction or maintenance, do not transfer a flammable compressed gas from one tank to another on any public roadway.

Poisons. You must protect yourself, other people, and property from harm. Remember that many products classed as poison are also flammable. If you think a leaking poison liquid or gas might be flammable, take the added precautions needed for flammable liquids or gases. Do not allow smoking, open flame, or welding. Warn others of the hazards of fire, inhaling vapors, or coming into contact with the poison.

A vehicle involved in a leak of Division 2.3 or Division 6.1 must be checked for stray poison before being used again.

If a Division 6.2 (infectious substances) package is damaged in handling or transportation, you should immediately contact your supervisor. Packages that appear to be damaged or show signs of leakage should not be accepted.

Radioactive materials. If a leak or broken package involves radioactive material, tell your dispatcher or supervisor as soon as possible. If there is a spill, or if an internal container might be damaged, do not touch or inhale the material.

Fire and Flames

Fire and flames can be a real hazard when you are hauling hazardous materials. Even the smallest spark can start a fire or cause an explosion. Never drive a placarded vehicle near open fires unless you can safely pass without stopping.

You might break down and have to use stopped vehicle signals. Use reflective triangles or red electric lights. Never use burning signals, such as flares or fusees, around a tank used for flammable liquid or flammable gas. This is true whether the vehicle is loaded or empty. It's also true for vehicles loaded with Division 1.1, 1.2, or 1.3 explosives, or flammable liquids or gases.

Don't smoke. When you are loading hazardous materials, keep fire away. Don't let people smoke nearby. Never smoke around explosives, oxidizers, or flammables.

General Emergency Procedures

As a professional driver, your job at the scene of an accident is to keep people away from the area. Limit the spread of material, if you can do so safely. Communicate the danger to emergency response personnel.

Follow these steps:

1. Check to see that your driving partner is OK
2. Keep shipping papers with you
3. Keep people far away and upwind
4. Warn others of the danger
5. Send for help
6. Follow your employer's instructions

Cargo Leaks

If you discover a cargo leak, identify the material by using shipping papers, labels, or package location. Don't touch any leaking material. Under the stress of handling an accident or leak, you may forget your cargo is hazardous. You could injure yourself this way.

Do not try to identify material or find the source of a leak by smell. Many toxic gases destroy one's sense of smell. They can injure or kill you even if they don't smell. Do not eat, drink, or smoke around a leak or spill.

If hazardous material is spilling from your vehicle, do not move it any more than safety requires.

You may move off the road and away from places where people gather if doing so results in greater safety. Only move your vehicle if you can do so without danger to yourself or others.

Never continue driving with hazardous material leaking from your vehicle to find a phone booth, truck stop, help, or similar reason. Remember that the carrier pays for the cleanup of contaminated parking lots, roadways, and drainage ditches. The costs are huge. So don't leave a lengthy trail of contamination. If hazardous material is spilling from your vehicle, park it. Secure the area and stay there.

Send someone else for help. When sending someone for help, give that person:

- a description of the emergency
- your exact location and direction of travel
- your name, the carrier's name, and the name of the community or city where your terminal is located
- the shipping name, hazard class, and ID number of the material, if you know them

This is a lot for someone to remember. It is a good idea to write it all down for the person you send for help. The emergency response team must know these things to find you and to handle the emergency. They may have to travel miles to get to you. This information will help them to bring the right equipment the first time, without having to go back for it. Never move your vehicle if doing so will cause contamination or damage the vehicle. Keep downwind and away from roadside rests, truck stops, cafes, and businesses.

Never try to repack leaking containers. Although it is sometimes legal to do so, it must be done right. Unless you have been told otherwise, don't try it. Call your dispatcher or supervisor for instructions. If you need to, call emergency personnel.

National Response Center

The National Response Center helps coordinate emergency response to chemical hazards. They are a resource to the local police and firefighters. The person in charge of a vehicle involved in an accident may have to phone the National Response Center. This call will be in addition to any made to police or firefighters.

You or your employer must phone when any of the following occur as a direct result of a hazardous materials incident:

- A person is killed.
- A person receives injuries requiring hospitalization.
- Estimated carrier or other property damage exceeds $50,000.
- The general public is evacuated for one or more hours.
- One or more major transportation arteries or facilities are closed or shut down for one hour or more.
- Fire, breakage, spillage, or suspected radioactive contamination occurs.
- Fire, breakage, spillage, or suspected contamination occurs involving shipment of etiologic agents (bacteria or toxins).
- A situation exists of such a nature (e.g., continuing danger to life exists at the scene of an incident) that, in the judgment of the carrier, should be reported.

 The phone number of the National Response Center is (800) 424-8802. It's available 24 hours a day, and is toll-free.

Those making the immediate telephone report should be ready to give:

- their names
- the name and address of the carrier they work for
- the phone number where they can be reached
- the date, time, and location of the incident
- the extent of injuries, if any
- the classification, name, and quantity of hazardous materials involved, if available
- the type of incident and nature of hazardous material involvement and whether a continuing danger to life exists at the scene

If a reportable quantity of hazardous substance was involved, the caller should give the name of the shipper. Also, the caller should give the quantity of the hazardous substance discharged.

Be prepared to give your employer the required information. Carriers must make detailed written reports within 30 days.

Chemtrec

The Chemical Transportation Emergency Center (CHEMTREC) in Washington also has a 24-hour toll-free line. CHEMTREC was created to provide emergency personnel with technical information about the physical properties of hazardous products. The National Response Center and CHEMTREC are in close communication. If you call either one, they will tell the other about the problem when appropriate.

TIP *CHEMTREC's phone number is (800) 424-9300.*

Special Rules for Transporting Explosives

Some locations require permits to transport Division 1.1, 1.2, and 1.3 explosives. States and counties may also require drivers to follow special routes.

The federal government may require permits for special hazardous materials cargo, like rocket fuel. Find out about permits and special routes for places you drive.

Even the most commonplace cargo can be dangerous if transported in a defective vehicle or by an unskilled driver. Cargo that isn't properly secured can spill out on the roadway. Or, the driver can allow an accident to happen, spilling the cargo. Property nearby can be damaged, and others on the roadway can be injured.

When the cargo itself is hazardous, the smallest mistake can lead to disaster. So if you haul "hazmat," you must be very well informed and very skilled.

PASS POST-TRIP

Instructions: For each true/false test item, read the statement. Decide whether the statement is true or false. If it is true, circle the letter A. If it is false, circle the letter B. For each multiple-choice test item, choose the answer choice—A, B, C, or D—that correctly completes the statement or answers the question. There is only one correct answer. Use the answer sheet provided on page 325.

1. The _____ must make sure a hazardous product is being transported in the proper packages.
   A. shipper
   B. carrier
   C. driver
   D. consignee

2. A written route plan is something you must apply for before making a trip with explosives.
   A. True
   B. False

3. The "Hazardous Materials Table" and the "List of Hazardous Substances and Reportable Quantities" are _____.
   A. different names for the same thing
   B. available only to police and fire departments
   C. published in a different form by each state
   D. the two main lists used by shippers, carriers, and drivers

4. Hazardous materials markings can be a label, tag, or sign.
   A. True
   B. False

5. Your load contains more than 1,000 pounds of mixed hazard classes from Placard Table 2. You have not loaded more than 5,000 pounds of any one class at any one place. You _____.
   A. must have a separate placard for each hazard class in your load
   B. must have a separate placard for each hazard class in your load that totals 1,000 pounds
   C. must have a separate placard for each hazard class in your load that totals more than 5,000 pounds no matter where you got it
   D. can use the "DANGEROUS" placard all by itself

6. Shipping papers for a hazardous materials shipment must_____.
   A. have a shipper's certification
   B. be kept in a door pocket or on the seat when the driver is not at the controls
   C. be marked differently from other cargo documents
   D. all of the above

7. When you haul radioactive materials, you must have your written plan with you.
   A. True
   B. False

8. Trailers used to haul Division 1.1, 1.2, or 1.3 explosives must _____.
   A. have padded walls and floors
   B. use a floor liner that does not contain iron or steel
   C. have a date of manufacture after January 1, 1983
   D. be equipped with an air ride suspension

9. The radioactive "transport index" refers to radiation that _____.
   A. is in the packages
   B. surrounds each package and passes through nearby packages
   C. can reach to the outside of the packages
   D. is harmless

10. In addition to a special license and training, you may need a special permit from the federal, state, or local government to haul hazardous materials.
    A. True
    B. False

CHAPTER 13 ANSWER SHEET

Test Item	Answer Choices				Self - Scoring
					Correct Incorrect
1.	A	B	C	D	☐ ☐
2.	A	B	C	D	☐ ☐
3.	A	B	C	D	☐ ☐
4.	A	B	C	D	☐ ☐
5.	A	B	C	D	☐ ☐
6.	A	B	C	D	☐ ☐
7.	A	B	C	D	☐ ☐
8.	A	B	C	D	☐ ☐
9.	A	B	C	D	☐ ☐
10.	A	B	C	D	☐ ☐

PART THREE

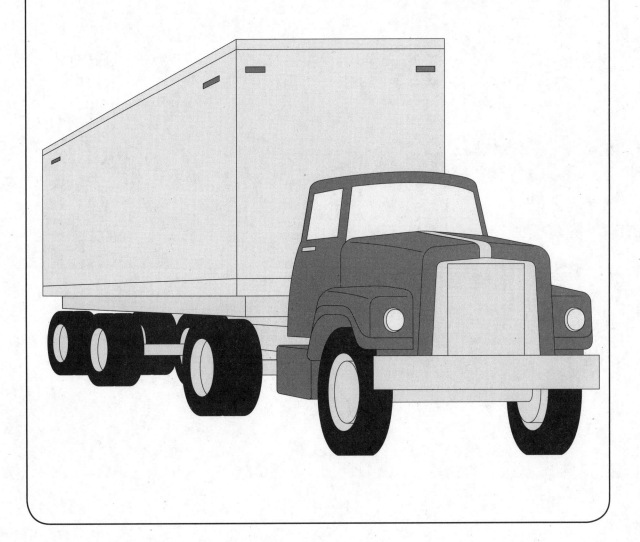

Chapter 14

Review

In this chapter, you will learn about:

- how to review what you've read
- general knowledge needed to get a CDL with no endorsements and the air brake restriction
- air brakes
- vehicle combinations
- tank vehicles
- hazardous materials

To complete this chapter, you will need:

- a pencil or pen
- a book of notes

You're almost at the end of this book. Just three more chapters to go! They'll help you put the finishing touches on the job of preparing to take the CDL tests.

You've read through a lot of pages and covered a lot of ground. If you plan to haul hazardous materials in a cargo tank with air brakes, you've probably read every chapter! What follows are questions on both the general and the endorsement knowledge areas. Answering the questions will help you measure how much you remember.

In this chapter, you'll also find tips on how to review what you've read if you find that necessary.

You can use this chapter in the way you think will work best for you. Take the tests first. Then use your test results to guide your review. Or, you can review your reading first, then take the tests.

HOW TO REVIEW

Quick—what was Chapter 4 about? What was the subject of Chapter 10?

You may recall Chapter 10 quite well, especially if you finished it recently. On the other hand, you may hardly be able to remember the subject of Chapter 4. If that's the case, it's not really that

surprising. It may be some time since you read Chapter 4. You've worked to absorb a lot of information since then. Before you take your CDL Knowledge Test, though, the information from Chapter 4 should be as fresh in your mind as that from Chapter 10. You may need to review.

You could read each chapter again from the start. You probably don't have time for this, though. Anyway, it's not necessary. Why review what you know? All you need to review is what you don't know as well.

How do you identify what you need to review? Try the systematic approach. Just as having a system helped you get through the material the first time, a system will help you do a quick and thorough review. Follow these steps:

- retake the PASS Pre-trip
- reread the chapter objectives
- scan the chapter
- read your notes
- retake the PASS Post-trip
- retake the diagnostic test
- take the Sample CDL tests

Retake the PASS Pre-trip

Take another look at the three questions that began each chapter in Part Two. If you know the material well, you should be able to answer each question correctly. If you can't, you should review the information tested by any question you missed.

Reread the Chapter Objectives

The chapter objectives list the major areas of information in the subject covered by the chapter. How confident do you feel in your knowledge of that area? Use your feelings to guide you to the part of the chapter you need to review.

Scan the Chapter

Mark for review any chapter that covers information you're still in doubt about. Look over the chapter main headings and subheadings. Pause to take stock of how familiar the words seem. Can't quite remember what a topic was about? Then plan to go over that part of the chapter in detail.

Pay special attention to any areas you marked with your highlighter. Those were areas that were new or hard for you at first. Make sure you feel confident about them now.

Take the time to reread thoroughly the parts of chapters that answer any test questions you may have missed.

If you have quite a lot of time left before you take your CDL Knowledge Test, scan Chapters 4 through 10 all the way through one more time. Start with the chapter that's the least fresh in your mind.

Also review completely any chapters you need to prepare for an endorsement.

Read Your Notes

As you read, you made notes of words that were new or ideas you wanted to remember. You wrote yourself a brief summary of each chapter. Use these notes to refresh your memory.

Retake the PASS Post-trip

Look at your answer sheets from the PASS Post-trip questions at the end of each chapter in Part Two. If you missed any the first time, review the part of the chapter covered by those questions. Then take the Post-trip again. You should be able to answer all of them easily and correctly. Try for a perfect score on the PASS Post-trips this time.

Retake the Diagnostic Test

Chapter 1 had a diagnostic test. Take the test again. Compare your score with the first time you took the test. Note which questions you missed, if any. They may be the same, or they may be different. Either way, they point to any area of information you could be stronger in before you take the CDL Knowledge Test.

Take the Sample CDL Tests

You can use the two review knowledge tests in this chapter before or after you review the other chapters. The results will pinpoint any areas of information you're still weak in.

Take the review tests under "actual" test conditions. Turn off the radio. Don't smoke, eat, or drink while taking the test. Put your books and dictionary away. You won't be able to use them when you take the actual CDL Knowledge Test.

Follow the instructions given.

For answers, please refer to pages 398–400.

SAMPLE TEST 1

Test Item	Answer Choices				Self - Scoring Correct	Incorrect
1.	A	B	C	D	☐	☐
2.	A	B	C	D	☐	☐
3.	A	B	C	D	☐	☐
4.	A	B	C	D	☐	☐
5.	A	B	C	D	☐	☐
6.	A	B	C	D	☐	☐
7.	A	B	C	D	☐	☐
8.	A	B	C	D	☐	☐
9.	A	B	C	D	☐	☐
10.	A	B	C	D	☐	☐
11.	A	B	C	D	☐	☐
12.	A	B	C	D	☐	☐
13.	A	B	C	D	☐	☐
14.	A	B	C	D	☐	☐
15.	A	B	C	D	☐	☐
16.	A	B	C	D	☐	☐
17.	A	B	C	D	☐	☐
18.	A	B	C	D	☐	☐
19.	A	B	C	D	☐	☐
20.	A	B	C	D	☐	☐

Cut here ✂

Test Item	Answer Choices				Self - Scoring Correct	Incorrect
21.	A	B	C	D	☐	☐
22.	A	B	C	D	☐	☐
23.	A	B	C	D	☐	☐
24.	A	B	C	D	☐	☐
25.	A	B	C	D	☐	☐
26.	A	B	C	D	☐	☐
27.	A	B	C	D	☐	☐
28.	A	B	C	D	☐	☐
29.	A	B	C	D	☐	☐
30.	A	B	C	D	☐	☐
31.	A	B	C	D	☐	☐
32.	A	B	C	D	☐	☐
33.	A	B	C	D	☐	☐
34.	A	B	C	D	☐	☐
35.	A	B	C	D	☐	☐
36.	A	B	C	D	☐	☐
37.	A	B	C	D	☐	☐
38.	A	B	C	D	☐	☐
39.	A	B	C	D	☐	☐
40.	A	B	C	D	☐	☐

Cut here

Test Item	Answer Choices				Self - Scoring Correct Incorrect	
41.	A	B	C	D	☐	☐
42.	A	B	C	D	☐	☐
43.	A	B	C	D	☐	☐
44.	A	B	C	D	☐	☐
45.	A	B	C	D	☐	☐
46.	A	B	C	D	☐	☐
47.	A	B	C	D	☐	☐
48.	A	B	C	D	☐	☐
49.	A	B	C	D	☐	☐
50.	A	B	C	D	☐	☐
51.	A	B	C	D	☐	☐
52.	A	B	C	D	☐	☐
53.	A	B	C	D	☐	☐
54.	A	B	C	D	☐	☐

Cut here

SAMPLE TEST 1

1. One purpose of Commercial Driver Licensing is to disqualify drivers who operate commercial motor vehicles in an unsafe manner.
   A. True
   B. False

2. All CDL skill tests must be taken in _____.
   A. a representative vehicle
   B. a company vehicle
   C. a vehicle with air brakes
   D. a rented vehicle

3. To meet the driver qualification requirements for a CDL, you must _____.
   A. have at least a year of CMV driving experience
   B. be at least 18 years old for intrastate driving or at least 21 years old for interstate driving
   C. take prescription medication
   D. have perfect eyesight

4. If you are convicted of a traffic violation in a state other than your home state, you must make notification _____.
   A. to your employer
   B. within 30 days of the date of conviction
   C. in writing
   D. all of the above

5. A driver who is disqualified _____.
   A. must always receive a jail sentence
   B. has forfeited bond
   C. may not drive a CMV
   D. may not drive a motor vehicle

6. When operating a CMV, which of the following first offenses will result in the loss of your CDL for one year?
   A. driving with an expired or suspended CDL
   B. excessive speeding, reckless driving, or any traffic offense that causes a fatality
   C. DUI, DWI, and driving while on drugs
   D. driving with a load of hazardous materials that are not properly secured

7. Once you get your CDL for a specific vehicle group, you may never change it to another group.
   A. True
   B. False

8. The only sure cure for fatigue is _____.
   A. narcotics
   B. amphetamines
   C. real sleep
   D. caffeine

9. Medical certificates must be renewed every _____.
   A. year
   B. six months
   C. three years
   D. two years or when specified by a physician

10. Prescription drugs are allowed if a doctor says the drugs will not affect safe driving ability.
    A. True
    B. False

11. You are required to stop at a railroad grade crossing when_____.
    A. the crossing is located in a city or town with frequent train traffic
    B. when there are no warning signals at the crossing
    C. the cargo being carried makes a stop mandatory under state or federal regulations
    D. when a flagman tells you that it's safe to proceed

12. The law requires CMVs to have _____.
    A. service brakes
    B. emergency brakes
    C. parking brakes
    D. all of the above

13. You may not use tires that _____ on your vehicle.
    A. are belted
    B. are flat or have leaks
    C. have a tread depth of more than 2/32 of an inch
    D. are treaded

14. To drive your vehicle in a way that would cause it to break down or cause an accident is not only unsafe, it's illegal.
    A. True
    B. False

15. Which of the following is a hazardous materials class?
    A. "Requires Placards"
    B. "Other Regulated Materials"
    C. "Dangerous"
    D. none of the above

16. The color of the turn signals facing forward must be _____.
    A. amber only
    B. white
    C. white or amber
    D. whatever the manufacturer selected

17. When the engine starts, the ammeter jumps to the charge side and flutters. This is a sign _____.
    A. that the ammeter is faulty
    B. that the generator needs repair
    C. of normal operation
    D. that the engine is warmed up

18. When you first start your vehicle, the warning lights should come on.
    A. True
    B. False

19. Your vehicle's battery box must have _____.
    A. at least two batteries
    B. enough fluid to work properly
    C. at least one loose ground wire
    D. a secure cover

20. Which of the following creates the braking force in a braking system?
    A. air pressure
    B. hydraulic pressure
    C. vacuum pressure
    D. friction

21. Rust around wheel nuts often means that _____.
    A. the nuts are loose
    B. the nuts are broken
    C. the nuts are about to break
    D. the vehicle needs painting

22. The engine speed that you use for shifting should be based on _____.
    A. information in your owner's manual
    B. the vehicle's GVWR
    C. the vehicle's mpg
    D. what feels right for the situation

23. When roads are wet, icy, or covered with snow, you should _____.
    A. use your retarder for braking as much as possible
    B. turn the retarder off
    C. use the maximum setting if the retarder has power adjustment settings
    D. be ready to turn the retarder on the minute you start to skid

24. Lane changes, turns, merges, and tight maneuvers require _____.
    A. lots of luck to perform without accidents
    B. frequent checking and rechecking, using your mirrors
    C. maintaining a constant speed
    D. power steering

25. Total stopping distance is _____ plus reaction distance plus braking distance.
    A. visual distance
    B. perception distance
    C. vehicle length
    D. trip distance

26. Dirty headlights give about _____ as much light as clean ones.
    A. 100 percent
    B. 90 percent
    C. 75 percent
    D. 50 percent

27. To prepare for winter driving, check that your antifreeze will do the job _____.
    A. by inspecting the color
    B. by checking the level in the reservoir
    C. using a special coolant tester
    D. by draining the system and replacing the antifreeze

28. In hot weather, tire inflation pressures can be higher than normal. You should _____.
    A. inspect the tires less often
    B. let air out of your tires when the pressure exceeds 105 psi
    C. not let air out of the tires when they are hot
    D. spin the tires to cool them off

29. If you try to downshift while coming down a mountain, you might _____.
    A. damage the clutch
    B. damage the engine
    C. get stuck in neutral
    D. lose traction

30. A vehicle marked at the rear with a red triangle having an orange center _____.
    A. is hauling hazardous materials
    B. is moving slowly
    C. is a farm vehicle
    D. makes frequent stops

31. When steering to avoid a crash _____.
    A. keep one hand on the steering wheel and the other hand free for shifting
    B. turn and apply the brake at the same time for more control
    C. use both hands on the steering wheel and don't brake while turning
    D. none of the above

32. The most common skid is _____.
    A. rear wheels losing traction due to excessive braking or acceleration
    B. steering tires sliding due to front-brake lock-up
    C. trailer wheels sliding out on curves
    D. caused by a front tire blow-out

33. The first step in correcting a drive-wheel braking skid is to let off the brake.
    A. True
    B. False

34. Poorly distributed cargo can _____.
    A. cause the vehicle to pull to one side
    B. make the vehicle feel like it's going to tip over
    C. make the vehicle harder to put into motion and stop
    D. all of the above

35. Bridge laws _____.
    A. are the same as the maximum legal axle weight
    B. control traffic on a bridge
    C. apply only to drawbridges
    D. can lower the maximum axle weight limit

36. Loading a device beyond its rating is _____.
    A. illegal and unsafe
    B. impossible
    C. the smart way to increase a payload
    D. a good way to test your equipment

37. If your vehicle is overloaded to the rear, it is _____.
    A. very stable
    B. likely to tip over
    C. likely to go into a jackknife
    D. likely to go into a front-wheel skid

38. The legal size and weight distribution limits _____.
    A. should be your guide under all conditions
    B. are not changed by adverse conditions
    C. may not ensure safe operations in bad weather
    D. are often changed by states from season to season

39. Blocking used to prevent cargo movement _____.
    A. is secured to the cargo compartment floor
    B. is secured to the cargo itself
    C. is secured to the cargo and the walls
    D. must be placed every 2 1/2 feet

40. You must inspect cargo within _____ miles after starting a trip.
    A. 10
    B. 25
    C. 50
    D. 100

41. The lighter your load, _____.
    A. the less traction your vehicle has
    B. the longer your stopping distance
    C. the longer your stopping time
    D. all of the above

42. You inspect your vehicle because _____.
    A. it's important for safety
    B. it's required by federal regulations
    C. it's required by state law
    D. all of the above

43. You should do a post-trip inspection _____.
    A. at the end of each run for each vehicle you operate
    B. on the last vehicle you operate each day
    C. on whichever vehicle the dispatcher selects
    D. on the same vehicle every day

44. Drive tires must have at least _____ of tread, and no less.
    A. 1/4 inch
    B. 2/32 inch
    C. 1/2 inch
    D. 4/32 inch

45. A serious hazard of exhaust system leaks is _____.
    A. excess smoke can enter the cab
    B. exhaust smoke can damage the cargo
    C. poisonous fumes can enter the cab
    D. excess smoke can damage the vehicle

46. An air suspension _____.
    A. should not fill evenly all the way around
    B. should fill before the air brake system reaches normal pressure
    C. should allow the vehicle to tilt with the load
    D. should not leak more than three psi in five minutes at normal pressure

47. Mud flaps _____.
    A. should touch the ground
    B. should have advertising on them
    C. must be made of high-grade rubber
    D. may be required by state law

48. During the pre-trip inspection, approach the vehicle and _____.
    A. note the vehicle's general condition
    B. check for body damage and major leaks
    C. see if the vehicle leans to one side
    D. all of the above

49. Normal clutch travel is_____.
    A. less than one or two inches
    B. about one or two inches
    C. more than two inches
    D. not something the driver can measure

50. During the walk-around part of the pre-trip inspection, you should _____.
    A. leave the vehicle key in the switch so you won't lose it
    B. pull the key out of the switch and put it on the dash so you'll know where it is
    C. leave the vehicle key in the switch so it will be handy for the engine test
    D. put the key in your pocket so no one can move the vehicle while you are under it

And Now a Few Extras

Instructions: Match the letter for the vehicle system condition in Column B with the gauge reading in Column A.

Column A Gauge Reading	Column B System Condition
51. 180 to 250 degrees Fahrenheit	A. normal cooling system temperature
52. 165 to 185 degrees Fahrenheit	B. normal transmission oil temperature
53. 30 to 75 psi	C. normal operating oil pressure
54. 5 to 15 psi	D. normal idle oil pressure

SAMPLE TEST 2

Test Item	Answer Choices				Self - Scoring Correct Incorrect	
1.	A	B	C	D	☐	☐
2.	A	B	C	D	☐	☐
3.	A	B	C	D	☐	☐
4.	A	B	C	D	☐	☐
5.	A	B	C	D	☐	☐
6.	A	B	C	D	☐	☐
7.	A	B	C	D	☐	☐
8.	A	B	C	D	☐	☐
9.	A	B	C	D	☐	☐
10.	A	B	C	D	☐	☐
11.	A	B	C	D	☐	☐
12.	A	B	C	D	☐	☐
13.	A	B	C	D	☐	☐
14.	A	B	C	D	☐	☐
15.	A	B	C	D	☐	☐
16.	A	B	C	D	☐	☐
17.	A	B	C	D	☐	☐
18.	A	B	C	D	☐	☐
19.	A	B	C	D	☐	☐
20.	A	B	C	D	☐	☐

Test Item	Answer Choices				Self - Scoring Correct Incorrect
21.	A	B	C	D	☐ ☐
22.	A	B	C	D	☐ ☐
23.	A	B	C	D	☐ ☐
24.	A	B	C	D	☐ ☐
25.	A	B	C	D	☐ ☐
26.	A	B	C	D	☐ ☐
27.	A	B	C	D	☐ ☐
28.	A	B	C	D	☐ ☐
29.	A	B	C	D	☐ ☐
30.	A	B	C	D	☐ ☐
31.	A	B	C	D	☐ ☐
32.	A	B	C	D	☐ ☐
33.	A	B	C	D	☐ ☐
34.	A	B	C	D	☐ ☐
35.	A	B	C	D	☐ ☐
36.	A	B	C	D	☐ ☐
37.	A	B	C	D	☐ ☐
38.	A	B	C	D	☐ ☐
39.	A	B	C	D	☐ ☐
40.	A	B	C	D	☐ ☐

Cut here

Cut here

Test Item		Answer	Choices		Self - Scoring	
					Correct	Incorrect
41.	A	B	C	D	☐	☐
42.	A	B	C	D	☐	☐
43.	A	B	C	D	☐	☐
44.	A	B	C	D	☐	☐
45.	A	B	C	D	☐	☐
46.	A	B	C	D	☐	☐
47.	A	B	C	D	☐	☐
48.	A	B	C	D	☐	☐
49.	A	B	C	D	☐	☐
50.	A	B	C	D	☐	☐
51.	A	B	C	D	☐	☐
52.	A	B	C	D	☐	☐
53.	A	B	C	D	☐	☐
54.	A	B	C	D	☐	☐

SAMPLE TEST 2

1. All drivers who need a CDL must take the _____.
   A. air brakes test
   B. combination vehicles test
   C. chauffeur's license test
   D. general knowledge test

2. A driver who meets the requirements of the CDL laws does not have to meet the qualifications set in FMCSR Part 391.
   A. True
   B. False

3. At least once a year, a driver must fill out a form listing all motor vehicle violations (other than parking) that occurred during the previous 12 months _____.
   A. even if there were no violations
   B. only if there was a conviction
   C. only if the bond was forfeited
   D. only if the carrier requires it

4. Your application for a CMV driving job must list the names and addresses of all employers who ever hired you.
   A. True
   B. False

5. Which of the following is a serious CMV driving violation?
   A. driving a representative vehicle
   B. reckless driving
   C. countersteering
   D. driving with a load of hazardous materials

6. Which of the following first offenses in a CMV will result in the loss of your CDL for life?
   A. leaving the scene of an accident involving a CMV you were driving
   B. using a CMV to commit a felony involving controlled substances
   C. driving any vehicle under the influence of alcohol or an illegal drug
   D. driving a CMV under the influence of alcohol or an illegal drug

7. You are pulled over for weaving between lanes. If you refuse to take a test for alcohol, you may be disqualified just as if you were DUI. This is because of the _____.
   A. higher standard of care law
   B. reasonable cause law
   C. implied consent law
   D. none of the above

8. If your employer gives you a road test, and you pass it, you don't have to take the CDL Performance Test.
   A. True
   B. False

9. Which of these statements about night driving is true?
   A. Most people are more alert at night than during the day because they drink more coffee.
   B. Most heavy vehicle accidents occur between midnight and 6 a.m.
   C. Most hazards are easier to see at night than during the day due to artificial lighting.
   D. If you become sleepy, it's best to push on and finish the long trip.

10. Which of these statements about alcohol is true?
    A. Some people aren't affected by drinking.
    B. A few beers have the same effects on driving ability as a few shots of whiskey.
    C. Coffee and fresh air can sober up a person.
    D. Exercise can help you burn off alcohol faster.

11. When making an emergency stop on the highway, you must put on your four-way flashers and _____.
    A. keep them on until you are underway again
    B. keep them on until you get your warning devices out
    C. hang a red flag from the rear of your vehicle
    D. blow your horn

12. Locked-up brakes are more often the result of panic stopping than faulty equipment.
    A. True
    B. False

13. Fatigued drivers should _____.
    A. drink coffee to combat sleepiness
    B. take a rest break until they are fully alert
    C. take medication
    D. do exercises to get energized

14. If your vehicle doesn't pass a roadside inspection _____.
    A. it can be declared "out of service"
    B. you cannot drive the vehicle, even to a repair station
    C. you may repair it on the spot, then resume your trip
    D. all of the above

15. Federal law requires _____.
    A. two white headlights, one at the left front and one at the right front of the tractor
    B. fog lamps for bad weather or dark roads
    C. stop lamps on projecting loads
    D. one red backup lamp

16. Trucks must carry a five-pound B:C fire extinguisher, unless they are placarded for hazardous materials.
    A. True
    B. False

17. Normal oil pressure while idling is _____.
    A. 165 to 185 degrees Fahrenheit
    B. 20 to 24 psi
    C. 30 to 75 psi
    D. 5 to 15 psi

18. Overcharging in a vehicle battery _____.
    A. means there isn't enough power to crank the engine
    B. leads to thickened battery fluid
    C. shortens battery life
    D. all of the above

19. Once a circuit breaker has opened, it must be replaced.
    A. True
    B. False

20. Tire inflation pressure should be checked _____.
    A. after the tire has warmed up
    B. when the tire is cold
    C. after the first 25 miles of a run
    D. every two hours or 100 miles of travel

21. Progressive shifting means _____.
    A. downshifting at lower rpm as you reach the lower gears
    B. downshifting at higher rpm as you reach the lower gears
    C. upshifting at lower rpm as you reach the higher gears
    D. upshifting at higher rpm as you reach the higher gears

22. You should downshift _____.
    A. when the tachometer or speedometer indicates it's necessary
    B. before starting down a steep hill
    C. before entering a sharp curve
    D. all of the above

23. Maintain good visual alertness by _____.
    A. always looking as far ahead as you can see
    B. fixing your eyes on the road right in front of you
    C. keeping a close watch on the road behind you at all times
    D. shifting your attention between the mirrors and the road ahead

24. Signaling other drivers when it is safe to pass is _____.
    A. a great aid to highway safety
    B. a courtesy professional drivers should extend to others
    C. an unsafe practice
    D. required

25. Traveling in the lane alongside other vehicles
    _____.
    A. is a good idea since they can always see you that way
    B. can cause you to be trapped when you need to change lanes
    C. will make traffic flow more smoothly
    D. will keep people from cutting in front of you

26. If another driver approaches you at night with high beams on, _____.
    A. get them to turn them off by flashing yours
    B. watching the approaching lights closely will show you where the vehicle is
    C. look slightly right in the right lane or at the right edge marking if there is one
    D. keep your high beams on until the other vehicle passes.

27. Ice on your radiator shutters or winterfront _____.
    A. will melt off after the engine warms up
    B. can cause the engine to overheat
    C. will help insulate the engine compartment
    D. is a minor winter inconvenience

28. In hot weather, bleeding tar can _____.
    A. damage your tires
    B. damage the road
    C. give your vehicle more traction
    D. make parts of the road very slippery

29. When going down a long grade, the gear selected by drivers of the latest, fuel efficient trucks is affected by _____.
    A. their streamlined shapes that reduce air drag
    B. their low friction parts
    C. their more powerful engines that allow you to climb hills in higher gears
    D. all of the above

30. Impaired drivers often _____.
    A. drive too fast, too slow, or change speed for no reason
    B. drive especially well because they are trying so hard
    C. are safer because they tend to drive slowly
    D. none of the above

31. You have to leave the road to avoid another vehicle. What should you not do?
    A. keep one set of wheels on the pavement
    B. stay on the shoulder until you can come to a stop
    C. turn widely to be sure of avoiding the other vehicle
    D. stay off the brake

32. The best way to stop a front-wheel skid is to _____.
    A. let the vehicle slow down
    B. brake hard
    C. turn sharply
    D. shift into neutral

33. Cargo that isn't properly secured can shift. This can _____.
    A. result in damaged cargo
    B. overturn the whole vehicle
    C. cause the cargo to spill out, resulting in a major highway hazard
    D. all of the above

34. State laws can regulate _____.
    A. the weight of your vehicle, but not the cargo
    B. the weight of your cargo, but not your vehicle
    C. both your vehicle and cargo weight
    D. neither vehicle nor cargo weight. Only the federal government can regulate these

35. Tire load is _____.
    A. the most weight a tire can carry safely
    B. the same as inflation pressure
    C. a measure of whether a tire is underinflated
    D. how much weight a tire puts on the road

36. An empty vehicle is _____.
    A. very stable
    B. likely to tip over
    C. likely to go into a jackknife
    D. likely to go into a rear-wheel skid

37. A vehicle with a high center of gravity is _____.
    A. very stable
    B. likely to tip over
    C. likely to go into a jackknife
    D. likely to go into a front-wheel skid

38. When securing cargo, you should use _____.
    A. no fewer than three tiedowns, even for the smallest load
    B. one tiedown for every ten feet of cargo
    C. as few tiedowns as possible, since they only increase your overall weight
    D. none of the above

39. You may need to tarp a load both to protect the cargo and _____.
    A. yourself
    B. others on the road
    C. the vehicle
    D. the shipper

40. If your load is sealed, _____.
    A. you are not responsible for its weight
    B. break the seal, inspect the load, and reseal it every three hours or 150 miles
    C. break the seal and inspect the load when you are within 25 miles of your destination
    D. you are still responsible for exceeding gross weight or axle limits

41. Dry bulk tanks _____.
    A. are designed to control the shifting of the load
    B. are stable in curves and turns
    C. tend to have a high center of gravity
    D. present the fewest driving problems of all vehicles

42. If a state official conducts a roadside inspection and finds your vehicle to be unsafe, _____.
    A. the official will offer to repair it for you
    B. your vehicle will be put "out of service"
    C. your vehicle will be taken away from you
    D. your license to drive will be taken away from you

43. When you stop during a trip, you should check _____.
    A. the engine oil and coolant
    B. the post-trip inspection report made by the vehicle's last driver
    C. your cargo securement devices
    D. your disc brakes

44. Steering axle tires must have at least _____ of tread, and no less.
    A. 1/4 inch
    B. 2/32 inch
    C. 1/2 inch
    D. 4/32 inch

45. In many states, your vehicle will be placed "out of service" if there are _____ broken or missing leaves in your leaf spring suspension.
    A. more than one tenth
    B. more than one fifth
    C. more than one fourth
    D. any

46. A bent _____ is a steering system defect.
    A. torque rod
    B. tie rod
    C. spring hanger
    D. shock absorber

47. Snow chains _____.
    A. are required by federal regulations
    B. may be required by states during the winter
    C. are standard equipment on heavy vehicles
    D. are so sturdy they never break

48. When you start your vehicle's engine, the _____ should come up to normal with a few seconds.
    A. oil pressure
    B. fuel pressure
    C. coolant temperature
    D. all of the above

49. During the pre-trip inspection, check the brake (stop) lights by _____.
A. applying the trailer hand brake if your vehicle has one, or getting someone to apply the treadle valve
B. bleeding the air out of the system so the tractor protection valve will act
C. fanning off air pressure so that the brakes will apply
D. pulling the red knob

50. Regardless of the vehicle you have, the seven-step vehicle inspection is all you need to do.
A. True
B. False

And Now a Few Extras

Instructions: Match the letter for the description in Column B with the number of the term in Column A that fits the description.

Column A Weight Term	Column B Definition
51. GVW	A. a rating given to the vehicle by its manufacturer, which is the maximum weight of a combination vehicle plus cargo
52. GCW	B. a rating given to the vehicle by its manufacturer. It is the maximum weight for a single vehicle plus its load
53. GVWR	C. the total weight of a single vehicle, plus its load
54. GCWR	D. the total weight of the power unit or tractor, plus any trailers, plus the cargo

ENDORSEMENT TEST REVIEWS

The following sections will give you a brush-up on the questions specifically geared toward the endorsement tests we discussed earlier. The four categories covered are: Air Brakes, Combination Vehicles, Tank Vehicles, and Hazardous Materials. Use the Answer Sheet starting on page 361 to record your answers. The correct answers can be found starting on page 399.

Air Brakes Questions

The following questions should serve as a good review for you as you prepare for the Air Brakes endorsement test. Answer sheets are provided starting on page 361. For answers, refer to page 399.

1. Fanning air brakes _____.
   A. cools them off
   B. increases the air pressure
   C. increases the braking power
   D. decreases the air pressure

2. When some brakes in the system are doing more work than others, _____.
   A. those brakes will develop more heat
   B. vehicle handling will be affected
   C. those brakes will lose stopping power first
   D. all of the above

3. If your air compressor stops working, _____.
   A. you will lose your brakes immediately
   B. the safety relief valve will close
   C. you'll have some air pressure left stored in the air tanks
   D. your brakes won't be affected at all

4. An air brake system safety relief valve opens at about _____.
   A. 20 to 45 psi
   B. 60 psi
   C. 120 psi
   D. 150 psi

5. Your air brake safety relief valve has opened several times. This means _____.
   A. the system is working normally
   B. your air pressure is probably low
   C. the air compressor has failed
   D. you should have the system repaired

6. If your vehicle has a front brake limiting valve control, it should be in the _____.
   A. "slippery" position on wet roads
   B. "normal" position only on dry roads
   C. "normal" position under all road conditions
   D. "slippery" position under all road conditions

7. The service brake system _____.
   A. is an inspection done by the company maintenance staff
   B. applies and releases the brakes when you use the brake pedal during normal driving
   C. is the pump and check valves that keep the air tank pressure serviced
   D. is only on vehicles with hydraulic brakes

8. Spring brakes are _____.
   A. applied by air pressure
   B. held in the released position by air pressure
   C. not affected by air pressure
   D. have nothing to do with the emergency brake system

9. Most large vehicles with air brakes have spring brakes _____.
   A. that are parking brakes
   B. that are emergency brakes
   C. that are part of the parking and emergency brake systems
   D. that will stop the vehicle in a shorter distance than the service brakes

10. You should avoid setting the parking brakes _____.
    A. on a steep grade
    B. if the air pressure is low
    C. if the brakes are hot
    D. in hot weather if the brakes are wet

11. Spring brakes should come on automatically when you pump air pressure down to _____ psi.
    A. zero
    B. 20 to 45
    C. 60
    D. 90

12. If your brakes are out of adjustment, you won't have good _____ braking power.
    A. service
    B. emergency
    C. parking
    D. all of the above

13. On newer vehicles with air brakes, the parking brake control is _____.
    A. a lever
    B. a square, red, push-pull control knob
    C. a diamond-shaped, yellow, push-pull control knob
    D. a round, blue, push-pull control knob

14. If a low air pressure warning device comes on, _____
    A. pull off the road as soon as it's safe to do so
    B. you can safely continue until you get to a service station
    C. you may continue, but remember to make note of it later on your post-trip inspection report
    D. turn it off so it won't distract you while driving

15. The tractor protection valve _____.
    A. works automatically or can be operated manually
    B. protects the tractor air tanks in case of air pressure loss
    C. separates the tractor air supply from the trailer air supply
    D. all of the above

16. The tractor protection valve _____.
    A. cannot be operated manually
    B. is the same thing as the trailer brake control
    C. is also called the breakaway valve
    D. all of the above

17. The trailer hand brake _____.
    A. lets you control the amount of air to the trailer brakes
    B. may be used as a parking brake
    C. exhausts the air supply from the trailer brakes
    D. exhausts the air supply from the tractor brakes

18. Which of the following statements about glad hands is true?
    A. The service brake glad hands are color-coded red.
    B. The service brake glad hands are color-coded blue.
    C. The spring brake glad hands are color-coded blue.
    D. The spring brake glad hands are color-coded yellow.

19. If the service air line breaks, _____ (unless you have a dual air system).
    A. the air pressure will drop slowly
    B. you should apply the spring brakes
    C. the spring brakes will come on
    D. the check valve will prevent the dry tank from losing its air supply

20. If a spring brake line ruptures, _____ (unless you have a dual air system).
    A. the air pressure will drop slowly
    B. you should apply the spring brakes
    C. the spring brakes will come on
    D. the check valve will prevent the dry tank from losing its air supply

21. If the line from the compressor to the main supply tank ruptures, _____.
    A. the air pressure will drop slowly
    B. you should apply the spring brakes
    C. the spring brakes will come on when the service pressure drops below 45 psi
    D. the check valve will prevent the dry tank from losing its air supply

22. Brake drums or discs _____.
    A. must not have cracks
    B. must have linings that fit loosely on the shoes
    C. must not have cracks longer than half the width of the friction area
    D. should be well-greased between the drum or disc and the lining

23. Slack adjusters _____.
    A. never need adjustment themselves
    B. should only be adjusted by someone with an S endorsement
    C. should travel at least two inches from where the push rod is attached
    D. can be adjusted manually or automatically

24. Air tanks should be drained_____.
    A. at the end of each day or shift
    B. to keep sludge from clogging brake system valves
    C. manually when automatic devices fail
    D. all of the above

25. To determine if slack adjusters need adjustment, _____.
    A. apply the service brakes and release the parking brakes
    B. apply both the service and parking brakes
    C. pull out the automatic slack adjustment control knob
    D. release both the service and parking brakes and pull hard on the slack adjuster. It should not move more than one inch

26. To test the brakes on a single air system, run the engine at a fast idle to charge the air system. Your gauge should show you _____.
    A. pressure building from 50 to 90 psi within three minutes
    B. pressure building from 85 to 100 psi within 45 seconds
    C. the compressor cutting out below 50 psi or above 70 psi
    D. the governor stopping below 100 psi or above 125 psi

27. To test the brakes on a dual air system, run the engine at a fast idle to charge the air system. Your gauge should show you _____.
    A. pressure building from 50 to 90 psi within three minutes
    B. pressure building from 85 to 100 psi within 45 seconds
    C. the compressor cutting out below 50 psi or above 70 psi
    D. the governor stopping below 100 psi or above 125 psi

28. Your vehicle's single brake system needs adjustment if _____.
    A. pressure builds from 50 to 90 psi within three minutes
    B. pressure builds from 85 to 100 psi within 45 seconds
    C. the compressor cuts out between 70 psi and 90 psi
    D. the governor stops between 100 and 125 psi

29. In a single vehicle with a fully charged air system, air pressure loss (after an initial drop) should be less than _____ per minute.
    A. 2 psi
    B. 3 psi
    C. 4 psi
    D. 5 psi

30. When checking a single vehicle for air leaks with the brakes on, air pressure loss (after an initial drop) should be less than _____ per minute.
    A. 2 psi
    B. 3 psi
    C. 4 psi
    D. 5 psi

Combination Vehicles/Doubles/Triples Questions

The following questions should serve as a good review for you as you prepare for this endorsement test. Answer sheets are provided starting on page 363. For answers, see page 399.

1. Before you begin coupling, _____.
   A. pre-position the fifth wheel
   B. check your path for hazards
   C. check the condition of the tractor frame
   D. put the spring brakes on

2. You can damage the landing gear by _____.
   A. aligning the tractor and trailer
   B. backing under the trailer at an angle
   C. positioning the trailer kingpin in the center of the fifth wheel V-slot
   D. backing under the trailer in a straight line

3. The trailer should be lowered so _____.
   A. the tractor strikes the nose of the trailer
   B. the trailer is well above the fifth wheel
   C. you can use the landing gear to make major adjustments
   D. it is raised slightly when the tractor is backed under it

4. Which of the following coupling steps comes first?
   A. apply the trailer brakes
   B. raise the landing gear
   C. put the transmission in reverse
   D. connect the air supply lines

5. When rolling up the landing gear to complete the coupling process, _____.
   A. begin in high gear to make the job go faster
   B. use low gear until the tractor is supporting the trailer
   C. leave the landing gear half-way down to speed the uncoupling process later
   D. use plenty of grease to make the job easier

6. If one of the glad hands on your blue air line is missing a seal, _____.
   A. it's a minor problem and you can proceed with the coupling
   B. you might end up with a service line air leak
   C. you might end up with an emergency line air leak
   D. you can take one off the red line and use that as a replacement

7. When uncoupling, if you park the trailer at an angle, _____.
   A. you'll save space in the yard
   B. you won't have to walk so far to get around the vehicle
   C. your job will be easier
   D. you could damage the landing gear

8. When uncoupling, it's a good practice to _____.
   A. couple the glad hands together
   B. couple the glad hands to dummy couplers
   C. hang the electrical cable with the plug end down
   D. all of the above

9. When you inspect the coupling, check to make sure that _____.
   A. there's a little slack in the air lines
   B. the landing gear handle is hanging free
   C. there's no slack in the air lines
   D. the electrical cable is unplugged

10. Inspect a converter dolly as if it were _____.
    A. cargo
    B. a trailer
    C. an axle with a fifth wheel
    D. an engine

11. A converter dolly may have its own air tank.
    A. True
    B. False

12. If you crossed your air lines on a new trailer with spring brakes, _____.
    A. you won't be able to release the trailer brakes
    B. you won't be able to release the tractor brakes
    C. you won't have any brakes at all
    D. you will have no way of knowing there's a problem

13. If you crossed the air lines on an old trailer, _____.
    A. you won't be able to release the trailer spring brakes
    B. you won't be able to release the tractor brakes
    C. you won't have any brakes at all
    D. you won't have any trailer brakes

14. To test the tractor protection valve, reduce the air pressure to _____. The valve should close automatically.
    A. zero
    B. 20 to 45 psi
    C. 60 psi
    D. 100 psi

15. To test the trailer emergency brake system, charge the trailer brakes. Then _____.
    A. press the treadle valve
    B. push in the blue, round knob
    C. pull the diamond-shaped, yellow knob
    D. pull the red, eight-sided knob

16. To test the trailer service brakes, _____.
    A. press the treadle valve
    B. use the trailer brake hand valve
    C. push in the blue round knob
    D. pull the red, eight-sided knob

17. To test for crossed air lines, _____.
    A. pull the red, eight-sided knob, then push it in and listen for brake movement air release
    B. pull the yellow knob out and push it in while listening for air flow
    C. turn the engine off, apply and release the brake pedal, and look for an air pressure drop
    D. turn the engine off, apply and release the trailer brakes with the hand valve, and listen for brake movement and air release

18. When you turn a corner, your trailer wheels follow a different path than your tractor wheels. This is _____.
    A. an alignment problem
    B. a steering problem
    C. a sign of a poorly coupled tractor-trailer
    D. normal

19. Trucks with double or triple trailers _____.
    A. have a dangerous "crack-the-whip" effect
    B. present a rearward amplification hazard
    C. can turn over in response to abrupt steering movements
    D. all of the above

20. If your vehicle starts to skid and you think it may jackknife, _____.
    A. put the tractor brakes on hard
    B. release the tractor brakes
    C. put on the trailer brakes
    D. turn the steering wheel sharply to the left, then back to the right

21. When coupling doubles, if the second trailer does not have spring brakes, _____.
    A. couple the second trailer to the first trailer before putting the tractor on
    B. drive the tractor close to the second trailer, connect the emergency line, charge the trailer air tank, and disconnect the emergency line
    C. you need a second tractor that can be left near the rear trailer
    D. put the rear trailer against a dock

22. When coupling doubles, start by _____.
    A. coupling the tractor to the second trailer
    B. backing the converter dolly under the second trailer
    C. coupling the converter tongue to the pintle hook on the first trailer
    D. putting the second trailer against a dock

23. On trailers used as doubles and triples you'll find shut-off valves (cut-off cocks) in the _____.
    A. service air lines
    B. emergency air lines
    C. service and emergency air lines
    D. parking brake system only

24. Should you unhook the pintle hook with the converter dolly still under a trailer?
    A. Yes. This is the standard procedure.
    B. Yes. This keeps the trailer from falling.
    C. No. This could damage the pintle hook.
    D. No. You could be injured if you do this.

25. Vehicles with double and triple trailers _____.
    A. have more parts that should be inspected
    B. need about the same amount of space and traffic gaps as other CMVs
    C. respond to steering corrections just like straight trucks
    D. are as stable as any other CMV

26. The dolly support _____.
    A. supports the dolly while it is being backed under a trailer
    B. supports the dolly air lines
    C. supports the spare tire carrier
    D. supports the tongue when the dolly isn't connected to anything

27. There is no way to check if the entire brake system in a triple is charged.
    A. True
    B. False

28. To make a right turn with a long vehicle combination, _____.
    A. swing to the left as you begin the turn
    B. use all the lanes you need before and during the turn
    C. guard against other vehicles trying to pass you on the left
    D. if you must use another lane to avoid running over the curb, swing wide as you complete the turn

29. Which has the most rearward amplification?
    A. a five-axle tractor-semitrailer
    B. turnpike doubles
    C. B-train doubles
    D. triples

30. You can prevent rollovers by keeping your load _____.
    A. low and centered between the trailer sides
    B. low and toward the right or curb side
    C. over the tractor drive wheels
    D. low and toward the left or road side

Tank Vehicles Questions

The following questions should serve as a good review for you as you prepare for the Tank Vehicles endorsement test. Answer sheets are provided starting on page 364. For answers, refer to page 400.

1. The best way to prevent rolling over when pulling a tanker through a curve is _____.
    A. travel at speeds below the posted limits
    B. travel faster than the posted speed limit
    C. get a running start going into the curve, then brake through the curve
    D. go into the curve at a slow speed, then speed up halfway through the curve

2. You will feel the effects of liquid surge less if you _____.
    A. turn slowly and carefully
    B. brake to a stop gradually
    C. haul thick liquids rather than thin ones
    D. all of the above

3. A smooth bore tanker is one that _____.
    A. is a long, hollow tube with no sharp corners
    B. is divided into compartments by bulkheads
    C. has baffles to control the surge factor
    D. can carry several different products at the same time

4. Dividers inside tankers that have openings at the top and bottom are called _____.
    A. bulkheads
    B. baffles
    C. barriers
    D. none of the above

5. Baffles in a tanker
    A. make this type of tanker easiest to clean
    B. control forward and back liquid surge
    C. eliminate side-to-side liquid surge
    D. allow you to haul several different products at the same time

6. Which is the most stable liquid tanker shape?
    A. van
    B. double conical
    C. cylindrical
    D. elliptical

7. The cladding in a tanker _____.
    A. cuts down on liquid surge
    B. cools down or heats up the cargo
    C. protects the tanker from corrosive cargo
    D. protects cargo from temperature extremes

8. What makes a bulk liquid tanker hard to handle?
    A. the cargo's high density
    B. the cargo's low center of gravity
    C. the cargo's high center of gravity
    D. friction

9. The amount of liquid to load into a tank depends on _____.
    A. the amount the liquid will expand in transit
    B. the weight of the liquid
    C. legal weight limits
    D. all of the above

10. Outage is _____.
    A. allowing for liquids to expand in transit
    B. the same as innage
    C. another name for a liquid level sensing device
    D. a dangerous practice when hauling liquid loads

11. When you have a liquid load, fanning the brakes will _____.
    A. cause the brakes to heat up and possibly fail
    B. increase the surge factor
    C. cause the vehicle to rock
    D. all of the above

12. Stop valves in a cargo tank _____.
    A. break off in an emergency
    B. close in an emergency
    C. can only be operated manually
    D. protect the tank and piping from rear-end collisions

13. A tanker that is "authorized" _____.
    A. has a DOT specification number
    B. may carry hazardous materials
    C. must be retested according to a specific schedule
    D. all of the above

14. Retest markings _____.
    A. must be stamped on the tank itself
    B. give the day, month, and year of the last retest
    C. show what type of test or inspection was performed
    D. are made by the driver

15. A retest marking reading "10-09, P, V, L" would tell you _____.
    A. that in October 2009, the cargo tank received and passed a pressure retest, external visual inspection and test, and a lining inspection
    B. that this is a DOT Specification 1009 tank carrying poisonous viscous liquid cargo
    C. that this tank passed a test on October 2009 and that it can carry various liquid loads
    D. nothing. This is not an official retest marking.

Hazardous Materials Questions

The following questions should serve as a good review for you as you prepare for the HazMat endorsement test. Answer sheets are provided staring on page 365. For answers, refer to page 400.

1. Requirements for the right kind of hazardous materials packages and the right way to load, transport, and unload bulk tanks are examples of _____.
   A. marking rules
   B. containment rules
   C. rules for communicating risk
   D. all of the above

2. Requirements to use the right shipping papers, package labels, and placards are examples of hazardous materials_____.
   A. marking rules
   B. containment rules
   C. rules for communicating risk
   D. all of the above

3. To assure that only safe drivers transport hazardous materials, drivers who transport hazmat must _____.
   A. have a CDL
   B. know the hazardous materials rules and pass a written test for the Hazardous Materials Endorsement
   C. complete special training for hauling certain materials
   D. all of the above

4. You are picking up a hazardous materials shipment and several packages are leaking. You should _____.
   A. wrap them in plastic
   B. be sure to wash out your truck when you finish unloading this shipment
   C. refuse this shipment
   D. note that the packages are leaking on your pre-trip inspection report

5. The person attending a vehicle placarded for hazardous materials _____.
   A. can be anyone you, as the driver, select
   B. must be the driver and no one else
   C. must have a perfect driving record
   D. must be qualified

6. A vehicle placarded for Division 1.1, 1.2, or 1.3 explosives _____.
   A. may never be left unattended under any condition
   B. may be left unattended in a safe haven
   C. may not be parked on the consignee's property
   D. may never be parked on private property

7. Which of these rules apply when hauling hazardous materials?
   A. Smoking is prohibited within 25 feet of a vehicle with hazardous materials.
   B. You may not park a vehicle with hazardous materials within 300 miles of a fire.
   C. You may only fuel a vehicle carrying hazardous materials at special fuel islands.
   D. When driving a vehicle with hazardous materials past a fire, you must stop and check the area before proceeding.

8. If you are hauling hazardous materials and have _____ tires on your vehicle, you must stop and check them every time you park.
   A. dual
   B. radial
   C. bias
   D. treaded

9. Hazardous waste may not be hauled in a vehicle that isn't marked with _____.
   A. the name or trade name of the private carrier operating the vehicle
   B. the motor carrier identification number
   C. the name of the person operating the vehicle, if that's not the carrier
   D. all of the above

10. If an X or an RQ is in the HM column of a shipping paper entry, the _____.
    A. material listed on that line is the largest part of the shipment
    B. entry refers to the materials that must be top loaded
    C. material listed on that line should be discarded
    D. shipment is regulated by hazardous materials regulations

11. Which of the following statements about hazardous materials is true?
    A. Contact with any hazardous material will kill you.
    B. Hazardous materials pose a risk to health, safety, and property during transportation.
    C. You must have a Hazardous Materials Endorsement on your CDL to drive a vehicle hauling any amount of hazardous material.
    D. A vehicle with any hazardous material on it must have placards.

12. On the Hazardous Materials Table, column 1 shows _____.
    A. how to ship an item
    B. to what shipping mode the information on that line applies
    C. how much of the material can be put in a vehicle without placarding
    D. what materials can be loaded together

13. On the Hazardous Materials Table, proper shipping names are shown _____.
    A. in plain type and alphabetical order
    B. in italic type and loading order
    C. so you know you must not transport that shipment
    D. so you know you can transport that shipment without placards

14. Which of these statements about shipping papers is true?
    A. Shipping papers may include hazardous and non-hazardous items.
    B. Hazardous items must be listed before non-hazardous items on the shipping paper.
    C. Reportable quantities will be identified differently from hazardous materials.
    D. all of the above

15. Hazardous waste shipments are identified by _____.
    A. the letters "HW" on the vehicle placards
    B. the word "waste" before the name of the material on the shipping papers
    C. the shipping name
    D. the name of the carrier

16. The Uniform Hazardous Waste Manifest must have _____.
    A. the quantity listed in pounds only
    B. an attached copy of the loading plan
    C. the signatures of everyone who released, transported, and received the shipment, with the dates of the transfer actions
    D. a record of the route taken

17. Tankers used to transport hazardous materials on vehicles must be marked with _____.
    A. the UN ID number of the product
    B. the date of manufacture
    C. the owner's PUCO or ICC number
    D. the CHEMTREC phone number

18. Which of the following statements about "Radioactive" labels is true?
    A. Yellow III indicates the highest level of radioactivity.
    B. Yellow II indicates the highest level of radioactivity.
    C. White I indicates the highest level of radioactivity.
    D. Black III indicates the highest level of radioactivity.

19. You are scheduled to transport 1,000 pounds each of non-flammable gas and pressurized liquid oxygen. Which of the following statements is true?
    A. You must refuse this load.
    B. You'll need two different types of placards.
    C. You'll need a liquid tanker with double-wall bulkheads.
    D. You'll need a "DANGEROUS" placard.

20. When a vehicle must be placarded, the placards must be _____.
    A. on the front, back, and both sides
    B. on both doors of the cab
    C. on the front and back, just above the license plate
    D. on each package

21. Tankers used to haul hazardous materials must _____.
    A. meet DOT specifications
    B. meet inspection, retest, and marking requirements
    C. have the month and year of the most recent test stamped into the metal of each tank or on a certification plate
    D. all of the above

22. Of the explosives, the least dangerous is _____.
    A. Division 1.1
    B. Division 1.2
    C. Division 1.3
    D. Division 1.6

23. Which of the following statements about flammable solids and fire is true?
    A. Friction can cause flammable solids to ignite.
    B. Flammable solids might retain heat from manufacturing or processing, and this heat can start a fire.
    C. Flammable solids may burn when combined with water.
    D. all of the above

24. Of the following hazardous material classes, which is the most dangerous?
    A. Division 6.1
    B. Division 6.2
    C. Division 6.3
    D. Division 6.4

25. ORM is regulated because _____.
    A. it's dangerous to transport because of the way it's packaged
    B. it is shipped only in large quantities
    C. it is an unknown material
    D. it is not properly labeled

26. Shipping papers _____.
    A. must have the shipper's certification
    B. must be marked differently from shipping papers for regular cargo
    C. must be carried in the vehicle so they are easy to get to in case of an accident, or for an inspection
    D. all of the above

27. Which of these statements about loading and unloading cargo tanks is false?
    A. The person attending the tanker must have a clear view of it.
    B. The person in attendance must be within 50 feet of the tanker.
    C. The person in attendance must be aware of the danger.
    D. The person in attendance must be authorized and able to move the vehicle.

28. Some hazardous materials may not be loaded together with other hazardous materials. The prohibited combinations are listed in the _____.
    A. Hazardous Materials Table
    B. List of Hazardous Materials and Reportable Quantities
    C. Segregation and Separation Chart
    D. Hazardous Waste Manifest

29. Do not load _____ with the engine running.
    A. explosives
    B. flammable liquids (unless they're needed to pump the product)
    C. compressed gases (unless they're needed to pump the product)
    D. all of the above

30. Which of the following statements about transporting hazardous materials is true?
    A. Trailers used to haul Division 1.1, 1.2, or 1.3 explosives must have a floor liner that doesn't contain iron or steel.
    B. You must never enter a tunnel when carrying cylinders of hydrogen or cryogenic liquids on your vehicle.
    C. You must stop before crossing a railroad track when hauling any amount of chlorine.
    D. all of the above

31. Which of the following statements about transporting chlorine is true?
    A. You must have a gas mask when you're hauling chlorine in a cargo tank.
    B. Before you couple or uncouple your tank, you must detach any loading and unloading connections.
    C. You must never leave the tank uncoupled from the tractor without chocking it first.
    D. all of the above

32. Which of the following statements about hauling hazardous materials with a cargo tank is true?
    A. The only time you should have liquid discharge valves open is during loading and unloading.
    B. For each shipment of flammable cryogenic liquid, you must make a written record of the cargo tank pressure and the outside temperature.
    C. You must complete your trip from start to finish within the one-way travel time.
    D. all of the above

33. You must control how far packages labeled "Radioactive II" or "Radioactive III" are located from people and cargo space walls by referring to _____.
    A. the transport index and the Separation Distances Table
    B. the Table of Hazardous Materials
    C. the List of Hazardous Materials and Reportable Quantities
    D. the Segregation and Separation Chart

34. Who is responsible for checking that the shipper has correctly named, labeled, and marked a hazardous materials shipment?
    A. the D.O.T.
    B. the carrier
    C. the shipper
    D. the manufacturer

35. Which of these statements about repairing vehicles with hazardous cargo is true?
    A. Devices that produce sparks may not be used to repair or maintain the cargo area or fuel system of your vehicle when you have a placarded load.
    B. While the work is being done, there must be someone able to move the vehicle immediately if that becomes necessary.
    C. The vehicle must be removed from the building as soon as the work is done.
    D. all of the above

36. When handling packages of explosives, you must _____.
    A. not use hooks or other metal tools
    B. keep bystanders 100 feet away
    C. roll the packages carefully with no sharp or jarring actions
    D. all of the above

37. Which of these statements about hazardous materials accidents is true?
    A. Open smoking packages of flammable solids if you can get to them.
    B. Remove broken and unbroken packages of flammable solids from the vehicle if it is safe to do so.
    C. Use a dry type of cleaner if a corrosive material leaks in your vehicle.
    D. all of the above

38. If hazardous material is leaking from your vehicle, _____.
    A. drive to a phone, truck stop, or service station as quickly as you can
    B. drive at as high a speed as is safe to spread the material thinly
    C. park the vehicle, secure the area, and send someone else for help
    D. get someone to watch the vehicle and go for help yourself since you know best what's needed

39. The markings on cargo tanks tell you about _____.
    A. the capacity
    B. the original test date
    C. the retest date
    D. all of the above

40. As the driver, your job at the scene of an accident is to _____.
    A. keep people away and upwind
    B. limit the spread of hazardous material, regardless of the risk to your safety
    C. make sure the shipping papers stay in the truck
    D. direct traffic

41. If someone is killed, injured and sent to the hospital, or total property damage exceeded $50,000 as a direct result of a hazardous materials incident, you or your employer must phone _____.
    A. your insurance agent
    B. the National Response Center
    C. the DOT
    D. CHEMTREC

42. A resource for local police and firefighters handling a hazardous materials incident is _____.
    A. the IMO
    B. the NRC
    C. the DOT
    D. the ICAO

43. If you are involved in a hazardous materials spill, you should call_____.
A. either the National Response Center or CHEMTREC, but you may not call both
B. either the National Response Center or CHEMTREC
C. the National Response Center, instead of local police or firefighters
D. the DOT

44. When you haul explosives, _____.
A. some locations may require you to have special permits
B. states may require you to follow special routes
C. counties may require you to follow special routes
D. all of the above

45. A vehicle involved in a leak of Division 2.3 or Division 6.1 must be _____.
A. sent to a decontamination station
B. checked for stray poison before being used again
C. washed down by CHEMTREC
D. never used again to haul food for humans

By now you should feel well-prepared for the knowledge part of your CDL test. If you've tackled every question in this book, you've had a lot of practice answering test questions. You can think of the CDL Knowledge Test as just another set of questions about material you're familiar with. And you can feel proud of the job you've done in this self-guided training course.

You know that you must take a Skills Test as well as a Knowledge Test to get your CDL. The next chapter will help you practice so you'll be ready for a typical CDL Skills Test.

ENDORSEMENT TESTS ANSWER SHEETS

Test Item	Answer Choices				Self - Scoring Correct Incorrect	
Air Brakes						
1.	A	B	C	D	☐	☐
2.	A	B	C	D	☐	☐
3.	A	B	C	D	☐	☐
4.	A	B	C	D	☐	☐
5.	A	B	C	D	☐	☐
6.	A	B	C	D	☐	☐
7.	A	B	C	D	☐	☐
8.	A	B	C	D	☐	☐
9.	A	B	C	D	☐	☐
10.	A	B	C	D	☐	☐

Cut Here

Test Item	Answer Choices				Self - Scoring
					Correct **Incorrect**
11.	A	B	C	D	☐ ☐
12.	A	B	C	D	☐ ☐
13.	A	B	C	D	☐ ☐
14.	A	B	C	D	☐ ☐
15.	A	B	C	D	☐ ☐
16.	A	B	C	D	☐ ☐
17.	A	B	C	D	☐ ☐
18.	A	B	C	D	☐ ☐
19.	A	B	C	D	☐ ☐
20.	A	B	C	D	☐ ☐
21.	A	B	C	D	☐ ☐
22.	A	B	C	D	☐ ☐
23.	A	B	C	D	☐ ☐
24.	A	B	C	D	☐ ☐
25.	A	B	C	D	☐ ☐
26.	A	B	C	D	☐ ☐
27.	A	B	C	D	☐ ☐
28.	A	B	C	D	☐ ☐
29.	A	B	C	D	☐ ☐
30.	A	B	C	D	☐ ☐

Cut Here

Test Item

Answer Choices

Combination Vehicles/
Doubles/Triples

Self - Scoring

Correct Incorrect

	A	B	C	D
1.	A	B	C	D
2.	A	B	C	D
3.	A	B	C	D
4.	A	B	C	D
5.	A	B	C	D
6.	A	B	C	D
7.	A	B	C	D
8.	A	B	C	D
9.	A	B	C	D
10.	A	B	C	D
11.	A	B	C	D
12.	A	B	C	D
13.	A	B	C	D
14.	A	B	C	D
15.	A	B	C	D
16.	A	B	C	D
17.	A	B	C	D
18.	A	B	C	D
19.	A	B	C	D
20.	A	B	C	D

Test Item	Answer Choices				Self - Scoring Correct Incorrect	
21.	A	B	C	D	☐	☐
22.	A	B	C	D	☐	☐
23.	A	B	C	D	☐	☐
24.	A	B	C	D	☐	☐
25.	A	B	C	D	☐	☐
26.	A	B	C	D	☐	☐
27.	A	B	C	D	☐	☐
28.	A	B	C	D	☐	☐
29.	A	B	C	D	☐	☐
30.	A	B	C	D	☐	☐
Tank Vehicles						
1.	A	B	C	D	☐	☐
2.	A	B	C	D	☐	☐
3.	A	B	C	D	☐	☐
4.	A	B	C	D	☐	☐
5.	A	B	C	D	☐	☐
6.	A	B	C	D	☐	☐
7.	A	B	C	D	☐	☐
8.	A	B	C	D	☐	☐
9.	A	B	C	D	☐	☐
10.	A	B	C	D	☐	☐

Cut Here

Test Item	Answer Choices				Self - Scoring Correct Incorrect
11.	A	B	C	D	☐ ☐
12.	A	B	C	D	☐ ☐
13.	A	B	C	D	☐ ☐
14.	A	B	C	D	☐ ☐
15.	A	B	C	D	☐ ☐
Hazardous Materials					
1.	A	B	C	D	☐ ☐
2.	A	B	C	D	☐ ☐
3.	A	B	C	D	☐ ☐
4.	A	B	C	D	☐ ☐
5.	A	B	C	D	☐ ☐
6.	A	B	C	D	☐ ☐
7.	A	B	C	D	☐ ☐
8.	A	B	C	D	☐ ☐
9.	A	B	C	D	☐ ☐
10.	A	B	C	D	☐ ☐

Cut Here

Test Item	Answer Choices				Self - Scoring	
					Correct	Incorrect
11.	A	B	C	D	☐	☐
12.	A	B	C	D	☐	☐
13.	A	B	C	D	☐	☐
14.	A	B	C	D	☐	☐
15.	A	B	C	D	☐	☐
16.	A	B	C	D	☐	☐
17.	A	B	C	D	☐	☐
18.	A	B	C	D	☐	☐
19.	A	B	C	D	☐	☐
20.	A	B	C	D	☐	☐
21.	A	B	C	D	☐	☐
22.	A	B	C	D	☐	☐
23.	A	B	C	D	☐	☐
24.	A	B	C	D	☐	☐
25.	A	B	C	D	☐	☐

Cut Here

Test Item	Answer Choices				Self - Scoring Correct Incorrect	
26.	A	B	C	D	☐	☐
27.	A	B	C	D	☐	☐
28.	A	B	C	D	☐	☐
29.	A	B	C	D	☐	☐
30.	A	B	C	D	☐	☐
31.	A	B	C	D	☐	☐
32.	A	B	C	D	☐	☐
33.	A	B	C	D	☐	☐
34.	A	B	C	D	☐	☐
35.	A	B	C	D	☐	☐
36.	A	B	C	D	☐	☐
37.	A	B	C	D	☐	☐
38.	A	B	C	D	☐	☐
39.	A	B	C	D	☐	☐
40.	A	B	C	D	☐	☐
41.	A	B	C	D	☐	☐
42.	A	B	C	D	☐	☐
43.	A	B	C	D	☐	☐
44.	A	B	C	D	☐	☐
45.	A	B	C	D	☐	☐

Cut Here

Chapter 15

CDL Skills Tests

In this chapter you will learn about:

- what the CDL Skills Tests cover
- how the CDL Skills Tests test your ability to inspect your vehicle
- how the CDL Skills Tests test your basic ability to control your vehicle
- how the CDL Skills Tests test your safe driving ability

To complete this chapter, you will need:

- the owner's manual for your vehicle
- the Federal Motor Carrier Safety Regulations pocketbook (or access to U.S. Department of Transportation regulations, Parts 383, 393, and 396 of Subchapter B, Chapter 3, Title 49, Code of Federal Regulations)
- a CDL preparation manual from your state Department of Motor Vehicles, if one is offered
- the driver's manual from your state Department of Motor Vehicles

By now you should feel well prepared for the CDL Knowledge Test. You've seen many sample, pre-trip, post-trip, and review test questions. Did you answer every one of them? Then you should feel less anxious about this type of test-taking.

As you know, that's not all there is to CDL testing. You'll have to take a Skills Test as well. The main skill areas tested are:

- pre-trip inspection
- basic vehicle control
- safe driving
- air brakes (to avoid the Air Brake Restriction)

At least some of the skills must be tested in "on-street" conditions. Some of them may be tested off the street, or with a truck simulator.

You must take your CDL tests in a vehicle that is "representative" of the one you plan to drive.

This means that your test vehicle must belong to the same group, A, B, or C, as your work vehicle. It doesn't mean that the two vehicles must be the same make or model. They simply must belong to the same group.

The representative vehicle may have to be equipped with air brakes. If you wish to drive a CMV with an air brake system, you must take your Skills Tests in a vehicle with air brakes. If you take the Skills Tests in a vehicle without air brakes, an air brake restriction is put on your CDL. This would also be the case if you don't pass the air brake part of the Skills Tests. This means you are not allowed to drive a CMV with an air brake system. This includes braking systems that rely in whole or in part on air brakes.

Refer to the table below to see what tests you must take for the different vehicle groups.

Required CDL Skills Tests

Vehicle Group	Skills Tests Required
Group A	Pre-trip Inspection
	Basic Control Skills
	Road Test
Group B	Pre-trip Inspection
	Basic Control Skills
	Road Test
Group C	Road Test

"Passing" on the Skills Tests depends on what's being tested. In all cases, anyone who breaks a traffic law automatically fails the test. So does anyone who causes an accident during the test.

Your state may excuse military drivers from having to take the Skills Tests. You must meet certain conditions, though. You must have a current, valid license, and have a good driving record. A "good driving record" means that, in the past two years, you have not had your license suspended, revoked, or canceled. You must not have been disqualified from driving in the past two years. Your driving record for the past two years must be free of traffic violations connected with an accident (other than parking violations).

Other restrictions may apply.

Let's assume, though, that you will be taking the Skills Tests. In this chapter, we'll give you an idea of what the tests will be like. You'll learn what the examiners will be looking for. And, we'll give you some tips on how to "practice" and prepare for the Skills Tests.

PRE-TRIP INSPECTION

Part of the Skills Test will be on your ability to inspect your vehicle. When you take the Knowledge Test, you'll answer questions about vehicle inspection. When you take the Skills Tests, you will actually do one.

Having read Chapter 9, you know what a "basic" inspection involves. Keep this in mind as well. You must take your Skills Tests in, and therefore inspect, a representative vehicle. If you don't want the Air Brake Restriction, your vehicle must have air brakes. You'll have to show you know how to inspect the air brake system. Then, if you want one or more of the endorsements, your vehicle inspection will include extra steps. These extra steps are covered in the chapters that deal with the endorsements. Be sure to read those chapters thoroughly.

Most states will have you do an inspection as part of the Skills Tests. The purpose is to see if you know whether the vehicle is safe to drive. You may actually only do certain parts of a complete inspection. You will have to describe anything you would do if you were inspecting your vehicle completely.

During the Inspection Test, you may also be asked to describe in some detail the different parts and systems on your vehicle. Or, the examiner may have you stop your inspection from time to time and ask you questions about your equipment.

As you examine your vehicle, tell the examiner what you are inspecting. Describe what you find. You're not likely to take a defective vehicle to the test. However, the examiner will want to know you would recognize a defect if you saw one. So as you inspect your vehicle, mention the defects likely to be found at the different sites. Explain how those defects impair safety.

The examiner may ask you for more details. For example, you might say, "The hoses are not cracked or rubbing." The examiner may tell you to point out just which hoses you are talking about. You may be asked to state the function of the hoses.

Or, you might point to the battery and state that it's in good shape. The examiner might ask you to name the part and ask how you know it's in good shape. You'll have to respond like this: "This is the battery. It's securely mounted, and it's not cracked. There's no corrosion. The cables are securely attached to the correct posts," and so on.

The examiner will not prompt you or give you hints. If you forget a step, the examiner will just assume you don't know how to do a complete inspection. For this reason, you may want to have an Inspection Aid with you. Usually, you may have a checklist with you when you take the Inspection Test. Check with your state to see if for some reason this is not allowed. Otherwise, feel free to use the checklist. You'll be less likely to overlook something if you check off items as you go.

You'll usually find just such Inspection Aids in the CDL manual your state provides. There will likely be one for straight trucks, and another for tractor-trailers. You'll also find Inspection Aids at the end of this section. Your state may allow you to use the aids it provides, but not the ones in this book. You could ask about this when you arrange to take your Skills Tests. Or, bring both with you just in case.

Inspection Test Site

You'll usually take the Inspection Test in a flat parking area or open lot. Examiners will avoid using a parking space on a street, or any place where traffic passes close by. If you do have to take the test in a busy area, don't get so involved taking the test that you forget to check for traffic and other hazards.

Safety Tips

Do keep your own safety in mind while you take the test. Never get under, in front of, or behind a vehicle when there's any chance at all that the vehicle might move. Don't brace your foot on top of the bumper when you pull the engine hood open.

Use care getting in and out of the vehicle. Know where the cab footholds are on your vehicle. Make sure they're free of dirt and grease. Then use them. Use the three-point stance.

Figure 15-1. The Three-Point Stance.

The Three-Point Stance means you use the steps or footholds and the hand-hold. Use either both hands and one foot, or both feet and one hand. That is, you use three limbs to enter the cab.

When you get out of the truck, climb out backwards, as if you were using a ladder. Never jump out of the cab.

If you are about to do something really dangerous, the examiner may stop you. For the most part, though, your safety is your responsibility, not the examiner's.

Test Instructions

The examiner may give you instructions for taking the Inspection Test. They could go something like this:

"Please conduct a thorough inspection of your vehicle. As you do the inspection, point to or touch the things you are inspecting. Explain what you are looking for.

"Start by inspecting the engine compartment. Then start the engine. After you have done the start-up checks, turn off the engine. Do the rest of the inspection.

"Start whenever you are ready."

Make sure you understand the instructions. Ask questions if you don't. The examiner will be prepared to answer them.

General Procedure

As a general procedure, you'll be expected to follow these main steps:

- inspect the engine compartment
- perform the in-vehicle checks
- shut down the engine
- perform the external inspection

For the external inspection, you may only have to do one side of the vehicle. Also, you may only have to explain some things once, even if they appear in several places on the vehicle. For example, each wheel has tread. The first time you inspect a tire, you'll have to explain correctly what you are looking for then as far as the tread is concerned. After that, the examiner may tell you that you only have to state that this is a tire and that you're inspecting the tread. You might not have to repeat the entire explanation. (If the examiner doesn't mention this, ask. Don't assume this will be the accepted procedure. Check and make sure.)

Usually, you can go about the inspection however you prefer. Just make sure you follow the main steps just listed. Of course, if the examiner gives you specific or different instructions, follow them to the letter.

Inspecting a cabover? You may not have to raise the cab to inspect the engine compartment. Most of the engine items can be checked without raising the cab. Ask the examiner if you should raise the cab. You may be told you don't have to. Of course, you must tell the examiner about any items that can't be checked without raising the cab. Remember, the examiner won't prompt you for this information.

The Inspection Test isn't timed. You can work at your own pace. You won't get any extra points for speed. In fact, if you rush, you might overlook something. Take whatever time you need to be thorough.

"Examining" the Examiner

As you've learned, the examiner will not prompt you or give you hints. The examiner may remind you about the main steps. But you won't hear anything like "You forgot the brakes" or "What about the tires?"

All the same, the examiner might give you unintended hints. The examiner might start walking toward the next area to inspect. Or, the examiner might look at a vehicle part you should inspect next.

During the inspection, if you go blank for a minute, examine the examiner. See where the examiner is standing or looking. That could be enough to jog your memory and enable you to go on.

Pre-trip Inspection Standards

In this section, we'll list the main items you should inspect and describe how to inspect them. We'll take them in the order they appear in the Inspection Test, then in their place in the seven-step inspection. You'll see what the examiner will be expecting. Don't forget, you may be asked to give far more detail about vehicle parts and systems than what we mention here. You may have to state what this part does and why you are inspecting it. You may have to explain how a defect impairs safety.

You'll find those details in Chapters 5 through 9, plus the chapters on air brakes and endorsements.

Inspect the engine compartment. Check the oil level, using the dipstick. The dipstick measures the amount of oil that lubricates the engine. The level should be above the "Add" mark.

Check the level of coolant, which cools the engine. If your vehicle has a sight glass, use it to view the coolant level. Or else, take off the radiator cap. The coolant level should be above "Low." Note that if your vehicle does have a sight glass, you'll be marked wrong for failing to use it.

TIP *If the engine is hot, don't open the radiator! That could be dangerous. Instead, describe the process of checking the coolant level.*

If you have power steering, check the fluid. The fluid is part of the hydraulic system that assists steering action to the front wheels. With the engine off, pull the dipstick. The level should be above the "Add" mark.

If you have an automatic transmission, check the fluid level. (You may have to do this with the engine running. Your owner's manual will tell you if that is the case.)

You may not actually have to open reservoirs and pull dipsticks. The examiner may just have you describe the process. Your description must be thorough!

Check these belts for tightness and excessive wear:

- alternator
- air compressor
- fan
- power steering
- water pump

Press down on the center of the belt. There should be no more than 3/4 inch slack. Belts should not be frayed or cracked. They shouldn't have loose fibers.

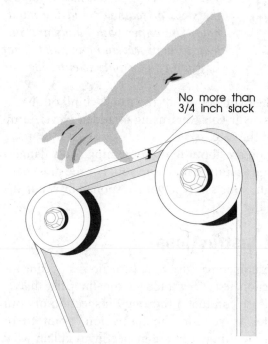

No more than 3/4 inch slack

Figure 15-2. Check belts for snugness.

Check the fluid level in the battery. It should be up to the filler rings in each cell. Make sure each cell has a vent cap. Check to see that the vents are not clogged. Check the battery mount. Check the hold-down bars to make sure the battery is snug. The battery box cover should be in place. The battery box itself must not be cracked or leaking. Look for battery cables that are frayed, worn, or cracked. Check to make sure the battery connections are tight.

Look for leaks in the engine compartment. Look for fluid puddles on the ground under the engine or drips on the engine's underside.

In-vehicle checks. Make sure the parking brake is on. The clutch and gearshift disengage the engine from the drive train. This keeps the vehicle from moving and reduces the load on the starting motor. Depress the clutch before turning on the starter. Keep it there until the engine reaches idling speed. Shift into "Park" if your transmission is automatic.

Adjust the seat and mirrors. You need a clear view of the traffic to the rear. The mirrors should not be cracked, loose, or dirty.

Turn on the starter. Release it as soon as it starts. Listen for any unusual engine noises.

Watch the gauges. The oil pressure should come up to normal within seconds after the engine is started. The warning light should go off. Engine oil temperature should show a gradual rise to normal. The ammeter and/or voltmeter should show the generator or alternator is working. The warning light should go off. The needle should flutter, then give normal readings.

Check for looseness in the steering linkages. Move the steering wheel back and forth. Free play should be about two inches for a 20-inch steering wheel. Test power steering with the engine on. Note the degree of free play that occurs before the front left wheel just barely moves. It should be less than five to 10 degrees.

Test for excessive clutch pedal travel. Depress the clutch until you feel a slight resistance. One to two inches of travel before you feel resistance is normal.

The horn warns other drivers or pedestrians of danger. Test it to make sure it works.

Check the windshield for cracks, dirt, and stickers that obscure your view. Make sure the windshield wipers and washers work. The blades should be secure on the wiper arm, and the rubber should not be worn.

You must have emergency equipment for use during a breakdown or at the scene of an accident. You should have:

- spare fuses, if your vehicle uses them
- three red reflective triangles
- a properly charged and rated fire extinguisher

Your emergency equipment may also include these items:

- snow chains
- tire-changing equipment
- spare lamps and bulbs
- electrical tape and tools
- flashlight
- pliers
- screwdriver
- wire
- tire pressure gauge

The heater/defroster warms the cab and prevents frost and mist from forming on the windshield. Make sure it works.

Shut down the engine. Put the transmission in the lowest forward gear.

Check to see that the lights are working. Make sure the parking brake is set. Turn on the headlights (on low beams) and the four-way flashers, and get out. Take the key with you. Go to the front of the vehicle.

Check that the low beams are on and both of the four-way flashers are working. Push the dimmer switch and check that the high beams work. Turn off the headlights and four-way hazard warning flashers. Turn on the parking, clearance, side-marker, and identification lights.

Don't forget to check the dashboard lighting indicators for signals, flashers, and headlight high beams. They should go on when the lights are on.

Turn on the right-turn signal, and start the "walk-around" part of the inspection.

External inspection. Check the tires. Tread depth should be 4/32 inch on front tires, 2/32 inch on the other tires. Use a tire pressure gauge to check inflation. You can use a tire billy or mallet to check that the tire isn't flat. You'll be marked wrong if you simply kick the tire.

The wheel rims keep the tires on the wheel. Check the wheel for damaged or bent rims. Rims shouldn't have welding repairs. Rust trails suggest that the rim is loose on the wheel.

Lug nuts hold the wheel on the axle. Check for missing or loose lug nuts. Rust trails suggest the nuts might be loose. Bolt holes should not be out-of-round or cracked.

Spacers or axle collars between dual wheels keep the wheels evenly separated. For duals, make sure the wheels are evenly separated and that the tires don't touch each other.

Check that the wheel hub oil seal isn't leaking. If your hub has a sight glass, use it to check the oil level.

Check the axle seals. Wheel/axle mountings should not be cracked, distorted, or leaking.

If mud flaps are required, make sure you have them. They should be firmly attached and free of damage.

Check the brake assembly. On some brake drums, you can see the brake linings from outside the drum. If this is the case on your vehicle, check that a visible amount of brake lining is showing. If a backing plate fully encloses the brake shoe assembly, you can't do this check. You can state what the linings do and what shape they should be in.

Brake shoes rub on the inside of the brake drum to slow the vehicle. There should be no cracks, dents, or holes in the brake shoes. Look for missing or loose bolts.

The brake chamber converts air pressure to mechanical force to operate the wheel brakes. Brake chambers should not be cracked or dented. They must be securely mounted.

The slack adjuster is the linkage from the brake chamber to the brake shoe. It works the brakes. Check for broken, loose, or missing parts. The angle between the push rod and the adjuster arm should be a little over 90 degrees when the brakes are released. It should be not less than 90 degrees when the brakes are applied. When you pull it by hand, the push rod should not move more than about an inch.

Brake hoses carry air or hydraulic fluid to the wheel brake assembly. Check for worn, cracked, or frayed hoses.

Check the suspension. Check all the brackets, bolts, and bushings used to attach the spring to the axle and the vehicle frame. Look for cracked or broken spring hangers. There should be no broken, missing, or loose bolts or axle mounting parts. Look for missing or damaged bushings.

Leaf or coil springs dampen wheel vibration from rolling on the road. Look for leaves that have broken or shifted. They should not contact the tires, rim, brake drum, frame, or body. Look for missing or broken leaves in the leaf spring. If you have coil springs, the coil should not be broken or bent out of shape.

You may have a torsion bar suspension on some rear tractor wheels. Here, a steel bar, rod, or arm assembly acts as a spring instead of a leaf or coil spring. Make sure the bar or arm is not cracked, broken, or missing.

Check for loose, broken, or leaking shock absorbers.

Check the steering linkage. This assembly transmits steering action from the steering box to the wheel. Connecting links, arms, and rods should not be worn or cracked. Joints and sockets must not be worn or loose. There should be no loose or missing bolts.

The steering box holds the parts that transform steering column action into wheel turning action. Look for missing nuts, bolts, and cotter keys. If you have power steering, check for leaks. Inspect the power steering hose for damage.

Check the fuel tank. It should be secure and free of leaks and damage. The caps should be on tight.

The frame supports the vehicle body or trailer platform over the wheels. It should not be cracked or bent. Look for loose, cracked, bent, broken, or missing cross members. There should be no cracks in the cargo compartment or trailer floor.

The header board prevents cargo from shifting forward and hurting the driver during a panic stop. If you're required to have one, it should be securely mounted and free of damage. It must be strong enough to hold back the cargo. If you have a canvas or tarp carrier, it must be securely mounted and lashed down.

The drive shaft transmits the power from the transmission to the drive axle. It must not be bent or cracked. Shaft couplings must look secure.

The exhaust system conducts combustion gases from the engine. All the outside parts should be securely mounted. Check for cracks, holes, or bad dents.

Check the lights and reflectors (parking, clearance, and identification, and turn signal lights). You should have all the required lights and reflectors, and they should be clean. Reflectors and clearance lights should be red at the rear, amber elsewhere. You shouldn't have any broken or missing reflectors or lights.

Running lights should be red at the rear. You must check them separately from signal, flasher, and brake lights.

Check that both brake lights come on when the brakes are applied. Each signal light should flash, and the four-way flashers must work.

Check the braking system. Test for leaks in the hydraulic brake system. Pump the brake pedal three times. Then apply firm pressure to the pedal and hold for five seconds. The pedal should not move.

Test how well the service brake works. Go about five mph. Push the brake pedal firmly. The brakes should apply evenly all around the vehicle. They should not pull to one side or the other. The brake pedal should not travel all the way to the floor before the brakes apply. Nor should the pedal give you a lot of resistance or take great effort to apply.

The parking brake should keep the vehicle from rolling when parked. Set the brake, then try to pull forward. The brakes should hold the vehicle back.

Air brake checks. If the air pressure is low, the low air pressure warning will sound immediately after the engine starts but before the air compressor has built up pressure. Test this by letting air pressure build to governed cut-out pressure (between 100 and 125 psi). The low air pressure warning should come on by the time the air pressure gets to 60 psi but not before it reaches 70 psi.

The air compressor maintains air pressure in the air brake system. With the engine off, you'll have to point to, touch, or press the belt to test for snugness. Press it at the center. It should not move more than 3/4 inch. Note that the belt is not frayed or cracked. There should be no loose fibers or signs of wear.

Your compressor may not be belt-driven. If that's the case, tell the examiner. Note that the compressor drive appears to be working and is not leaking.

Your air system should have no leaks.

Perform these brake system checks. Let air pressure build to governed cut-out pressure. That should occur between 100 and 125 psi. With the engine still idling, step on and off the brake, reducing the air pressure. The compressor should cut in at about 100 psi. The air pressure should begin to rise.

With the engine off, the wheels chocked, and the parking brake released, apply the foot brake. Air pressure should not drop more than three pounds in one minute on a single vehicle. The drop should be no more than four pounds in one minute for combination units.

Fan off the air pressure by rapidly applying and releasing the foot brake. The low pressure warning alarm should come on before air pressure drops below 60 psi.

Continue to fan off the air pressure. At about 20 to 40 psi on a tractor-trailer, the tractor protection valve should close (pop out). On other vehicles, the spring brake push-pull valve should pop.

Combination vehicle checks. The catwalk is a platform at the rear of the cab for the driver to stand on when connecting or disconnecting trailer lines. It should be solid, securely bolted to the tractor frame, and clear of loose objects.

Air and electrical lines carry air and power to the trailer. Check that the air hoses are not cut, cracked, chafed, or worn. Steel braid should not show through. Listen for leaks. The lines must not be tangled, crimped, or pinched. They should not drag against tractor parts. Electrical line insulation must not be cut, cracked, chafed, or worn. The electrical conductor must not be showing through. Neither the air nor electrical lines should be spliced or taped.

Air and electrical connectors connect air supplies and electrical power to the trailer. Make sure trailer air connectors are sealed and in good condition. Glad hands should be locked in place and free of damage. There must not be any air leaks you can hear. The trailer electrical plug must be firmly seated and locked in place.

The fifth wheel platform holds the fifth wheel skid plate and the locking jaws mechanism. Check for cracks or breaks.

Locking pins hold a sliding fifth wheel in place on the slider rails. Look for loose or missing pins in the slide mechanism. If the slider is air-powered, there should be no leaks. Check that the fifth wheel is not so far forward that the tractor frame will strike the landing gear during turns.

The release arm releases the fifth wheel locking jaws. Then the trailer can be uncoupled. The release arm should be in the engaged position. The safety latch locks the locking jaws closed. It should be engaged.

The kingpin attaches the trailer to the tractor. The apron provides a surface for resting the trailer on the fifth wheel. The kingpin must not be bent. The apron should lie flat on the fifth wheel skid plate. Check that the part of the apron you can see is not bent, cracked, or broken.

The landing gear supports the front end of the trailer when the trailer is not coupled to the tractor. The landing gear should be fully raised. Look for missing parts. The support frame should not be bent or damaged. You must have a crank handle and it must be secured. If the landing gear is power-operated, there must be no air or hydraulic leaks.

Inspection Test Scoring

The examiner uses a form to score the Inspection Test. Most states will score the Inspection Test as follows. You'll get a point for every item you inspect correctly. Of course, larger vehicles have more items that must be inspected. So, you'll need more points to pass. The lowest "passing" score is 39 items correctly inspected. This applies to inspecting a straight truck with two axles but without air brakes. If you're inspecting a three-axle tractor coupled to a two-axle trailer with air brakes, you would need to inspect 80 percent of the items correctly to pass.

Speaking of air brakes, if you don't perform the air brakes check during the engine start procedure, you automatically fail this part of the Inspection Test.

States do have the option to score the Inspection Test differently from the way we've just described. You may certainly ask how your Inspection Test will be scored. Try not to worry about the scoring. Focus instead on doing a thorough inspection.

You could fail the Inspection Test and be allowed to take the rest of the Skills Tests. The examiner would decide if the vehicle is safe to operate. Then the examiner could choose to continue with the skills testing. If you pass these other tests, you would have to retake the Inspection Test some other time before you would get your CDL.

COUPLING AND UNCOUPLING

Group A vehicle drivers may be tested on coupling and uncoupling. States may give you an extra Knowledge Test. They may also test you on the process during the Skills Test. Here's an example of the type of performance some examiners will be looking for.

Uncoupling

Set the tractor/truck and trailer parking brakes. Chock the wheels. Make sure you're working on solid ground. Lower the landing gear.

Release the locking jaws. Release the tractor brakes. Pull ahead until the fifth wheel clears the apron. Set the tractor/truck parking brakes.

Disconnect and secure the air and electrical lines. Pull forward about 16 feet.

Doubles and triples. Unhook and secure the air lines. Unhook electrical lines and safety chains. Release the pintle hook. Raise the eye of the tongue off the hook and clear the hook. Pull forward about 16 feet.

Coupling

Back up and align the fifth wheel jaws with the kingpin. Back slowly, until the fifth wheel touches the trailer apron without flattening. Secure the tractor.

Check that the fifth wheel coupler (locking jaws) is open. Compare the height of the fifth wheel with the trailer. Connect the air and electrical lines. Apply and release them several times. Check to see whether air is releasing from the quick release valve.

Apply the tractor brakes. Back under the trailer until the fifth wheel coupler engages and locks.

Check the coupler. Pull against the trailer with the trailer brakes on.

Raise the landing gear. Secure in low range.

Doubles and triples. Back slowly, so the truck hook is in line with the trailer eye. Stop with the eye over the hook. Lower the eye and secure the hook.

Raise the landing gear or remove supports. Hook up air lines. Hook up the electrical lines and safety chains. Remove the chocks.

Scoring the Coupling and Uncoupling Skills Test

There are some mistakes you could make in coupling and uncoupling that could result in immediate failure of this part of the Skills Test. Here's what some states consider serious uncoupling mistakes:

- not setting the trailer brakes before leaving the cab to lower the landing gear
- not chocking the wheels when your vehicle doesn't have spring brakes
- failing to lower the landing gear before unlocking the fifth wheel
- failing to shut off the air lines and setting the tractor parking brakes before disconnecting the air lines

- not disconnecting the air and electrical lines before separating the tractor and trailers

The following coupling mistakes could cause immediate failure in some states:

- flattened fifth wheel when coupling the units
- not setting the tractor parking brake before getting out of the cab to connect the air and electrical lines
- not setting the trailer brakes before backing under the trailer to engage the fifth wheel coupler
- failing to chock the wheels before backing under the trailer to engage the fifth wheel coupler when the vehicle doesn't have spring brakes
- not checking the coupling by pulling forward with the trailer brakes on or the wheels chocked
- failing to raise the landing gear before moving

BASIC CONTROLS SKILLS TEST

The next part of the test covers your basic vehicle control skills. FMCSR Part 383 says CDL applicants must have (and be able to show they have) these skills:

- ability to start, warm up, and shut down the engine
- ability to put the motor vehicle in motion and accelerate smoothly, forward and backward
- ability to bring the vehicle to a smooth stop
- ability to back the vehicle in a straight line, and check the path and clearance while backing
- ability to position the vehicle to make left and right turns, then make the turn
- ability to shift as needed and choose the right gear for speed and highway conditions
- ability to back along a curved path
- ability to observe the road and the behavior of other vehicles, especially before changing speed and direction

Most states will test these abilities with a "skills test." This may also be called the "range test" or "static test." It will usually involve going through a series of exercises on a course laid out just for this purpose.

Skills Test

Typical exercises are:

- straight line backing
- offset back left
- offset back right
- conventional parallel park
- sight-side parallel park
- 90 degree alley dock

Note: All states are required to have applicants conduct at least three of the above maneuvers during the skills test. States can also add maneuvers at their discretion. For example, some states continue to include the backward serpentine in their skills test. Check with your local DMV to get specific maneuvers.

Often, you'll be asked to do four of these. The easier exercises will likely be given first. Then you'll do the harder ones.

Three factors are scored as errors. They are:

- pullup
- encroachment
- wrong final vehicle position

Pullup. Anytime you stop and reverse direction to get a better position, it's scored as a pullup. A pullup is an error. Stopping without changing direction does not count as a pullup.

Encroachment. Touching or crossing an exercise boundary or cone with any part of the vehicle is an encroachment. Think of the boundaries or cones as walls. If your vehicle would touch or punch a hole in the "wall," that's an encroachment.

You're scored for an encroachment when you cross a boundary from the right side over to the wrong side. This is different from how pullups are scored. A pullup is a pullup no matter which side you're on.

Like pullups, encroachments are errors.

Final position. For some exercises, the final position of your vehicle counts. You could lose points or be marked for an error if your vehicle is not in the correct final position. Later in this section, we'll describe the exercises in greater detail.

Step 1: Vehicle overview for general condition

Step 2: Engine compartment checks
Fluid levels and leaks
Hoses and belts
Battery
Windshield washer
Wiring

Step 3: Inside the cab checks
Parking brake on
Gauge readings
Warning lights
Controls
Emergency equipment
Optional equipment

Step 4: Check lights
Headlights (low and high beams)
Four-way flashers
Parking lights
Clearance lights
Side marker lights
Identification lights
Right-turn signal

Step 5: Walk-around inspection
   A. Left front side: wheel and tire, suspension, brakes, axle, steering, side marker lamp and reflector, door glass, latches and locks, mirrors
   B. Front of cab: axle, steering system, windshield, lights and reflectors
   C. Right front side: door glass, latches and locks, wheels and tires, suspension, brakes, axle, steering, condition of bed or body, side marker lamp and reflector
   D. Right side: fuel tank, exhaust system, frame and cross members, wiring, spare tire, lights and reflectors, mirrors
   E. Cargo securement, curbside doors
   F. Right rear: wheels and tires, suspension, brakes, axle, lights and reflectors
   G. Rear: lights and reflectors, license plate, mudflaps, condition of bed or body, rear doors
   H. Cargo securement
   I. Left rear and left front: wheels and tires, suspension, brakes, steering, axle, lights and reflectors, battery, condition of bed or body

Step 6: Check signal lights

Step 7: Brake check

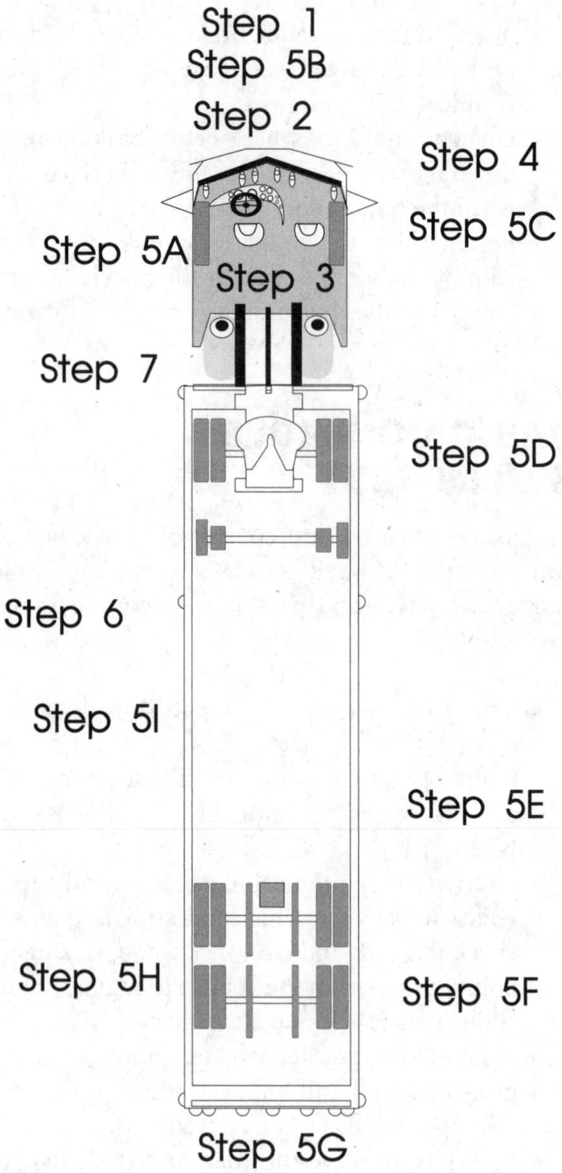

Figure 15-3. Vehicle inspection aid for straight trucks.

Step 1: Vehicle overview for general condition

Step 2: Engine compartment checks
Fluid levels and leaks
Hoses and belts
Battery
Compressor
Windshield washer
Wiring

Step 3: Inside the cab checks
Parking brake on
Gauge readings
Warning lights
Controls
Emergency equipment
Optional equipment

Step 4: Check lights
Headlights (low and high beams)
Four-way flashers
Parking lights
Clearance lights
Side marker lights
Identification lights
Right-turn signal

Step 5: Walk-around inspection
  A. Left front side: wheel and tire, suspension, brakes, axle, steering, side marker lamp and reflector, door glass, latches and locks, mirrors
  B. Front of cab: axle, steering system, windshield, lights and reflectors
  C. Right front side: door glass, latches and locks, wheels and tires, suspension, brakes, axle, steering, coupling to trailer, landing gear, side marker lamp and reflector
  D. Right side: fuel tank, exhaust system, frame and cross members, wiring, spare tire, lights and reflectors, mirrors
  E. Cargo securement, trailer doors
  F. Right rear: wheels and tires, suspension, brakes, axle, lights and reflectors
  G. Rear: lights and reflectors, license plate, mudflaps, coupling between trailers, landing gear, rear doors
  H. Cargo securement
  I. Left rear and left front: wheels and tires, suspension, brakes, steering, axle, lights and reflectors, battery, coupling to trailer

Step 6: Check signal lights

Step 7: Brake check

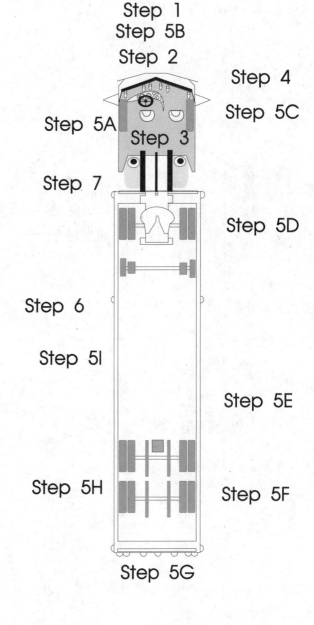

Figure 15-4. Vehicle inspection aid for tractor-trailers with air brakes.

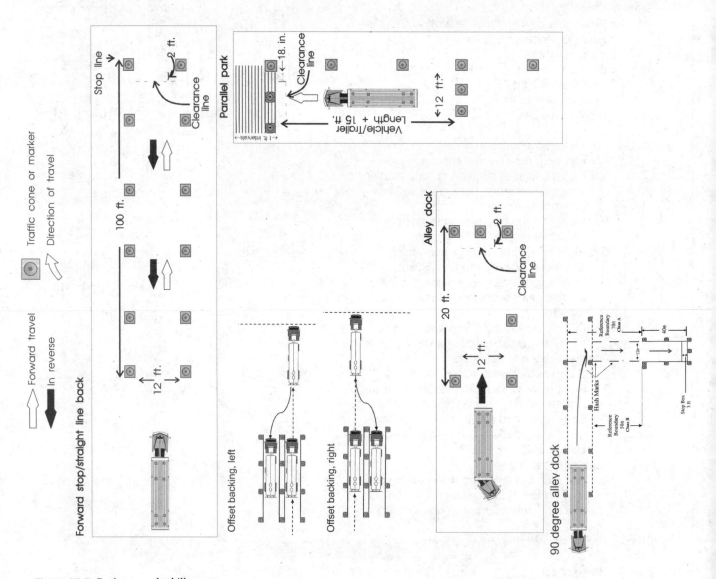

Figure 15-5. Basic controls skills test course.

Test Site

The test course will usually be on an open, paved area away from traffic and clear of overhead obstructions. As safe as it is, you should make a point of checking for hazards before you begin. Remember, one of the points of the test is to see if you know how to check for hazards, especially when backing.

Figure 15-5 shows you what most test courses look like.

Boundaries may be marked with paint or chalk lines, cones, yellow rope, or tape.

Test Instructions

The examiner may give you instructions for taking the Skills Tests. They could go something like this:

"Try not to go over any exercise lines or hit any of the cones. It is better to do a pullup than to go over a boundary. The base of each cone or marker marks the exercise boundary. So if a wheel passes over the base of a cone, that counts as going over the boundary.

"You'll get directions for each exercise as we come to it. When you complete each exercise, tap your horn.

"If you see me raise my arm straight up with the palm out, stop the vehicle."

Of course, if the examiner gives you specific or different instructions, follow them to the letter.

Make sure you understand the instructions. Ask questions if you don't. The examiner will be prepared to answer them.

Test Exercises

Forward stop/straight line back. This exercise tests two skills. It tests your ability to bring the vehicle to a smooth stop. And, it tests your ability to back the vehicle in a straight line, checking the path and clearance while backing. You'll show these abilities by first driving through a lane or alley. You'll stop as close as possible to a stop line at the end. Next, you'll back down and out of the alley. You must not touch any boundaries while you do this.

For the forward stop, aim to stop with your front bumper as close as possible to the line at the end of the alley. You must come within two feet of the stop line, but don't go past the line. You may stop only once. Don't pull ahead once you've stopped. Don't lean out of the window or open the door to see better. Use your mirrors.

For the straight line back, you'll drive forward. You'll position the vehicle so the rear of the vehicle is about even with the stop line. The examiner will usually signal when you're in the right position to begin. Then you'll back straight down the alley.

Don't touch either side boundary. Stop with your front bumper about even with the end of the alley.

Figure 15-6 shows a typical forward stop/straight line back exercise layout.

Alley dock. This exercise tests your ability to back along a curved path. You'll drive by the alley so the entrance is on your left. Then you will try to back into the alley. You must stop within two feet of the rear of the alley without going past the end boundary. Don't cross any boundary lines while you're backing.

Back serpentine

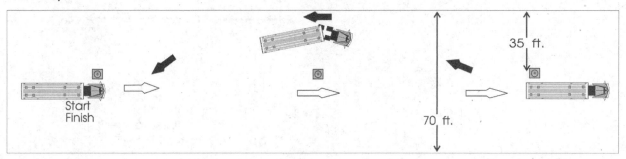

Figure 15-6. Back serpentine.

You may not make a 90-degree turn in front of the alley and then back straight in. In fact, you can't set up at more than a 45-degree angle. Pretend that there's a curb across from the alley that prevents you from doing this.

Figure 15-7 pictures the alley dock exercise.

Measured right turn. Some states include this exercise in the skills test. This measures your ability to position the vehicle for, then make, a right turn. You will start about 30 to 50 feet from a marker, as if you were approaching an intersection. You will drive forward. Then you will make a right turn around a cone or marker. Make the right rear wheels come as close to the cone as possible without touching it. In trying to stay clear of the cone, don't make the mistake of turning too wide. Missing the cone by more than six inches is an error in some states.

Parallel parking. There are two parallel parking exercises, the sight-side and the conventional. For the sight-side, you'll park in a space that's on your left. The space will be 10 feet longer than the length of your straight truck (or 10 feet longer than your trailer if your vehicle is a tractor-semitrailer). You will drive past the space, as if it were a parking space on the street. Then you'll back into it.

Try to get as close as possible to the rear and the "curb" of the space. Get within 18 inches of the rear boundary, but don't go over it. Don't cross any lines or hit any markers. Be careful parallel-parking with a trailer. When you jack the trailer, you can easily put the rear of the tractor over the "curb" boundary. This would be marked as an encroachment. If your vehicle is a straight truck, get your vehicle completely into the space. When parallel-parking a tractor-trailer, you only have to get the trailer in the space.

The parking space is on the right in the conventional parallel parking exercise. Otherwise, the exercise is the same as the sight-side parallel park.

Figure 15-9 shows the parallel park layout.

Backward serpentine. This measures your ability to back along a curved path. You'll start at the head of a row of three cones (see Figure 15-10). You'll line up near the first cone and drive forward along the right side of the line of cones. In other words, the cones will be on your left. Drive forward and stop when the rear of your vehicle is past the third cone. Then you will back up so your vehicle will wind around the three cones. You'll pass the third cone so the cone is on your left. Then pass the second cone so the cone is on your right. Pass the first cone so the cone is on your left. End up in the position you started from.

As with the other exercises, don't cross any boundaries or touch any markers. Each pullup counts as an error.

 In reverse

⦿ Traffic cone or marker

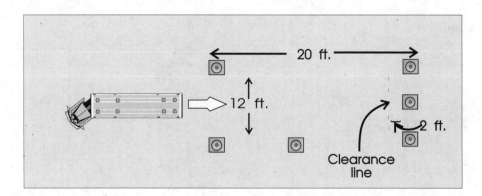

Figure 15-7. Typical alley dock course layout.

Off-set back left

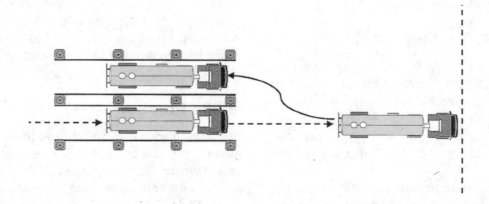

Off-set back right

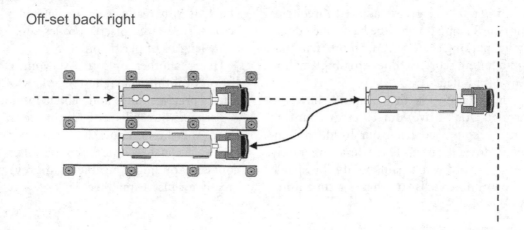

Figure 15-8. Off-set back left and right.

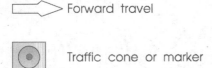 Forward travel

Traffic cone or marker

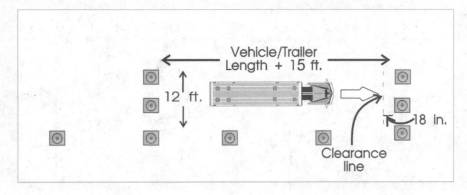

Figure 15-9. Typical parallel park exercise layout.

You can see that some of the exercises test the same skill. In some states, you will only do some, not all, of the exercises.

Skills Test Scoring

Most states will count each pullup and each encroachment as an error. You will also lose points for an incorrect final position. For example, stopping too short of or over the end line would be an error. Depending on which tests you take, you could make a total of 12 to 16 errors and still pass.

Other states start you out with 100 points. You lose a point for every error. After the Skills Test and the Road Test, you must end up with 75 points to pass.

In still other states, you are allowed three tries for the forward stop/straight line back, alley dock, and parallel parking exercises. Any more than that results in a failing score. So does crossing any line or driving over a flag.

You can certainly ask the examiner how the test is scored. Don't drive yourself crazy counting points and keeping score in your head, though. Just try to do your best. These tests are rarely timed. You usually won't gain points for speed. Unless the examiner tells you there's a time limit,

take it slow and easy. Go as slowly as you need to so you don't make mistakes.

Even if you don't pass the Skills Test, the examiner may decide to give you the Road Test. If you have passed the Inspection Test and the Road Test, you'll have to repeat and pass the Skills Test to get your CDL.

Skills Practice

If you own your own vehicle and have a learner's permit, you can practice in it whenever you have time. Where can you practice? Try a church or school yard or store parking lot. Any paved, open area will do. Get permission from the owner, manager, principal, or whoever's in charge of the area first, of course. Mark off your boundaries with chalk. If you really want to get fancy and have "cones," use large plastic bottles or jugs. Or, get some lengths of plastic pipe.

Here's another option. Ask your DMV for a list of local third-party testers. Many of these also offer a practice range and vehicle for a fee.

Perhaps your skills are badly in need of improvement. You should know that many private schools and vocational/technical schools offer "refresher courses." You might be able to improve your skills greatly for just a small fee.

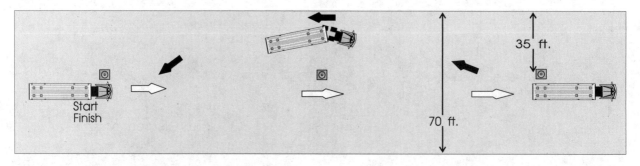

Figure 15-10. Typical backward serpentine course layout.

SAFE DRIVING SKILLS TEST

The last part of the test covers your ability to drive safely in traffic.

FMCSR Part 383 states that CDL applicants must have (and be able to show they have) these skills:

- ability to use proper visual search methods
- ability to give the right signals when changing speed or direction
- ability to adjust speed to the shape and condition of the road, the weather and visibility, the traffic, the cargo, and the driver's own condition
- ability to choose a safe gap for changing lanes, passing, and crossing and entering traffic
- ability to position the vehicle before and during a turn to keep other vehicles from passing on the wrong side
- ability to position the vehicle before and during a turn to prevent problems caused by off-tracking
- ability to maintain a safe following distance based on road conditions, visibility, and vehicle weight
- ability to adjust speed and following distance, brake, and change directions based on the weather conditions and stay in control

Most states will test this with a Road Test.

Test Site

You'll drive a pre-set route. This will usually include:

- four left and four right turns
- traffic lights, stop signs, and uncontrolled intersections
- a straight section of a city business street
- two intersections where you must stop
- a through intersection
- a multi-lane section of roadway
- moderate traffic
- one uncontrolled railroad crossing
- a curve to the left or right
- a section of expressway or two-lane rural or semi-rural road
- a downgrade and an upgrade
- parking on an upgrade and a downgrade
- an underpass, low clearance, or bridge

Later in this section, you'll learn the skills such a route tests.

Test Instructions

The examiner may give you instructions for taking the Road Test. They could go something like this:

"I will give you instructions as we go along. I'll always give directions for turns and so forth as far in advance as possible.

"Along the way I will point out a location. I'll ask you to pretend it is the top of a steep hill that you will be going down. I'll ask you to go through the motions of what you would do if it were a real hill. Tell me what you are doing as you do it. The fact that there is this pretend hill doesn't mean there won't be real hills on the route we'll take."

(If you are going for the hazmat endorsement, the examiner will likely ask you to pretend there's a railroad crossing on the route. At the pretend crossing, you should do whatever you would do at a real crossing. You should tell the examiner what you are doing.)

"There will be no trick directions to get you to do something illegal or unsafe."

Of course, if the examiner gives you specific or different instructions, follow them to the letter.

Make sure you understand the instructions. Ask questions if you don't. The examiner will be prepared to answer them.

Along the way, the examiner will mark the test form. This doesn't mean you have done something wrong. It could mean you did something right. The examiner will probably tell you not to worry about this. That's good advice. Don't let the examiner distract you. Pay attention to your driving.

Test Exercises

Each feature of the test route is designed to challenge one or more safe driving skills. Let's look at some of the challenges this route presents.

Left and right turns. Turns test your ability to give the right signals when changing speed or direction. They also test your ability to position the vehicle before and during a turn to keep other vehicles from passing on the wrong side. You must also prevent problems caused by off-tracking in the turn.

The examiner will watch you approach the turn, stop if necessary, enter the turn, and complete the turn. Slow far enough in advance as you approach a turn. You should slow to 15 mph when you approach a blind intersection. Downshift if necessary.

Don't forget to signal that you are slowing down. Start signaling that you intend to turn when you are 100 feet from making the turn.

Don't pull up too closely to the vehicle in front of you. You should be able to see that vehicle's rear wheels.

When there's a stop light, stop sign, or yield sign, you will be marked wrong if you fail to stop. (Stopping because a vehicle blocks you partway through the turn isn't marked.) If you must stop, stop at the stop line, or within five feet of it. Don't stop out in the intersection, over the stop line, or past the sidewalk. Don't stop past the stop sign or other intersection marker. If there's a vehicle in front of you, don't pull up too closely to it. You should be able to see that vehicle's rear wheels.

If you do stop, come to a full stop. Don't roll or coast through the intersection. Your wheels should be straight ahead, not turned to the left or right.

Check for traffic. Look to the right, to the left, and to the rear. Use your mirrors. Make eye contact with other drivers and pedestrians.

Yield to vehicles having the right of way. Yield whenever it results in greater safety. Don't stop unnecessarily. That just confuses other drivers.

Start the turn from the proper lane. For a right turn, start from the right-most lane, but not over the marking on the left side of the lane unless it's necessary. Try to keep vehicles from coming up on your right. Turn without going over the curb or swinging too wide.

Start a left turn from the left-most lane, but not over the lane markings.

Turn left and right properly (review Chapter 7 if you don't recall the correct way to turn). Proceed at a smooth, even speed that allows you to maintain control over the vehicle. Don't stop in the middle of the turn (unless you have to for safety's sake), and don't change gears. (You may downshift if you need to before you start to turn. You may change gears again to get started up from a full stop.)

Keep both hands firmly on the wheel. Don't turn the wheel with your palms.

Complete the turn by turning into the proper lane. For a right turn, you'll lose points if your rear wheels touch or come up over the curb. You'll also lose points if you turn too wide and end up in the lane of oncoming traffic. An unnecessary button-hook will also be marked as an error. You should complete a right turn in the right-most lane.

For a left turn, you'll lose points if you end up in the right lane, or in the lane of oncoming traffic.

You'll be marked for an error if you cause traffic to back up behind you on any turn.

Cancel your turn signal. Accelerate smoothly. If you've made a left turn and are in the left lane, check your mirrors and move to the right lane as soon as it's clear.

Traffic lights, stop signs, and uncontrolled intersections. Obey all traffic signs. When signs tell you to stop, come to a full stop. Don't roll or coast through the stop.

At uncontrolled intersections, check for traffic. Look to the right, to the left, and to the rear. Use your mirrors. Make eye contact with pedestrians and other drivers.

Yield to vehicles having the right of way. Yield whenever it results in greater safety. Don't stop unnecessarily. That just confuses other drivers.

A straight section of a city business street. Watch for hazards at the side of the road and from entrances onto the street. Slow for hazards as soon as you see them. Plan at least seven to 15 seconds ahead for lane changes.

Travel in the right lane if it's clear. Choose the center lane if the right lane is obstructed by tree branches, utility poles, and so forth. You would also choose the center lane if the traffic in the right lane causes you to continually change out of it. If there is a lot of traffic entering and exiting the right lane, choose the center lane so you can travel at a safe speed without a lot of braking and shifting.

Center your vehicle in the lane. Don't drift to the right or left.

Keep up with the flow of traffic. Choose a speed that lets you travel without lots of slowing, stopping, and speeding up. Keep a following distance of one second for every 10 feet of vehicle length at 40 mph. Add a second for speeds over 40 mph. Avoid riding behind larger vehicles that might block your view.

The examiner may tell you to change lanes. First, check the traffic to the front and rear. Give extra attention to your blind spot.

Signal for the lane change.

Don't tailgate the vehicle ahead of you while you wait for a gap in the lane. Don't change lanes until you have a large enough gap to do so safely.

CDL Basic Control Skills Practice Test Performance Record

Instructions: Have an observer rate your performance by checking a box each time you make a pullup or encroachment. An incorrect final position counts as 1 error. The observer should mark NO ERRORS if your performance was error-free.

Test Item	No Errors	Errors				
		1	2	3	4	5
Offset Backing						
Pullups						
Encroachments						
Clearance						
Alley Dock						
Pullups						
Encroachments						
Final Position (within 2 ft. of rear dock)						
Forward Stop/Straight Line Backing						
Pullups						
Encroachments						
Final Position (within 2 ft. of stop line)						
Serpentine						
Pullups						
Encroachments						
Parallel Park (Conventional)						
Pullups						
Encroachments						
Final Position (within 18 in. of curb)						
Parallel Park (Sight Side)						
Pullups						
Encroachments						
Final Position (within 18 in. of curb)						

Figure 15-11. Sample skills test form.

CDL Basic Control Skills Practice Test
Self-Scoring Form

Instructions: Use the information from the Performance Record to score how well you did on the Basic Control Skills Practice Test.

Test Item	Self - Scoring	
	Pass	No Pass
Offset Backing		
Pullups (Falling = more than 4)	☐	☐
Encroachments (Falling = driving over marker or cone)	☐	☐
Clearance (Falling = wheel less than 1 ft. or more than 3 ft. from marker)	☐	☐
Alley Dock		
Pullups (Falling = more than 4)	☐	☐
Encroachments (Falling = backing over marker or curb)	☐	☐
Final Position (Falling = more than 2 ft. from, or over rear dock)	☐	☐
Forward Stop/Straight Line Backing		
Pullups (Falling = more than 3; more than 4 for truck and trailer)	☐	☐
Encroachments (Falling = jackknife)	☐	☐
Final Position (Falling = more than 2 ft. from, or over stop line)	☐	☐
Serpentine		
Pullups (Falling = more than 5)	☐	☐
Encroachments (Falling = backing over marker or curb)	☐	☐
Parallel Park (Conventional)		
Pullups (Falling = more than 3)	☐	☐
Encroachments (Falling = backing over marker or curb)	☐	☐
Final Position (Falling = further than 18 in. from curb)	☐	☐
Parallel Park (Sight Side)		
Pullups (Falling = more than 3)	☐	☐
Encroachments (Falling = backing over marker or curb)	☐	☐
Final Position (Falling = further than 18 in. from curb)	☐	☐

Blend smoothly with the traffic in the "new" lane. Don't turn sharply or abruptly. Maintain your speed. Move to the center of the lane. Regain your following distance in the new lane of traffic. Keep a traffic gap to the front and the rear.

Cancel the signal when you complete the lane change.

Two intersections where you must stop. Check for traffic to the right, to the left, and to the rear. Use your mirrors. Make eye contact with pedestrians and other drivers. Take your foot off the accelerator. Slow far enough in advance as you approach a turn. You should slow to 15 mph when you approach a blind intersection. Downshift if necessary. Don't forget to signal that you are slowing.

Don't pull up too closely to the vehicle in front of you. You should be able to see that vehicle's rear wheels.

Stop at the stop line, or within five feet of it. Don't stop out in the intersection, over the stop line, or past the sidewalk. Don't stop past the stop sign or other intersection marker. If there's a vehicle stopped in front of you, don't pull up too closely to it. You should be able to see that vehicle's rear wheels.

Come to a full stop. Don't roll or coast through the intersection.

A through intersection. Check for traffic to the right, to the left, and to the rear. Use your mirrors. Make eye contact with pedestrians.

Yield to pedestrians and other vehicles in the intersection.

Don't change lanes, or gears, in the intersection. Be in the right gear so you can proceed slowly enough without stalling. Don't allow the vehicle to drift back. Don't lug or rev the engine. Don't disrupt the flow of traffic.

Keep your hands on the wheel.

A multi-lane section of roadway. This tests your ability to choose a safe gap for changing lanes, passing, and crossing and entering traffic. Check traffic to the front and rear, especially your blind spot. Signal your intent to change lanes. Don't tailgate while you're waiting to change lanes. Wait for enough of a gap before you make your move.

Blend smoothly with the traffic in the "new" lane. Don't turn sharply or abruptly. Maintain your speed. Move to the center of the lane. Regain your following distance in the new lane of traffic. Keep a traffic gap to the front and the rear.

Cancel the signal when you complete the lane change.

Moderate traffic. Don't drive faster than needed to stay with the flow of traffic. Driving too slowly isn't any safer than driving too fast. Keep the proper following distance for your vehicle, speed, and driving conditions.

Be visually alert. Scan "the big picture," and check your mirrors often. Anticipate problems other drivers might present.

One uncontrolled railroad crossing. Your test route may have a real railroad crossing or a pretend one. When there's no signal, you must stop and check for trains. Activate the four-way flashers as you approach the crossing. Stop within 50 feet and no closer than 15 feet. Look to the left and right. Roll down the window and listen for a train.

For the hazardous materials endorsement, you should stop no closer than 15 feet from the nearest rail. You may have to set the parking brake and shift into neutral. Look and listen for a train before crossing the track.

Don't change gears, stop, or brake while on the track. Don't pass or change lanes. Travel at the posted speed limit, or slower if that would be safer. Don't pass another vehicle while crossing the track. Keep your hands on the steering wheel.

Of course, if the test route includes a controlled railroad crossing, you should obey the signals.

Review the regulations that cover railroad crossings. This is covered in Chapter 5.

Turn off the flashers when you have completed the crossing.

A curve to the left or right. Reduce your speed before you enter the curve. Maintain your speed in the curve. Stay in your lane. Don't let any wheels cross into the next lane. Make continual traffic checks. Make an extra effort to keep track of following vehicles when coming out of the curve.

A section of expressway or two-lane rural or semi-rural road. This tests your ability to merge, change lanes, and exit. When you enter the expressway, check the traffic. Check to the front and the rear. Pay special attention to your blind spot.

Signal as soon as the traffic on the expressway can see your signal.

Don't tailgate, and don't slow up traffic following you. Merge without stopping. Don't exceed the ramp speed.

Accelerate to the flow of traffic in the acceleration lane. Don't turn sharply onto the expressway lane. Don't lug or rev the engine. Move to the center of the right-most driving lane. Cancel your signal.

When you change lanes, check traffic to the front and rear, especially your blind spot. Signal your intent to change lanes. Don't tailgate while you're waiting to change lanes. Wait for enough of a gap before you make your move.

Blend smoothly with the traffic in the "new" lane. Don't turn sharply or abruptly. Maintain your speed. Move to the center of the lane. Establish your following distance in the new lane of traffic. Keep a traffic gap to the front and the rear.

Cancel the signal when you complete the lane change.

To exit the expressway, start with a traffic check. Pay special attention to the right blind spot. Turn on your right-turn signal.

Make a smooth change into the exit (deceleration) lane. Don't make a sharp turn. Enter the exit lane at its beginning.

Slow in the deceleration lane. When you're on the ramp, travel at or below the posted speed. Don't accelerate in the curve. Use the correct downgrade or upgrade shifting and braking procedure for graded curves.

Don't tailgate on the ramp. Cancel your signal when you're done exiting.

A downgrade and an upgrade. The upgrade on your route will challenge your ability to change gears to maintain your speed. Don't lug the engine. Stay in the right-most lane. Put on your four-way flashers if you are going much slower than the flow of traffic. Check the traffic to the left and the rear often. Use your mirrors.

The downgrade will test your ability to select a safe speed for going down the hill. It will test your use of the brakes. Select a lower gear before you start down the grade. Use a gear that will require little if any braking to control the vehicle. Don't rev the engine. You should absolutely not be in neutral.

Check your brakes before you head down the grade. Press on the treadle. Feel the brakes apply. Listen for a release of air.

Select a speed that is not too fast for the weight of the vehicle, the length or steepness of the grade, the weather, and road conditions. Once you have this "safe" speed, apply the brake hard enough to feel a definite slowing. When your speed has been reduced to five mph below the "safe" speed, release the brakes. This application should last for about three seconds. Once the speed has increased to the "safe" speed, repeat the procedure.

Use a steady, light braking. Don't fan, pump, or ride the brakes. Maintain an even speed.

Don't ride the clutch, rev the engine, or coast.

Increase your following distance. Stay in the right-most lane. Keep both hands on the wheel. Check the traffic to the left and the rear often. Use your mirrors.

Parking on an upgrade and a downgrade. As you approach the grade, check the traffic to the sides and the rear, using your mirrors. Turn on your right-turn signal. Get into the right-most or curb lane. Don't block any driveways or fire hydrants. Take your foot off the accelerator. Use steady braking as you slow. Downshift smoothly. Do not coast.

On the upgrade, the challenge is to park so the vehicle doesn't roll. Block the wheel against the curb. When you get going again, you'll have to start up against the pull of gravity.

To park, bring the vehicle parallel to the curb. Turn off the turn signal. Turn on the four-way flashers. Put the parking brake on. Put the transmission in neutral. Release the foot brake. Don't rest your foot on the clutch.

To start up again, check the traffic to the left and right. Pay special attention to the left side. Make eye contact with other drivers and pedestrians. Turn off the four-way flashers and turn on the left-turn signal.

Release the parking brake. Put the vehicle in gear. Don't turn the wheel before the vehicle moves. Don't stall the engine as you start up the grade.

Check the traffic to the left. Don't stall the engine as you accelerate. Blend smoothly with the traffic. Don't turn sharply into the traffic. Match your rpm to the road speed.

An underpass, low clearance, or bridge. An underpass or low clearance will test your ability to judge if it's safe to continue. You should be able to determine what the clearance is from the signs. You should know your vehicle height and be able to decide if you can make it under the underpass.

Before you cross a bridge, you'll have to decide if it can support your vehicle's weight. You'll have to know your vehicle's weight. Then you'll have to determine the bridge's weight limit from the signs.

After you drive under an overpass, the examiner may ask you to state what the posted clearance or height was. After you go over a bridge, the examiner may ask you to state what the posted weight limit was. If the test route does not have a bridge or an overpass, you may be asked about another traffic sign. Be prepared to identify any traffic sign that may be on the route. Be prepared to explain to the examiner the meaning of the sign.

 Not all test routes will have a grade. You may be asked to describe in words the process of driving up and down grades. This is something that you can easily practice without a vehicle.

Straight-line backing. You may be asked to back your vehicle in a straight line. Before you begin, turn on the four-way flashers and tap your horn. Keep straight and back slowly. Turn your head and check your mirrors often throughout the backing. You will be allowed to pull forward five times if needed to complete the maneuver.

Ninety-degree angle backing. Turn on the four-way flashers and tap the horn. Back slowly, turning your head and checking the mirrors throughout. You will be allowed to pull forward five times if needed to complete the maneuver.

General driving behavior. Throughout the test, you should demonstrate safe driving practices in general. Don't shift in turns and intersections. Downshift and upshift smoothly, without grinding the gears. You'll have to double-clutch if your vehicle doesn't have synchromesh gears. Don't over-rev or lug the engine. Don't coast with the clutch in. Don't ride the clutch to control your speed. Don't "snap" or "pop" the clutch.

Use the brakes correctly. Don't ride them, pump them, or fan them. Brake smoothly with steady pressure.

Use the correct steering techniques. Keep both hands on the wheel. Grip the wheel properly. Don't turn the wheel with your palms. Don't understeer or oversteer.

Obey all traffic signals. Pull over to the right and stop if you hear a siren or see the flashing lights of emergency vehicles. Drive within the speed limit or below it if conditions demand it. Don't drive any slower than safety requires. Don't make any unnecessary stops.

Obey all traffic signs. Be alert for and obey signs that ban heavy vehicles from certain roads or bridges.

Be alert for, and avoid, traffic hazards. When you check for traffic, move your head right or left. That way the examiner can clearly see you are checking. Avoid bumping other vehicles, objects, pedestrians, animals, and so on. Never put the vehicle up on a curb or sidewalk.

Stay centered in the correct lane. Don't wander across lane lines. Don't stop over the stop line. Start and complete the turn and lane change from the proper lane.

If you're taking your test at night, turn on your lights as soon as the law, or need for additional lighting, requires them. Use your high beams only to improve visibility. Be careful not to blind other drivers. Use the windshield washers and wipers and the heater/defroster if the weather calls for them.

Wear your seat belt.

Finally, obey the basic rules of the road that all drivers must follow. It would be a good idea to read your state's manual for automobile drivers. Review the rules about signs, road markings, right of way, and so forth. These rules apply to all road users, from the little compact car driver to the driver of an 18-wheeler. It is expected that you know and will obey these rules.

Also remember that states may have added requirements for CMV drivers beyond the federal regulations. It's your responsibility to inform yourself of these and fulfill them when you take your tests.

Road Test Scoring

Different states score the Road Test different ways. You may start out with a certain number of points, say 100. You lose a point for every error. To pass, you must end up with enough points left, like 75.

Or, you may earn a point for each mistake. If you earn too many points, 25 for example, you fail the test.

As we've already mentioned, your Road Test score might be combined with your Skills Test score.

You could make a mistake on a particular route test feature. Or, you could earn an error point for a general driving mistake, like grinding the gears. As you know, disobeying a traffic law and causing an accident are two mistakes that result in immediate failure. You might also fail the test if you get into a near accident, even if you manage to maneuver your way out of it.

If you're concerned about how the test will be scored, ask the examiner. Then put it out of your mind and concentrate on your driving.

Safe Driving Skills Practice

Practice these basic safe driving skills as you go about your normal routine. Perhaps you've gotten in the habit of stopping over the stop line or crowding the vehicle ahead of you. Do you neglect to signal when you change lanes? If your driving has gotten a little sloppy, start tightening up. Start driving according to the standards the Road Test will use. When you've finished, take a moment to assess your own performance.

Start now to get in the habit of doing things by the book. Then it will come naturally to you when you take the test. You won't have to think about it.

Well, that's it. There's not much left to do now but take the CDL tests. Have you met all your **PASS** mini-goals? Did you take all the sample, pre-trip, and post-trip tests? Have you worked on doing a thorough vehicle inspection? Are you practicing your basic vehicle control and safe driving skills?

If so, you've done a lot of work to prepare for your CDL tests. And, you did it on your own. No one stood over your shoulder and told you what to do. You can feel good about that. (Not everyone has that stick-to-it-iveness!) At this point, taking the CDL tests should be a piece of cake!

Chapter 16 is the last chapter in this book. In it you'll find tips to make your Test Day go smoothly.

Chapter 16

Final Tips for the Test

When you have finished this chapter, you will have learned:

- how to prepare your vehicle for the CDL tests
- how to arrange to take the CDL tests
- tips for taking the CDL tests
- retaking the CDL tests

To complete this chapter, you will need:

- the owner's manual for your vehicle
- the Federal Motor Carrier Safety Regulations pocketbook (or access to U.S. Department of Transportation regulations, Parts 383, 393, and 396 of Subchapter B, Chapter 3, Title 49, Code of Federal Regulations)
- a CDL preparation manual from your state Department of Motor Vehicles, if one is offered
- the driver's manual from your state Department of Motor Vehicles

Before we go any further, we'd like to congratulate you on getting this far. This was not the shortest, or most entertaining, book you could read in your spare time. You probably didn't become a CMV driver because you thought it would give you the chance to do lots of reading! But you did it. Now all that's left to do is to take the tests. Here's some final tips that should make Test Day a little easier on you.

PREPARE YOUR VEHICLE

There are a few things you can do with your vehicle that will help you during the test. These are:

- have the right vehicle
- have the vehicle in good working order
- be familiar with the vehicle

Have the Right Vehicle

This is just a reminder. You must take your CDL Skills Tests in a "representative vehicle." So, if you want a Group A CDL, you must take the tests in a Group A vehicle. If you don't want an air brake restriction, you must have air brakes on your vehicle.

States do not provide a vehicle for you to use in the tests. You must bring your own.

Have the Vehicle in Good Working Order

You know you will inspect your vehicle as part of the CDL tests. That would be a bad time to find out that it had some safety defect. Better to take plenty of time, well in advance, to have your vehicle thoroughly checked out. Get your vehicle in as close-to-perfect working order as your money will allow. That way you won't have to worry about the condition of your vehicle while you're taking the tests.

It wouldn't hurt to spruce up your vehicle, either. Tighten up loose fixtures. Repair minor flaws. Wash the vehicle. No, you won't be scored on whether your truck is clean or not. A clean vehicle will be easier to inspect. It will reflect well on you. Also, it will give you a feeling of confidence.

Inspection Equipment

Equip yourself with the following items to use during the Vehicle Inspection Test:

- flashlight
- tire pressure gauge
- tire tread depth gauge

A flashlight will help you see into the dark corners of your engine compartment. You'll be better able to see under the vehicle and in the cargo compartment. The pressure and tread depth gauges will give you the best tire measurements. Your state may not require that you inspect your tires this accurately. It isn't wrong to do so, though, and it is more professional.

Be Familiar with the Vehicle

You want to be completely familiar with the vehicle in which you'll take your tests. If it's a new vehicle, or one you don't drive often, try to get some time with it before Test Day. Learn where all the controls are. Get the feel of how they respond. Know which of the vehicle's noises are normal and which are signs of trouble. Get used to how the vehicle handles.

Are there some controls you don't use often, or at all? At least find out what they do and how to work them. The examiner could ask you to explain every lever and switch in the cab. "I don't know what that does. I never use it," is not a good response.

Read the owner's manual. You'll find many useful facts about what's normal for your vehicle and how best to operate it.

ARRANGING FOR THE CDL TESTS

The process of taking the CDL tests differs from state to state. You can get the most current details from your local state DMV Office. You can find the addresses of the main state DMV offices listed in the back of this book. Or, check your city telephone listings. Many have websites.

The following is an example of what happens in some states.

License Fees

Yes, it does cost money to apply for and test for a CDL. Fees vary from state to state.

Testing Sites

The written test will almost always be given at driver license stations. Not all such stations are large enough or well enough equipped for the Skills Test. These may sometimes be given at another site. Or, a third party may be authorized to give the Skills Tests.

Your local DMV can tell you where the Skills Tests may be taken. If you don't know the area, you may want to drive out there in advance of Test Day. At least find it on a map. No point in getting lost on the way to the test!

If you're curious and want to check out the test site ahead of time, call the DMV first. Make sure there's no problem with simply visiting the test site.

Testing Appointments

You can almost always take the Knowledge Test on a walk-in basis. Many states, however, give the Skills Test only by appointment. Be sure to find out if this is the case in your state.

Age Limits

Most states require CDL applicants to be at least 21 years old. Drivers between 18 and 21 years of age may sometimes apply for a Restricted CDL. This limits them to intrastate driving.

Instruction Permits

Instruction permits are mandatory. A CDL permit allows you to drive the type of commercial vehicle for which you are seeking to obtain your license. Whenever you drive your CMV during this period, you must have with you someone who holds a CDL for the same, or higher, vehicle group. This person must sit beside you as you drive.

To obtain the Class A permit, you must pass the permit Knowledge Test along with the Knowledge Test pertaining to the license you wish to obtain. You must also pass the Knowledge Test for the type of CDL and endorsements you want.

The fee for instruction permits will vary from state to state. Check with your local DMV for pricing.

Time Limits

The CDL tests can take anywhere from 45 minutes to one-and-a-half hours. Check with your DMV to see if this is the case in your state.

It's not a great idea to show up at a driver's license station at 4:55 P.M. wanting to take the CDL tests. That only forces the DMV staff to work overtime. That is not a good use of your tax dollars.

Keep the length of the testing sessions in mind when planning for test day. Will you need time off from work? Will you need to find a babysitter? If so, arrange to have enough time so you don't feel rushed while taking the tests.

Other Necessary Documents

Don't forget, there are a few other tests you must take besides those for your CDL. You must get a DOT physical if your current medical card has expired. All states require a medical examiner's card. You will have to show proof of the medical examiner's card to the DMV to be issued a CDL.

Most applicants will be expected to bring photo identification or a driver's license and a Social Security card.

States may require you to have other documents, licenses, and certificates as well. This is especially true for drivers seeking endorsements, especially the Tank Vehicle and Hazardous Materials endorsements. Check with your local DMV.

Special Services

Some states offer an oral Knowledge Test, instead of a written one. If you're interested in this option, be sure to ask about it ahead of time. You may need an appointment to take an oral rather than a written Knowledge Test.

TAKING THE TEST

Once you're in your seat with your pencil (or wheel) in hand, preparation time is over. Now's the time for the proof. The best thing you can do is relax and stay focused on the tests. Here's just a few more tips that may come in handy.

Knowledge Test

Read the instructions. If there are no instructions and you're unsure of how to proceed, ask the examiner.

Scan the entire test once. Immediately answer all the questions you're 100 percent certain about. Then go back and tackle the harder ones.

Sometimes you may find the answer to one question in another question. If you get stuck, look for other questions on the same subject. That may be just enough to jog your memory.

Once you've answered a question, don't change your answer. Your first impulse is probably correct. Change an answer only if you are positive your first answer was wrong.

If you change an answer, erase the old one cleanly. Mark the new answer clearly. Sometimes tests are scored by machines. Such machines can read even faint pencil marks. If it appears you selected two answer choices, you'll be marked wrong. This is true even if one of your choices was correct.

Be careful marking the answer form. Don't put the "right" answer in the "wrong" space. Just before completing the test, check and recheck your answers.

Don't leave any blanks. If you're not sure of an answer, use intelligent guessing to narrow your choices. When you absolutely don't know the answer, use wild guessing. There's still a slim chance you'll choose the right one. A question left blank can only be marked "wrong," however.

Relax, and let your thoughts come. When you feel yourself tensing up or getting tired, try these stress-relievers:

Neck rolls. Slowly, allow your head to droop forward, as if you were going to rest your chin on your chest. Then, just as slowly, raise your head back up. Slowly, tilt your head to the right, as if you could touch your right ear to your right shoulder. Raise your head back up. Slowly, drop your head back and look up at the ceiling. Bring your head back to the face-forward position. Last, tilt your head to the left. Repeat the whole series five times.

Next, keeping your neck straight, swivel just your head to the right and look along the line of your right shoulder. Swivel slowly back and to the left. Return to the face-forward position. Then repeat the movement, beginning at the left this time. Do this five times also.

Deep breathing. When people concentrate hard, they often hold their breath. That cuts off the supply of oxygen to the brain. Do that for too long and you'll become fatigued. You'll find it hard to think clearly.

Every now and then, make it a point to take a deep breath. Breathe in slowly and deeply through your nose. Breathe out just as slowly through your mouth.

Muscle relaxer. Some people hunch their shoulders when they're working hard. This is fatiguing. Here's how to break the tension. Take a deep breath. Shrug your shoulders up tight, as if you were trying to touch your ears with them. Hold that position for a count of five. Then release it all at once. Feel the tension flow out of your shoulders as you exhale.

If you've tensed up your whole body, you can do something similar. Take a deep breath. Try clenching and tensing all your muscles. Clench your jaw. Hunch up your shoulders. Squeeze the muscles in your buttocks, thighs, and calves. Curl up your toes.

Hold the tension for a few seconds, then release all the tension. Feel yourself relax as you breathe out.

Try any of these stress-relievers any time you feel tense, tired or distracted.

The Skills Test

Listen carefully to the instructions. Make sure you understand them. If you don't, ask the examiner for more details. Doing so will not cost you any points. Not understanding what you're being asked to do might.

If you get the chance, squeeze in some stress-relieving exercises between the Inspection, Skills, and Road Tests. Don't forget to breathe!

RETAKING THE TEST

Although you will likely pass the CDL tests the first time, you may be wondering what happens if you don't.

You can't take the Skills Tests until you pass the Knowledge Test. Some states will allow you to take other parts of the Skills Tests even if you fail one part. Other states will end the testing session if you fail one section.

Few states will limit the number of retakes you can have. Most will charge a fee for each retake, though. You should check with your local DMV to find out if this is the case in your state.

If you do fail a test the first time, don't give up. A few people, even well-prepared ones, do. Their nervousness about taking the test is so great that it impairs their performance.

The same people find they pass the test easily on the second try. Taking the test the first time gave them the practice and familiarity with test-taking they needed to pass.

Why worry about retaking the test? You're going to pass it the first time. We just know it.

Thank you for letting us help you prepare for the CDL tests. Please accept our best wishes for your success.

Appendix A

Answer Keys

CHAPTER 4

Pre-trip

1. B
2. B
3. B

Post-trip

1. B (Page 37)
2. B (Page 45)
3. B (Page 47, 49)
4. B (Page 37, 45)
5. B (Page 37)
6. B (Page 50)
7. B (Page 47)
8. B (Page 50)
9. B (Page 42)
10. B (Page 47)

CHAPTER 5

Pre-trip

1. B
2. A
3. B

Post-trip

1. B (Page 59)
2. A (Page 60)
3. B (Page 61)
4. B (Page 59)
5. C (Page 63)
6. A (Page 91)
7. B (Page 93)
8. C (Page 90)
9. A (Page 81)
10. B (Page 90)

CHAPTER 6

Pre-trip

1. B
2. B
3. A

Post-trip

1. A (Page 117)
2. B (Page 116)
3. D (Page 117)
4. B (Page 118)
5. D (Page 118)
6. C (Page 118)
7. D (Page 119)
8. B (Page 123)
9. C (Page 125)
10. B (Page 130)

CHAPTER 7

Pre-trip

1. A
2. B
3. A

Post-trip

1. A (Page 145)
2. C (Page 148)
3. A (Page 153)
4. C (Page 154)
5. B (Page 156)
6. A (Page 157)
7. D (Page 159)
8. A (Page 162)
9. B (Page 165)
10. A (Page 171)

CHAPTER 8

Pre-trip

1. B
2. B
3. B

Post-trip

1. C (Page 179)
2. D (Page 180)
3. B (Page 180)
4. D (Page 180)
5. B (Page 181)
6. A (Page 181)
7. B (Page 181)
8. B (Page 182)
9. D (Page 185)
10. C (Page 187)

CHAPTER 9

Pre-trip

1. A
2. A
3. B

Post-trip

1. A (Page 197)
2. D (Page 198)
3. B (Page 198)
4. B (Page 199)
5. D (Page 199)
6. B (Page 200)
7. B (Page 200)
8. B (Page 201)
9. A (Page 202)
10. A (Page 203)

CHAPTER 10

Pre-trip

1. B
2. B
3. B

Post-trip

1. B (Page 215)
2. C (Page 218)
3. B (Page 219)
4. C (Page 222)
5. D (Page 222)
6. C (Page 224)
7. B (Page 224)
8. C (Page 226)
9. C (Page 238)
10. A (Page 238)

CHAPTER 11

Pre-trip

1. B
2. A
3. A

Post-trip

1. C (Page 245)
2. C (Page 245)
3. A (Page 246)
4. C (Page 246, 247)
5. B (Page 248)
6. A (Page 251)
7. D (Page 245)
8. A (Page 261)
9. A (Page 257)
10. C (Page 261)

CHAPTER 12

Pre-trip

1. A
2. A
3. A

Post-trip

1. A (Page 267)
2. C (Page 268)
3. A (Page 268)
4. A (Page 268)
5. A (Page 269)
6. C (Page 271)
7. D (Page 271)
8. B (Page 271)
9. C (Page 272)
10. B (Page 278)

CHAPTER 13

Pre-trip

1. B
2. A
3. A

Post-trip

1. A (Page 284)
2. A (Page 285)
3. D (Page 291, 293)
4. A (Page 296)
5. D (Page 300, 303, 305)
6. D (Page 284)
7. A (Page 318)
8. B (Page 317)
9. B (Page 317)
10. A (Page 322)

CHAPTER 14: SAMPLE CDL TESTS

Sample Test 1

1. A (Page 37)
2. A (Page 41)
3. B (Page 45)
4. D (Page 42)
5. C (Page 43)
6. C (Page 43)
7. B (Page 47)
8. C (Page 65)
9. D (Page 63)
10. A (Page 61)
11. C (Page 69)
12. D (Page 82)
13. B (Page 88)
14. A (Page 91)
15. B (Page 94)
16. A (Page 102)
17. C (Page 117)
18. A (Page 118)
19. D (Page 125)
20. D (Page 126)
21. A (Page 137)
22. A (Page 145)
23. B (Page 148)
24. B (Page 149)
25. B (Page 153)
26. D (Page 159)
27. C (Page 161)
28. C (Page 163)
29. C (Page 165)
30. B (Page 167)
31. C (Page 170)
32. A (Page 172)
33. A (Page 172)
34. D (Page 179)
35. D (Page 181)
36. A (Page 181)
37. D (Page 182)
38. C (Page 182)
39. A (Page 186)

40. C (Page 187)
41. D (Page 190)
42. D (Page 196)
43. A (Page 196)
44. D (Page 199)
45. C (Page 200)
46. D (Page 202)
47. D (Page 203)
48. D (Page 205)
49. B (Page 207)
50. D (Page 207)

and a few extras:

51. B (Page 118)
52. A (Page 117)
53. C (Page 117)
54. D (Page 117)

Sample Test 2

1. D (Page 44)
2. B (Page 44)
3. A (Page 60, 61)
4. B (Page 42)
5. B (Page 41)
6. B (Page 43)
7. C (Page 46)
8. B (Page 46)
9. B (Page 65)
10. B (Page 66)
11. B (Page 69)
12. A (Page 83)
13. B (Page 63)
14. D (Page 92)
15. A (Page 102)
16. A (Page 116)
17. D (Page 117)
18. C (Page 121, 123)
19. B (Page 125)
20. B (Page 135)
21. D (Page 146)
22. D (Page 146)
23. D (Page 148)
24. C (Page 150)
25. B (Page 156)
26. C (Page 158, 160)
27. B (Page 162)
28. D (Page 163)

29. D (Page 165)
30. A (Page 167)
31. C (Page 170)
32. A (Page 172)
33. D (Page 179)
34. C (Page 180)
35. A (Page 181)
36. D (Page 181)
37. B (Page 182)
38. D (Page 185)
39. B (Page 186)
40. D (Page 187)
41. C (Page 190)
42. B (Page 196)
43. C (Page 197)
44. B (Page 199)
45. D (Page 202)
46. B (Page 203)
47. B (Page 204)
48. A (Page 207)
49. A (Page 209)
50. B (Page 209)

and a few extras:

51. C (Page 180)
52. D (Page 180)
53. B (Page 180)
54. A (Page 180)

Air Brakes Questions

1. D (Page 218)
2. D (Page 218)
3. C (Page 222)
4. D (Page 222)
5. D (Page 222)
6. C (Page 222)
7. B (Page 223)
8. B (Page 223)
9. C (Page 224)
10. C (Page 224)
11. B (Page 224)
12. D (Page 225)
13. C (Page 225)
14. A (Page 226)
15. D (Page 226)
16. C (Page 226)
17. A (Page 228)

18. B (Page 232)
19. C (Page 228, 233)
20. C (Page 228, 233)
21. D (Page 233)
22. C (Page 233)
23. D (Page 233)
24. D (Page 234)
25. D (Page 233)
26. A (Page 236)
27. B (Page 236)
28. C (Page 236)
29. A (Page 236)
30. B (Page 235)

Combination Vehicles/ Doubles/Triples Questions

1. B (Page 244)
2. B (Page 244)
3. D (Page 245)
4. D (Page 245)
5. B (Page 246)
6. B (Page 245)
7. D (Page 246)
8. D (Page 250)
9. A (Page 250)
10. C (Page 251)
11. A (Page 251)
12. A (Page 245)
13. D (Page 245)
14. B (Page 252)
15. D (Page 252)
16. B (Page 252)
17. D (Page 252)
18. D (Page 257)
19. D (Page 259)
20. B (Page 264)
21. B (Page 248)
22. B (Page 248)
23. C (Page 250)
24. D (Page 250)
25. A (Page 250)
26. D (Page 248)
27. B (Page 252)
28. D (Page 259)
29. D (Page 259)
30. A (Page 261)

Tank Vehicles Questions

1. A (Page 268)
2. D (Page 268)
3. A (Page 268)
4. B (Page 270)
5. B (Page 271)
6. D (Page 271)
7. C (Page 271)
8. C (Page 271)
9. D (Page 271)
10. A (Page 271)
11. D (Page 272)
12. B (Page 273)
13. D (Page 274)
14. C (Page 274)
15. A (Page 276)

Hazardous Materials Questions

1. B (Page 282)
2. C (Page 282)
3. D (Page 282, 283)
4. C (Page 284)
5. D (Page 284)
6. B (Page 284)
7. A (Page 285)
8. A (Page 286)
9. D (Page 287)
10. D (Page 289)
11. B (Page 281)
12. B (Page 291)
13. A (Page 291)
14. D (Page 293)
15. B (Page 293)
16. C (Page 295)
17. A (Page 295)
18. A (Page 296)
19. D (Page 299, 300)
20. A (Page 301)
21. D (Page 295, 303)
22. D (Page 307)
23. D (Page 307)
24. A (Page 308)
25. A (Page 308)
26. D (Page 308)
27. B (Page 308)
28. C (Page 308)
29. D (Page 309, 312)
30. D (Page 309, 312)
31. D (Page 313)
32. D (Page 309, 312, 313)
33. A (Page 310, 311, 313)
34. B (Page 284)
35. D (Page 314)
36. A (Page 309)
37. B (Page 314)
38. C (Page 321)
39. D (Page 303, 314)
40. A (Page 320)
41. B (Page 321)
42. B (Page 321)
43. B (Page 322)
44. D (Page 322)
45. B (Page 320)

Appendix B

Driver's License Administration by State

Visit the web site in your state for useful resources such as: state-specific commercial driver license regulations, licensing requirements, license offices and hours, and applications and fees.

Alabama Department of Public Safety
https://www.alea.gov/dps
(334) 242-4400

Alaska Division of Motor Vehicles
http://doa.alaska.gov/dmv/akol/index.htm
(907) 269-5551

Arizona Department of Transportation Motor Vehicle Division
https://azdot.gov/motor-vehicle-services
(602) 255-0072

Arkansas Department of Finance and Administration Office of Driver Services
http://www.arkansas.gov/dfa/driver_services/Pages/default.aspx
(501) 371-5581

California Department of Motor Vehicles
https://www.dmv.ca.gov/portal/
(800) 777-0133

Colorado Department of Revenue Division of Motor Vehicles
http://www.colorado.gov/revenue/dmv/
(303) 205-5600

Connecticut Department of Motor Vehicles
http://www.ct.gov/dmv/taxonomy/ct_taxonomy.asp?DLN=41638&dmvNav=|41638|
In Connecticut: (860) 263-5700; from elsewhere: (800) 842-8222

Delaware Division of Motor Vehicles
https://www.dmv.de.gov/
Georgetown (302) 853-1000 or 1004; New Castle (302) 326-5000 or 5005; Dover (302) 744-2500; Wilmington (302) 434-3200

District of Columbia Department of Motor Vehicles
http://dmv.dc.gov
(202) 737-4404

Florida Division of Driver Licenses
http://www.flhsmv.gov/ddl/dlclass.html
(850) 617-2000

Georgia Department of Driver Services
https://dds.georgia.gov/
In the Metro Atlanta Area: (678) 413-8400, (678) 413-8500, or (678) 413-8600; from elsewhere: (866) 754-3687 or (404) 657-9300

Hawaii Department of Transportation Motor Vehicle Safety Office
http://hidot.hawaii.gov
(808) 692-7650

Idaho Transportation Department
http://www.itd.idaho.gov/dmv/DriverServices/CDL.htm
(208) 334-8736

Illinois Secretary of State Commercial Driver's License
http://www.ilsos.gov/facilityfinder/facility
In-state: (800) 252-8980; from elsewhere: (217) 785-3000

Indiana Bureau of Motor Vehicles
http://www.in.gov/bmv/2854.htm
1-888-692-6841

Iowa Department of Transportation Motor Vehicle
Division Office of Driver Services
http://www.iowadot.gov/mvd/ods/cdl.htm
(800) 925-6469

Kansas Department of Revenue
http://www.ksrevenue.org/vehicle.html
(785) 296-3621

Kentucky Transportation Cabinet
*http://drlic.kytc.ky.gov/cdl/commercial_driving_
license.htm*
(502) 564-1257

Louisiana Department of Public Safety Office of
Motor Vehicles
*https://web01.dps.louisiana.gov/omvfaqs.nsf?Open
Database&Start=1&Count=1000&Expand=1*
(225) 379-1200

Maine Department of the Secretary of State Bureau
of Motor Vehicles
http://www.maine.gov/sos/bmv/licenses/index.html
(207) 624-9186

Maryland Department of Transportation Motor
Vehicle Administration
https://mva.maryland.gov/Pages/default.aspx
(410) 768-7000

Massachusetts Registry of Motor Vehicles
*https://www.mass.gov/orgs/massachusetts-registry-of-
motor-vehicles*
In the 339, 617, 781, or 857 area codes or from out-
of-state: (857) 368-8000; from elsewhere:
(800) 858-3926

Michigan Department of State
*http://www.michigan.gov/sos/0,4670,7-127-1627---,
00.html*
(888) SOS-MICH; (888) 767-6424

Minnesota Department of Public Safety Driver and
Vehicle Service Division
https://dps.mn.gov/divisions/dvs/Pages/default.aspx
(651) 297-3298

Mississippi Department of Public Safety
http://www.dps.state.ms.us/driver-services/
(601) 987-1212

Missouri Department of Revenue
http://dor.mo.gov/drivers/
(573) 526-2407

Montana Department of Justice Driver Services
http://www.mdt.mt.gov/travinfo/drivlic.shtml
(406) 444-4536

Nebraska Department of Motor Vehicles
*http://www.dmv.state.ne.us/examining/licpermits.
html#top*
(402) 471-3861

Nevada Department of Motor Vehicles
http://www.dmvnv.com/nvdl.htm
Las Vegas: (702) 486-4368; Reno/Sparks/Carson
City: (775) 684-4368; Rural Nevada (Toll Free):
(877) 368-7828; hearing impaired: (775) 684-4904

New Hampshire Department of Safety Division of
Motor Vehicles
*http://www.nh.gov/safety/divisions/dmv/
driver-services/index.htm*
(603) 227-4020

New Jersey Motor Vehicle Commission
https://www.state.nj.us/mvc/
In NJ (888) 486-3339; from elsewhere:
(609) 292-6500; hearing impaired: (609) 292-5120

New Mexico Department of Taxation and Revenue
Motor Vehicle Division
http://www.mvd.newmexico.gov/
(888) MVD-INFO; (888) 683-4636

New York Department of Motor Vehicles
http://dmv.ny.gov/
From area codes 212, 347, 646, 718, 917, and 929:
1-212-645-5550 or 1-718-966-6155
From area codes 516, 631, 845, and 914:
1-718-477-4820
From area codes 315, 518, 585, 607, and 716:
1-518-486-9786

Telecommunications Device for the Deaf:
1-800-368-1186
from outside the State of New York:
1-518-473-5595

North Carolina Division of Motor Vehicles
*http://www.ncdot.org/dmv/driver_services/
commercialtrucking/requirements.html*
(919) 715-7000

North Dakota Department of Transportation
Drivers License Division
http://www.dot.nd.gov/public/licensing.htm
(701) 328-2600

Ohio Department of Public Safety Bureau of
Motor Vehicles
http://bmv.ohio.gov/driver_license.stm
(614) 752-7500; hearing impaired: 614-752-4559

Oklahoma Department of Public Safety
http://www.dps.state.ok.us/dls/
(405) 425-2424

Oregon Motor Vehicles Division
https://www.oregon.gov/odot/dmv/pages/index.aspx
Salem Metro Area: (503) 945-5000; Portland
Metro Area; (503) 299-9999; Bend: (541) 388-6322;
Medford: (541) 776-6025; Roseburg: (541) 440-3395;
Eugene: (541) 686-7855; hearing impaired:
(503) 945-5001

Pennsylvania Department of Transportation Driver
and Vehicle Services
http://www.dmv.state.pa.us/
In-state: (800) 932-4600; from elsewhere:
(717) 412-5300; hearing impaired (within state):
(800) 228-0676; hearing impaired (from elsewhere):
(717) 412-5380

Rhode Island Division of Motor Vehicles
http://www.dmv.ri.gov/
(401) 462-5813

South Carolina Department of Motor Vehicles
*http://www.scdmvonline.com/DMVNew/default.
aspx?n=commercial_driver_licenses*
(803) 896-5000

South Dakota Department of Public Safety Driver
Licensing Program
*http://www.state.sd.us/dps/dl/Commercial/
Commercial.htm*
(605) 773-3178

Tennessee Department of Safety
http://state.tn.us/safety/driverlicense/cdlalt.htm
(615) 253-5221, (866) 849-3548; hearing impaired:
(615) 532-2281

Texas Department of Public Safety
https://www.dps.texas.gov/
(512) 424-2600

Utah Department of Public Safety Driver License
Division
http://publicsafety.utah.gov/dld/
(801) 965-4437

Vermont Department of Motor Vehicles
https://dmv.vermont.gov/
(802) 828-2000

Virginia Department of Motor Vehicles
*http://www.dmv.state.va.us/webdoc/citizen/drivers/
applyingcdl.asp*
(804) 497-7100

Washington State Department of Licensing
http://www.dol.wa.gov/driverslicense/cdl.html
(360) 902-3900; TTY: (360) 664-0116

West Virginia Division of Motor Vehicles
*http://www.transportation.wv.gov/dmv/Pages/default.
aspx*
In state: (800) 642-9066; from elsewhere:
(304) 558-3900

Wisconsin Division of Motor Vehicles
*http://www.dot.wisconsin.gov/drivers/drivers/apply/
types/cdl.htm*
(608) 264-7049

Wyoming Department of Transportation Driver
Services
https://www.dot.state.wy.us/driverservices
(307) 777-4800

Appendix C

Motor Transport Associations

Alabama
Alabama Trucking Association
P.O. Box 242337
Montgomery, AL 36124-2337
(877) 277-TRUK
www.alabamatrucking.org

Alaska
Alaska Trucking Association
3443 Minnesota Dr.
Anchorage, AK 99503
(907) 276-1149
www.aktrucks.org

Arizona
Arizona Trucking Association
7500 West Madison St.
Tolleson, AZ 85353
(602) 850-6000
www.arizonatrucking.com

Arkansas
Arkansas Trucking Association
P.O. Box 3476
Little Rock, AR 72203

1401 West Capitol Ave.
Little Rock, AR 72201
(501) 372-3462
www.arkansastrucking.com

California
California Trucking Association
4148 E. Commerce Way
Sacramento, CA 95834
(916) 373-3500
www.caltrux.org

Colorado
Colorado Motor Carriers Association
4060 Elati St.
Denver, CO 80216
(303) 433-3375
www.cmca.com

Connecticut
Motor Transport Association of Connecticut
60 Forest St.
Hartford, CT 06105
(860) 520-4455
www.mtac.us/

Delaware
Delaware Motor Transport Association
445 Pear St.
Dover, DE 19904
(302) 734-9400
https://delawaretrucking.org/

Florida
Florida Trucking Association
350 E. College Ave.
Tallahassee, FL 32301
(850) 222-9900
www.fltrucking.org

Georgia
Georgia Motor Trucking Association
2060 Franklin Way, Suite 200
Marietta, GA 30067
(770) 444-9771
www.gmta.org

Hawaii
Hawaii Transportation Association
P.O. Box 30166
Honolulu, HI 96820
(808) 833-6628
www.htahawaii.org

Idaho
Idaho Trucking Association
3405 East Overland Rd., Suite 175
Meridian, ID 83642
(208) 342-3521
www.idtrucking.org

Illinois
Illinois Trucking Associations
7000 Adams Suite 130
Willow Brook, IL 60527
(630) 654-0884
www.iltrucking.org

Indiana
Indiana Motor Truck Association
1 North Capitol Ave., Suite 460
Indianapolis, IN 46204
(317) 630-4682
http://www.intrucking.org/

Iowa
Iowa Motor Truck Association
717 East Court Ave.
Des Moines, IA 50309
(515) 244-5193
www.iowamotortruck.com

Kansas
Kansas Motor Carriers Association
2900 S.W. Topeka Blvd.
P.O. Box 1673
Topeka, KS 66601
(785) 267-1641
www.kmca.org

Kentucky
Kentucky Motor Transport Association
617 Shelby St.
Frankfort, KY 40601
(502) 227-0848
http://kmta.net/

Louisiana
Louisiana Motor Transport Association, Inc.
4838 Bennington Ave.
P.O. Box 80278
Baton Rouge, LA 70898
(225) 928-5682
http://www.louisianatrucking.com/

Maine
Maine Motor Transport Association
P.O. Box 857
142 Whitten Rd.
Augusta, ME 04332
(207) 623-4128
www.mmta.com

Maryland
Maryland Motor Truck Association, Inc.
3000 Washington Blvd.
Baltimore, MD 21230
(410) 644-4600
www.mmtanet.com

Massachusetts
Massachusetts Motor Transportation Association
12 Post Office Square
6th Floor
Boston, MA 02109
(617) 695-3512
www.mass-trucking.org

Michigan
Michigan Trucking Association
1131 Centennial Way
Lansing, MI 48917
(517) 321-1951
www.mitrucking.org

Minnesota
Minnesota Trucking Association
2277 Highway 36 West, Suite 302
Roseville, MN 55113
(651) 646-7351
www.mntruck.org

Mississippi
Mississippi Trucking Association
825 North President St.
Jackson, MS 39202
601-354-0616
www.mstrucking.org

Missouri
Missouri Trucking Association
102 East High St.
Jefferson City, MO 65101
(573) 634-3388
www.motrucking.org

Montana
Motor Carriers of Montana
501 N. Sanders #201
Helena, MT 59601
(406) 442-6600
www.mttrucking.org

Nebraska
Nebraska Trucking Association
1701 K St.
P.O. Box 81010
Lincoln, NE 68501
402-476-8504
www.nebtrucking.com

Nevada
Nevada Motor Transport Association
8745 Technology Way, Suite E
Reno, NV 89521
(775) 673-6111
http://nevadatrucking.com/

New Hampshire
New Hampshire Motor Transport Association
19 Henniker St.
Concord, NH 03302
(603) 224-7337
http://www.nhmta.org/

New Jersey
New Jersey Motor Truck Association
160 Tices Lane
East Brunswick, NJ 08816
(732) 254-5000
https://njtrucks.wildapricot.org/

New Mexico
New Mexico Trucking Association
4700 Lincoln NE
Albuquerque, NM 87109
(505) 884-5575
www.nmtrucking.org

New York
New York State Motor Truck Association
7 Corporate Drive
Clifton Park, NY 12065
(518) 458-9696
www.nytrucks.org

North Carolina
North Carolina Trucking Association
4000 Westchase Blvd., Suite 210
Raleigh, NC 27607
(919) 281-2742
https://www.nctrucking.com/

North Dakota
North Dakota Motor Carriers Association, Inc.
1937 East Capitol Ave.
Bismarck, ND 58501

P.O. Box 874
Bismarck, ND 58501
(701) 223-2700
www.ndmca.org

Ohio
Ohio Trucking Association
21 E. State St., Suite 900
Columbus, OH 43215
(614) 221-5375
http://ohiotrucking.org/

Oklahoma
Oklahoma Trucking Association
3909 N. Lindsay Ave.
Oklahoma City, OK 73105
(405) 525-9488
http://oktrucking.org/

Oregon
Oregon Trucking Associations
4005 SE Naef Rd.
Portland, OR 97267
(503) 513-0005
www.ortrucking.org

Pennsylvania
Pennsylvania Motor Truck Association
910 Linda Lane
Camp Hill, PA 17011
717-761-7122
www.pmta.org

Rhode Island
Rhode Island Trucking Association
660 Roosevelt Ave.
Pawtucket, RI 02860
(401) 729-6600
www.ritrucking.org

South Carolina
South Carolina Trucking Association
P.O. Box 50166
Columbia, SC 29250

2425 Devine St.
Columbia, SC 29205
(803) 799-4306
www.sctrucking.org

South Dakota
South Dakota Trucking Association
P.O. Box 89008
Sioux Falls, SD 57109
(605) 334-8871
www.southdakotatrucking.com

Tennessee
Tennessee Trucking Association
4531 Trousdale Dr.
Nashville, TN 37204
(615) 777-2882
www.tntrucking.org

Texas
Texas Trucking Association
700 East 11th St.
Austin, TX 78701
(800) 727-7135
www.tmta.com

Utah
Utah Trucking Association
4181 West 2100 South
Salt Lake City, UT 84104
(801) 973-9370
www.utahtrucking.com

Vermont
Vermont Truck & Bus Association, Inc.
P.O. Box 3898
Concord, NH 03302
(802) 479-1778
www.vtba.org

Virginia
Virginia Trucking Association
1707 Summit Avenue, Suite 110
Richmond, VA 23230
(804) 355-5371
www.vatrucking.org

Washington
Washington Trucking Associations and WTA
Services, Inc.
2102 Carriage Dr. SW, Bldg. F
Olympia, WA 98502
(800) 732-9019, (253) 838-1650
https://watrucking.org/

West Virginia
West Virginia Trucking Association
2006 Kanawha Blvd. E
Charleston, WV 25311
(304) 345-2800
http://www.wvtrucking.com/

Wisconsin
Wisconsin Motor Carriers Association
P.O. Box 44849
Madison, WI 53744
(608) 833-8200
www.witruck.org

Wyoming
Wyoming Trucking Association
555 N. Poplar St.
Casper, WY 82601

P.O. Box 1175
Casper, WY 82602
(307) 234-1579
www.wytruck.org

National
American Trucking Associations (ATA)
950 North Glebe Rd., Suite 210
Arlington, VA 22203
(703) 838-1700
www.trucking.org

Canadian Trucking Alliance
555 Dixon Rd.
Toronto, ON M9W 1H8
(416) 249-7401
www.cantruck.ca

Appendix D

Transportation Agencies

Federal Motor Carrier Safety Administration

http://www.fmcsa.dot.gov

Motor Carrier Safety Service Centers

Eastern Service Center
802 Cromwell Park Drive, Suite N
Glen Burnie, MD 21061
Phone: (443) 703-2240
Fax: (443) 703-2253
Connecticut, DC, Delaware, Massachusetts,
Maryland, Maine, New Hampshire, New Jersey,
New York, Pennsylvania, Puerto Rico, Rhode
Island, Virginia, Vermont, West Virginia,
Virgin Islands

Midwestern Service Center
4749 Lincoln Mall Dr., Suite 300A
Matteson, IL 60443
Phone: (708) 283-3577
Fax: (708) 283-3579
Iowa, Illinois, Indiana, Kansas, Michigan,
Minnesota, Missouri, Nebraska, Ohio, Wisconsin

Southern Service Center
1800 Century Blvd., Suite 1700
Atlanta, GA 30345
Phone: (404) 327-7400
Fax: (404) 327-7349
Alabama, Arkansas, Florida, Georgia, Kentucky,
Louisiana, Mississippi, North Carolina,
Oklahoma, South Carolina, Tennessee

Western Service Center
Golden Hills Office Centre
12600 West Colfax Ave., Suite B-300
Lakewood, CO 80215
Phone: (303) 407-2350
Fax: (303) 407-2339
Alaska, American Samoa, Arizona, California,
Colorado, Guam, Hawaii, Idaho, Northern
Mariana Islands, Montana, New Mexico, North
Dakota, Nevada, Oregon, South Dakota, Texas,
Utah, Washington, Wyoming

Appendix E

FMCSR Part 395

Federal Motor Carrier Saftey Regulations (FMCSR) Part 395 covers the Hours of Service of Drivers. It requires drivers to keep a Record of Duty Status. The subjects covered by this Part include the following:

- maximum driving and on-duty time
- travel time
- driver's record of duty status
- adverse driving conditions
- emergency conditions
- relief from regulations
- drivers declared "out of service"

Driving is different from almost every other profession. You usually don't work a set schedule. You don't stop driving just because the clock says it's 5:00 P.M. You don't stop driving just because Friday is about to turn into Saturday. You don't park the truck just because you've been on-duty for eight hours.

Instead, you work the number of hours your schedule or run requires, within the limits set in FMCSR Part 395. These limits are called the "hours of service regulations." The regulations state that you can work just so many hours at a time. Your legal day is the number of hours the regulations permit you to work before you must stop to rest. The number of hours you can work and the kind of work you can do on any day depends on the hours you worked the day before, up to eight days in the past.

You record the hours you worked and the kind of work you did in a log book. This log is called your "record of duty status." It helps you stay within the hours of service limits.

The first section in FMCSR Part 395 defines terms used throughout the rest of the Part. They are as follows:

On-duty time. From the moment you begin your work day until the time you are relieved of your duties, you are on-duty. If you are required to be ready for work, then you are on-duty. Even if you are just sitting at the terminal waiting for your truck to be loaded, you are on-duty. All of the time you spend working but not driving is called on-duty (not driving) or (ON) time. Here are some examples of on-duty not driving time:

- waiting to be dispatched
- inspecting or servicing your vehicle
- time spent as a co-driver (unless you are in the sleeper berth)
- loading and unloading, including supervising and taking care of paper work
- repairing, getting help for, or waiting with a disabled vehicle
- complying with the requirements of alcohol- and drug-testing programs
- providing a service for the carrier
- any other work that you are paid for by the trucking company or another job

You are on-duty (driving) or (D) when you are at the vehicle's controls. This is also referred to as "driving time" or simply "driving."

The on-duty and off-duty times add up to 24, the number of hours in an ordinary day.

Seven consecutive days. This is seven days, one right after the other. It can be any seven days in succession. (A "day" is any 24-hour period. A "24-hour period" is also defined in this section.)

Eight consecutive days. This is eight days, one right after the other.

Twenty-four-hour period. This is a period of 24 hours, one right after the other. You start counting based on what time it was at the terminal from which you are normally dispatched. That is, you don't add or subtract hours as you cross time zones.

Sleeper berth. When this term is used, it means that area of the truck meeting the specifications in

FMCSR Part 393. You'll see later that time spent in the sleeper is logged separately from ON or D (driving). You'll also see you can't log time as spent in the sleeper if you're napping on the seat. You must be "officially in" the "official sleeper."

Driver-salesman. Some drivers who work for private carriers are called "driver-salesmen." They sell products as well as transport them. Driver-salesmen drive within 100 miles of the terminal where they report for work. They do not spend more than 50 percent of their ON time driving.

Multiple stops. All stops made in any one village, town, or city may be counted as one stop.

Automatic on-board recording device. Your vehicle may have a device that completes your log for you. You may hear this called an ELD—an electronic logging device. It must record at least engine use, road speed, miles driven, the date, and time of day.

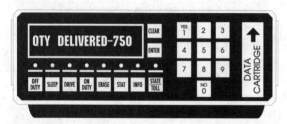

Figure E-1. An electronic logging device or on-board computer.

MAXIMUM DRIVING AND ON-DUTY TIME

FMCSR Part 395 sets the limits for how long you may drive. These rules limit the amount of total work that you may do and the total number of hours that you may spend driving. After 10 straight hours of off-duty time, you may drive for no more than 11 hours. After 10 straight hours of off-duty time, you may be on-duty for no more than 14 hours. Of the 14 hours, 11 may be spent driving.

As you might expect, there are exceptions. Drivers may extend the 14-hour-on-duty period to 16 hours on-duty under the following conditions:

- they are released from duty at the location to which they normally report for work for the five previous duty tours and

- they return to that normal work-reporting location and are released from duty within 16 hours and

- they haven't used this exception in the previous seven days except following a 34-hour restarting of a 7/8 day period.

This "restart" period allows drivers to begin a new weekly cycle if they take 34 consecutive (one right after the other) hours off.

In no case may the total number of hours that you spend driving exceed 11 hours.

These are rules for drivers of cargo-carrying vehicles. There are different rules for drivers who carry passengers.

You must get at least 10 hours of rest off-duty. You may split this into two very specific rest periods. In other words, you can combine one period of at least eight consecutive hours of rest in the sleeper berth with another rest period of at least two consecutive hours. You may take this second rest period in the sleeper berth, or you may be off-duty, or it may be a combination of these two. Note that for the sleeper berth time, you must be in the sleeper berth. You cannot be in the passenger seat and log that as sleeper berth time. The driving time immediately before and after these rest periods must not exceed 11 hours. Also, the total on-duty time before and after these rest periods must not include any driving beyond the 14th hour.

You may not return to driving without taking

- at least 10 consecutive hours off-duty or
- at least 10 consecutive hours in the sleeper berth or
- a combination of at least eight consecutive hours of sleeper berth and at least two consecutive hours of off-duty time

Drivers at natural gas or oil wells may rest elsewhere other than the sleeper berth but must get the same amount of rest.

If your company does not work every day of the week, you must not drive after being on-duty for 60 hours in seven straight days. This is often called the 60 hours/7-day rule. If your company does work every day of the week, you must not drive after being on-duty for 70 hours in a period of eight straight days. You may hear this referred to as the 70 hours/8-day rule. Drivers in Alaska are sometimes exempt from these rules. They may not drive more than 15 hours after being off-duty for eight hours. They may not drive after being on-duty for 20 hours or more after an eight-hour off-duty period. They may not drive after being on-duty for 70 hours in seven consecutive days for a carrier that doesn't work every day. They may not drive after being on-duty for 80 hours in eight consecutive days for a carrier that does work every day.

Drivers in Alaska may encounter adverse driving conditions. These drivers may drive as long as they need to in order to complete the run. After that, these drivers must be off-duty for at least 10 consecutive hours before driving again.

Driver-salesmen are exempt from these rules if driving time doesn't go over 40 hours in seven consecutive days.

Drivers who only work to make deliveries from retail stores to consumers are exempt from December 10 through December 25. They are limited to driving within 100 miles of where they normally report for work.

For some drivers of oil-well servicing vehicles, on-duty (ON) time does not always include waiting time. They must carefully account for all their waiting time.

TRAVEL TIME

Sometimes you may have to travel for your motor carrier. You are not driving or doing any other kind of work. Perhaps you're just flying to another state to collect a broken-down tractor. This is still counted as on-duty time unless you take 10 straight hours off when you reach the destination. If you do take the off-duty time, then you'll be considered off-duty for the entire time. That's the travel time plus the 10 straight hours off.

LOG BOOKS

You must keep a "record of duty status." Most people in the industry call this a "log book" or "log." A log book shows how you spent your time for each 24-hour period. You can use a handwritten log. If you do, you must make two copies. You must record the information on a form that meets the requirement of FMCSR Part 395. This is referred to as a "grid." There are examples of different types of grids in the FMCSR. You may also use Forms MCS-59, MCS-139, and 193A. These are older forms that still meet the requirements of FMCSR Part 395.

You may also record your hours with an electronic logging device (ELD). If ELDs are used on the trucks you are to drive, your carrier will make sure you know how to use them. You should still know how to do a log by hand. ELDs can break down. You will have to do your log by hand until the ELD is repaired.

You will log four different types of duty status. They are:

- off-duty or OFF
- sleeper berth or SB (but only if the sleeper berth is used)
- driving or D
- on-duty (not driving) or ON

Note that you don't record sleeper berth time if your vehicle doesn't have a sleeper.

Every time you change from one duty status to the other, you log it. You log where you were when this change occurred. You can show this with the highway number, the nearest milepost, the name of the nearest town, city or village, and the state abbreviation. If you are at a truck stop or service station, you can put the name of the business in place of the milepost. If you are at an intersection of two highways, put the number of the two roads instead of the milepost.

The log calls for other information besides change of duty status and location. You must also record:

- the date
- the total miles you drove the day you fill out the log
- the truck or tractor and trailer number
- the name of your motor carrier
- the driver's signature or certification
- the starting time of this 24-hour period
- the motor carrier's main office address
- remarks (the information about where your change of duty status took place)
- the co-driver's name, if there's a co-driver
- total hours
- shipping document numbers, or the name of the shipper and the freight

If your work isn't properly logged, both you and your motor carrier may be prosecuted. Your log must meet the following requirements:

Entries to be current. At all times, your log must be current to your last change of duty status.

Entries made by driver only. The log reflects the driver's duty status. If you are the driver, your handwritten log must be in **your** handwriting. It must be readable.

Date. Put the month, day, and year for the start of each 24-hour period.

Total mileage driven. "Total mileage driven" is NOT the total of the miles the truck moved. Don't include miles driven by the co-driver. (The co-driver's mileage goes on the log the co-driver completes.) Record the total miles driven by the driver.

Vehicle identification. Look for tractor numbers on the front of the cab or on the door. You can find trailer numbers on the front panel or the rear doors. If you cannot find the vehicle identification numbers, use the license plate number.

Name of carrier. This is the name of the motor carrier you are doing this run for. It may be different from the motor carrier you usually work for. You may work for more than one carrier in a 24-hour period. In that case, show both names. Follow each name with the start and finish time worked for that carrier. Don't forget to include AM and PM with these start and finish times.

Signature/certification. When you, as the driver, sign the log, you are saying everything is true and correct. You must use your legal signature, not your nickname or initials.

Twenty-four-hour period. Use the time in effect at your home terminal. Don't try to account for changes in time as you cross time zones. Start the 24-hour period, seven and eight consecutive days with the time your motor carrier specifies for your home terminal.

Main office address. If your motor carrier has more than one terminal, the address of the main office is the one called for here.

Recording days off-duty. You may have two or more consecutive 24-hour periods off-duty. You may record all of them on one log.

Total hours. Total all the hours spent in each duty status. Record the total for each to the right of the grid. They should add up to 24. If they don't, you've made an error somewhere. Either you added wrong or you forgot to record something.

Shipping document numbers. If you do not have a shipping document number to put here, use the name of the shipper and what's being shipped.

GRID PREPARATION

You must fill out the grid as described in FMCSR Part 395. You can use the form horizontally or vertically. The following description assumes you're using the form horizontally.

The four rows in the grid are for the length of time spent in each duty status. The columns mark off time segments. Each block in each row is one hour. Each long line in the middle of each block is the half hour (30 minutes). The short lines are quarter hours (15 minutes). The first one is one-fourth (15 minutes after the hour) and the second is three-fourths (45 minutes after the hour). In other words, each hour is broken into fourths. Each line within each box is one-fourth hour. You record your duty status to the nearest quarter of an hour.

To record the length of time spent in each duty status, start at the vertical line that represents the start time, to the nearest quarter hour. Put your pencil on that vertical line in the row for the duty status you are about to begin. Draw an unbroken horizontal line. The line stops at the quarter hour marker that stands for the time when you next switch duty status. Draw an unbroken vertical line up or down along the line that represents the time you switch status. Stop when you get to the row for the next duty status. Then start a new unbroken horizontal line in the new duty status.

This is easier to do than to describe! Figure E-2 may clear up any confusion you may have at this point.

Whenever you change duty status, you must record the location where the change took place. This is done in the remarks section. You recall there are three ways to describe the locations where the duty status change took place.

FILLING OUT THE LOG

You will have to fill out an original and one copy of each log. You must send the original log to your carrier within 13 days after you complete the form. If you drive for more than one carrier, you must send each carrier a complete copy of your log.

You must keep the copies of your logs for the past seven straight days with you.

EXEMPTIONS

Most drivers have to keep logs just as we've described. Some drivers are exempt. To be exempt, they must meet all these conditions:

- they work within a 100 air-mile radius of where they normally report for work
- they return to where they normally report for work within 12 consecutive hours
- they have eight consecutive hours off-duty between each 12-hour on-duty shift
- they don't exceed 10 hours driving time after eight consecutive hours off-duty
- their motor carrier keeps a record of their time as described in FMCSR Part 395

Some drivers in Hawaii are also exempt from logging requirements. So are

- some drivers transporting agricultural commodities
- some drivers involved in ground water well-drilling operations
- some transporters of construction materials and equipment
- some utility service drivers

While you may use an electronic logging device, you do need to know how to complete a paper log should the device fail.

EMERGENCIES

You may exceed the maximum driving time allowed under certain conditions. You may drive for two extra hours if you find yourself driving under adverse conditions that prevent you from completing your run within the allowed 11 hours. Adverse driving conditions do not mean a route without a café! Instead, adverse conditions include:

- snow
- sleet
- fog
- highways covered with snow or ice

Also included are unusual road or traffic conditions if the dispatcher didn't know about them when you began your run. If you meet such conditions and can't safely complete your run in the maximum allowed driving time, you may drive longer. You may drive a CMV no more than two more hours to finish the run or reach a place of

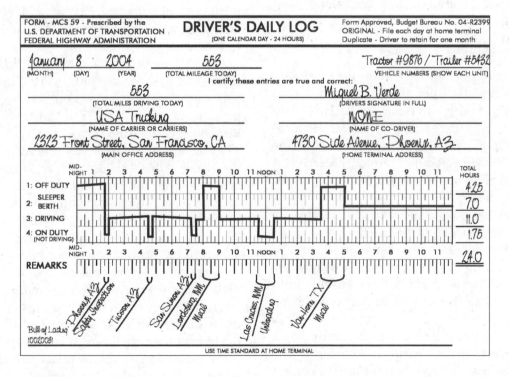

Figure E-2. A sample of a completed log.

safety. Still, drivers of property-carrying CMVs may not drive

- for more than 14 hours total, following 10 consecutive hours off-duty
- after being on-duty after the end of the 14th hour after coming on-duty, following 10 consecutive hours off-duty

At least 10 consecutive hours off-duty must separate each 11 hours on-duty. Drivers of property-carrying CMVs must not exceed 11 hours maximum driving time, following 10 consecutive hours off-duty.

If you work in Alaska and you encounter adverse driving conditions, you may continue to drive for two more hours. Even still, you may not drive more than 13 hours altogether, following 10 consecutive hours off-duty. Nor may you drive after you have been on-duty after the end of the 14th hour after coming on-duty, following 10 consecutive hours off-duty.

If an emergency comes up that causes a delay in your run, you may complete your run. This is only if the run could have been finished within the usual time limit had there not been an emergency.

The hours of service rules are also set aside when a driver provides transportation to help in a natural disaster, like an earthquake or flood.

PENALTIES

If you don't keep track of your time properly or don't obey the rules about how much time you spend working, you can be declared "out of service." Any agent of the Federal Highway Administration (FHWA) may inspect your log book and declare you out of service. These agents include highway patrol officers, DOT inspectors, and weigh masters.

You may be put out of service for disobeying the hours of service regulations. This would be the case if you work more hours than FMCSR Part 395 allows. When that happens, you must go off-duty immediately. You may not go back on-duty until you have had all the off-duty or sleeper time you are supposed to have.

You can be put out of service for not having a current record of duty status. If your log is current for the past six days, but not for the present day, you are given a chance to bring it up to date.

Otherwise, you are put out of service. Then you must be off-duty for the number of consecutive hours that the regulations require. The hours may be spent as sleeper berth time. Your carrier cannot require or permit you to drive during this period. You must get a copy of the out-of-service form to your carrier within 24 hours.

ELECTRONIC LOGGING DEVICES (ELDs)

As you have seen, you may use ELDs to do your log. This device will show how much time you spent in each duty status. It will show your start time for that 24-hour period. Electronic logging devices must be able to produce a printed log (printout) for officials who want to check your hours. The driver must sign any such printouts. ELDs must store information for the past seven consecutive days. They must record the same information as the handwritten log. This includes not only the four duty-status types. It also includes the date, total mileage today, vehicle numbers, carrier name, main office address, and so on.

If handwritten entries are needed, they must be made by the driver, in the driver's own handwriting.

Should your ELD fail, you must make note of it. As the driver, you will have to recreate any lost or missing records for the past seven days. You will have to continue to do your log by hand until the ELD is repaired.

On your vehicle, you must have instructions telling you how to store information in the ELD. You must have instructions on how to get information out of the ELD. You must have enough blank log grids for the current run.

As the driver, you must still get each day's log to the carrier within 13 days. With an ELD, there may be a way to do this electronically. Before you send your log to your carrier, you must check that all the entries the ELD has made are correct. In sending the log to the carrier you certify that this is a true and correct record of your duty status.

Even though your vehicle has an ELD, the FHWA may require you or your motor carrier to return to handwritten logs. One reason for this is if the carrier's drivers set a pattern of exceeding the hours of service limits. It can also be required if the motor carrier or any of the drivers tamper with the electronic logging devices.

Glossary

The following section contains definitions and explanations of terms used in your text and in the industry.

A

AC Alcohol concentration.

Agent A person authorized to transact business for and in the name of another. For instance, a driver becomes an agent of the trucking company when signing for freight.

Air bag suspension Trailer suspension system using air bags for a greater cushion than conventional leaf spring.

Air brakes Brakes that use compressed air instead of fluid.

Air cleaner Cleans air that enters the engine, sometimes called air filter.

Air compressor Builds up and maintains required air pressure in the brake system reservoir.

Air filter restriction gauge Gauge on a truck dashboard that registers the flow of air through the air filter element and registers the condition of the filter.

Air line Carries compressed air from one part of the air brake system to another, and from tractor to trailer.

Air-over-hydraulic brakes Brakes on a vehicle that have a regular hydraulic system assisted by air pressure.

Air reservoir Storage tank for compressed air.

Air shutter Mounted on the front of the radiator to control air entering the radiator and thereby to control engine temperature.

Alcohol concentration (AC) A measure of how much alcohol is in a person's blood or breath. You may also see the term "blood alcohol concentration" (BAC) used.

Alcohol (or alcoholic beverage) Beer, wine, distilled spirits (liquor).

Alternator Produces electricity for the battery and electric power to operate the lights, radio, and other items when the engine is running.

Ammeter Measures the amount of current flowing in an electrical current.

Amphetamines "Bennies," "speed," or "pep pills." Stimulant drugs.

Application pressure air gauge Indicates pressure being applied by brakes during brake operation.

Approved Approval issued or recognized by the Department of Transportation.

Articulate Refers to the ability to move or bend, as a tractor-trailer articulates when it turns.

Atmospheric gases Argon, krypton, neon, nitrogen, oxygen, and xenon.

Axle The bar that connects opposite wheels.

Axle temperature gauge Indicates the temperature of lubricant in an axle.

Axle weight How much weight the axle (or set of axles) transmits to the ground.

B

Baffle Dividers in a tanker that keep the load from shifting.

Battery Converts chemical energy into electricity.

Bias A type of tire with diagonal body plies, and narrow plies under the tread going in the same direction.

Blocking Supports that keep cargo from shifting during transport.

Brake drum The rotating unit of the brake that is attached to the wheel.

Brake lining Material attached to the brake shoe that creates friction.

Brake shoe Non-rotating part of the brake that contacts the brake lining and supplies braking force.

Braking distance Perception time plus reaction time plus brake lag.

Bridge formula Determines how much weight is put on any point by any group of axles.

Bridge law Regulates how much weight can be put on a bridge at any one point.

Btu British thermal unit. It's a measurement of heat energy.

Bulk packaging A packaging, including a transport vehicle or freight container, with a capacity greater than 119 gallons (liquid), 882 pounds (solid), or 1,000 pounds water capacity (gas).

C

C or Celsius or Centigrade It's one scale used for measuring temperature.

Cab Part of the vehicle where the driver sits.

Cabover Vehicle with most of its engine under the cab.

Carbon monoxide A poisonous gas in exhaust fumes.

Carboy A glass, plastic, or metal bottle or container holding several gallons of liquid. Carboys are often cushioned in a wooden box.

Cargo Freight.

Cargo tank Any bulk liquid or compressed gas packaging. Cargo tanks can be portable or permanently attached to any motor vehicle. They can be loaded or unloaded without being removed from the motor vehicle. Packaging built to specifications for cylinders are not cargo tanks.

Carrier A person who transports passengers or property by land or water as a common, contract, or private carrier, or by civil aircraft.

CDL Commercial Driver's License.

Centrifugal force The natural force that pulls objects in motion outward from the center of rotation.

Check valve Seals off one part of the air brake system from another.

Chock A block placed in front of a tire to keep the vehicle from moving.

Circuit A closed electrical path.

Circuit breaker Breaks and opens an electrical circuit in an overload.

Clean bore A tanker without compartments or baffles.

Combination vehicle One that articulates. It has joined sections that can move independently of each other.

Combustible Able to ignite and burn.

Combustible liquid A liquid with a flash point at or above 140 degrees and below 200 degrees Fahrenheit.

Commerce Trade, traffic, or transportation in the United States (or any area controlled by the United States). The trade can be between a place in a state and a place outside the state, even outside the United States.

Commercial Driver's License CDL

Commercial vehicle Commercial vehicles are vehicles with gross vehicle weights of 26,001 pounds or more; trailers with gross vehicle weights of 10,001 pounds or more; vehicles that transport hazardous materials that require placards, and buses designed to carry 16 or more people (including the driver).

Compressed gas Compressed gases are held under pressure. When in the container, compressed gases have an absolute pressure of more than 40 psi (that is, 40 psia) at 70 degrees Fahrenheit. Or, if measured at 130 degrees Fahrenheit, they have a pressure of 104 psia. A liquid flammable material with a vapor pressure over 40 psia at 100 degrees Fahrenheit is also a compressed gas.

Conductor A material through which electricity flows.

Consignee A company or person who has agreed to purchase a specific load of freight and to whom a shipment is delivered.

Consignor The shipper.

Consumer commodity A material that is packaged and distributed for retail sale for use by individuals. The material is meant for personal care or household use. This term also includes drugs and medicines.

Controlled intersection An intersection with a traffic light or signs that control traffic.

Controlled substance A substance listed on Schedules I through V of the Code of Federal Regulations.

Converter gear or dolly A coupling device; one or two axles and a fifth wheel used to couple a tractor to a semitrailer.

Convex mirror "Spot mirror" that shows a wider area than a flat mirror of the same size.

Coolant The liquid that reduces engine heat.

Corrosive material A corrosive is a solid or liquid that harms human skin on contact.

Couple Connect two sections of a vehicle combination.

Cryogenic liquid A coolant gas. It will boil at minus 130 degrees Fahrenheit when it's at one atmosphere, absolute (1 psia).

Cylinder A pressure vessel designed for pressures higher than 40 psia. A portable tank or cargo tank is not considered a cylinder.

D

Deceleration To slow down, to come to a stop (the opposite of acceleration).

Department of Transportation (DOT) The federal department that establishes the nation's overall transportation policies.

Designated facility A hazardous waste treatment, storage, or disposal facility that has been named on the manifest by the generator (creator of the waste).

Differential A system of gears that permits each wheel to turn at different speeds on the same axle, as occurs when going around a curve.

Disc brakes Brakes that function by causing friction pads to press on either side of a disc rotating along with the wheel.

Disc wheel A single unit that combines a rim and a wheel. The most common type is the Budd wheel.

Disqualification Suspension or complete loss of driving privileges.

Dock A platform where trucks are loaded and unloaded.

DOD The U.S. Department of Defense.

Double-clutching Shifting the gears of a truck transmission without grinding them (that is, shifting from the present gear to neutral, then shifting from neutral to the desired gear).

Driveaway-towaway operation A hauling operation in which the cargo is one or more vehicles, with one or more sets of wheels on the roadway.

Drive axle An axle that transmits power to the wheels and actively pulls the load.

Driver's record of duty status Form submitted by driver to carrier for each 24-hour period of work.

Drive shaft A heavy-duty tube that connects the transmission to the rear end assembly of the tractor.

Drive train A series of connected mechanical parts for transmitting motion.

Drive wheel The wheel to which power is transmitted to drive the vehicle.

Duals A pair of wheels and tires mounted together on the same side of one axle.

DUI Driving a commercial motor vehicle while under the influence of alcohol.

Dummy coupler A fitting used to seal the opening in an air brake hose connection (glad hands) when the connection is not in use. Sometimes called a dust cap.

Dunnage The material used to protect or support freight in trucks.

E

Eighteen-wheeler A combination of tractor and trailer with five axles and dual wheel combinations totaling 18 rims and tires.

Electrical system Consists of the starter, wiring, battery, generator, alternator, and generator regulator that work together to crank the engine for starting.

Employee Anyone who operates a CMV either directly for or under lease to an employer.

Employer Anyone who owns or leases a CMV or assigns employees to operate one.

Endorsement A special allowance added to your CDL that permits you to drive certain vehicles or haul certain cargo.

Engine retarder An auxiliary device added to the engine to reduce its speed. Also referred to as an engine brake.

En route On the way.

EPA The U.S. Environmental Protection Agency.

Etiologic agents Microorganisms (germs). They cause disease.

Explosives Explosives are chemical compounds, mixtures, or devices that explode.

F

F or Fahrenheit It's another scale for measuring temperature.

Fan belt A belt attached to the engine block that drives the fan.

Federal Motor Carrier Safety Regulations (FMCSR) Govern the operation of trucks and buses operated in interstate or foreign commerce by common, contract, and private motor carriers.

Felony A crime that is punishable by death or a prison term of more than one year.

Fifth wheel Part of the locking device used to connect a semitrailer and a tractor.

Filling density Refers to how full tankers and cylinders should be loaded with bulk liquids and compressed gases.

First aid Immediate and temporary care given to the victim of an accident or sudden illness until the services of a medical professional can be obtained.

Fishyback Transporting motor truck trailers and containers by ship.

Flammable A substance that easily bursts into flames (also"inflammable").

Flammable compressed gas. Compressed gas is said to be flammable if a mixture of 13 percent or less will form a flammable mixture when combined with air or it passes the UN Manual of Tests and Criteria tests for flammability.

Flammable liquid A liquid with a flash point below 100 degrees Fahrenheit.

Flammable solids A flammable solid is a solid material, other than an explosive, that can cause fire.

Flash point The lowest temperature at which a substance gives off flammable vapors. These vapors will ignite if they come in contact with a spark or flame.

Flatbed A truck or trailer without sides or a top.

FMCSR Federal Motor Carrier Safety Regulations.

Foot brake valve Foot-operated valve that controls the amount of air pressure delivered to or released from the brake chambers.

Freight container A reusable container with a volume of 64 cubic feet or more. It's designed and built so it can be lifted with its contents intact. It's mainly used to hold packages (in unit form) during transport.

Friction Resistance between the surfaces of two objects.

Fuel pump In a vehicle, the pump that moves fuel from the fuel tank to the engine.

Fuel tank A tank other than a cargo tank used to transport flammable or combustible liquid. It's also a tank that holds compressed gas used to fuel the transport vehicle it's attached to, or equipment on that vehicle. This is different from a tank semitrailer that transports fuel as cargo.

G

GCWR Gross combination weight rating.

Generator In a vehicle, the device that changes mechanical energy into electrical energy to run the lights and battery.

Glad hands Air hose brake system connections between the tractor and trailer.

Governor (air) Device to automatically control the air pressure being maintained in the air reservoirs. Keeps the air pressure between 90 and 120 psi and prevents excessive air pressure from building up.

Groove On a tire, the space between adjacent tread ribs.

Gross combination weight (GCW) The total weight of the power unit or tractor, plus any trailers, plus the cargo. (This would also apply to a vehicle towing another vehicle as cargo.)

Gross combination weight rating (GCWR) Gross combination weight rating is defined as the maximum gross combination weight stated by the manufacturer for a specific combination of vehicles, plus the load.

Gross vehicle weight (GVW) The total weight of a single vehicle, plus its load.

Gross vehicle weight rating (GVWR) The maximum gross vehicle weight stated by the manufacturer for a single vehicle, plus its load.

Gross weight The weight of a packaging, plus the weight of its contents.

GVWR Gross vehicle weight rating.

H

Hand throttle A manually set throttle in a tractor that is used to maintain a certain engine speed.

Hazardous material A substance or material, including a hazardous substance, that the Secretary of Transportation judges to pose an unreasonable risk to health, safety, and property when transported in commerce.

Hazardous substance A material listed in the Appendix to Part 172.101.

Hazardous waste Any material that is subject to the Hazardous Waste Manifest Requirement of the U.S. Environmental Protection Agency.

Hazardous waste manifest A document (shipping paper) on which all hazardous waste is identified and that must accompany each shipment of waste from the point of pickup to its destination.

Highway A public roadway that carries through traffic.

Hydraulic brakes Brakes that depend on the transmission of hydraulic pressure from a master cylinder to the wheel cylinders.

I

ICAO International Civil Aviation Organization.

Ignition point The temperature at which a flammable substance will ignite or burn.

Ignition switch Allows current to flow from the battery to the starter motor.

Import To receive goods from a foreign country.

Inertia The tendency of an object to remain in the same condition (that is, to stay still or keep moving).

Injector In a diesel engine, the unit that sprays fuel into the combustion chamber.

Insulated tanker Truck or trailer designed to haul cargo at controlled temperatures, whether cooled or heated.

Intermodal container A freight container designed and built to be used in two or more modes of transport.

Intermodal portable tank or IM portable tanks A specific class of portable tanks designed mainly for international intermodal use.

Intersection The place where two roadways intersect.

Intrastate traffic Traffic within one state only.

Irritating material When exposed to air or flames, it gives off dangerous fumes.

J

Jacking Changing the direction of the trailer.

Jackknife When a trailer forms a "V" with the tractor instead of being pulled in a straight line, usually resulting from a skid.

Jake brake (slang) The Jacobs engine brake that is used as an auxiliary braking device on a tractor.

K

Kingpin Hardened steel pin on a semitrailer that locks into the fifth wheel for coupling.

L

Landing gear Supports the front end of a semitrailer when not attached to a tractor.

Leaf spring suspension Conventional type of suspension system.

Limited quantity The maximum amount of a hazardous material for which there is a specific labeling and packaging exception.

Load lock A pole-shaped piece of equipment used to hold cargo in place during transport by way of a tension spring.

Log book A book carried by truck drivers that contains daily records of hours, route, and so forth, which is required by the Department of Transportation.

Low air warning device A mechanical warning of brake failure by means of a buzzer, a flashing red light, or a small red flag that drops into the driver's line of vision.

M

Manifest A document describing a shipment or the contents of a vehicle or ship.

Marker lights Clearance or running lights.

Marking Putting the descriptive name, instructions, cautions, weight, or specification marks required by regulations on outer containers of hazardous materials.

Mixture A material composed of more than one chemical compound or element.

Mode Any of the following transportation methods: rail, highway, air, or water.

Motor carrier A business that transports passengers or goods by a motorized vehicle.

Motor vehicle A vehicle, machine, tractor, trailer, or semitrailer, or any combination of these. This term assumes the vehicle is driven or drawn by mechanical power, is on the highway, and is transporting passengers or property. It doesn't refer to vehicles used on rail. It also doesn't include a trolley bus running on a fixed overhead electric wire and carrying passengers.

Mph Miles per hour.

N

Name of contents The proper shipping name specified in Part 172.101 or 172.102.

Net weight A measure of weight referring only to the contents of a package and that does not include the weight of any packaging material.

Non-bulk packaging A package smaller than the capacities stated for bulk packaging.

Non-liquefied compressed gas A gas that is completely in the form of vapor at 70 degrees Fahrenheit.

N.O.S. "Not otherwise specified."

O

Oil pressure gauge Instrument that measures the pressure of engine lubricating oil.

Operator A person who controls the use of an aircraft, vessel, or vehicle.

Organic peroxide A chemical.

ORM Other Regulated Materials.

Oscillatory sway Side-to-side motion at the top of a vehicle.

Other Regulated Materials "Other Regulated Materials" are those that don't meet any of the other hazardous materials class definitions but that are dangerous when transported in commerce.

Outage The amount by which a packaging falls short of being full of liquid. Usually expressed as a percent of volume. Outage allows for liquids to expand as they warm.

Out-of-service driver Driver declared out of service by a government representative because of violations.

Out-of-service vehicle A vehicle that cannot pass the government safety inspection and is declared out of service until the problems are corrected.

Outside container The outermost enclosure used to transport a hazardous material. This is other than a freight container.

Overpack An enclosure that is used by a single consignor to provide protection or convenience in the handling of a package or to combine two or more packages. Does not include a freight container.

Oversized vehicle Any vehicle whose weight and/or dimensions exceed a state's regulations.

Oxidizer Oxidizers cause other materials to react with oxygen.

P

Packaging The assembly of one or more containers and any other part needed to comply with the minimum packaging requirements.

Piggyback Intermodal transportation system wherein trailers or containers are carried by rail.

Pigtail Cable used to transmit electrical power to the trailer.

Pintle hook Coupling device at the rear of the truck for the purpose of towing trailers.

Placards Signs placed on a tanker or trailer during shipment to identify hazardous material and that must be visible from all angles.

Ply ratings The number of ply cords in the sidewall and tread of a tire (from two for cars up to as many as 20 for large off-the-road trucks).

Point of origin The terminal at which a shipment is received by a transportation line from the shipper.

Poison A substance that is so harmful to humans that it is dangerous to transport. It can kill when swallowed or inhaled, or when absorbed by the skin.

Pole trailer Trailer composed of a single telescopic pole, a tandem rear wheel unit, and a coupling device used to join the trailer to a tractor.

Pollution The presence of toxic substances that have entered the air, water, or soil and that poison the natural environment.

Portable tank Any packaging (except a cylinder having a 100-pound or less water capacity) over 110 U.S. gallons capacity. This definition includes tanks designed to be loaded into or on or temporarily coupled to a transport vehicle or ship. These tanks are equipped with skids, mounting, or other parts that make mechanical handling possible.

Pot torches Safety equipment used on a highway to warn traffic of an obstruction or hazard.

Power Takeoff (PTO) A device mounted on the transmission or transfer case, used to transmit engine power to auxiliary equipment.

Preferred route (or preferred highway) A highway for shipment of highway route controlled quantities of radioactive materials. These routes are named by state routing agencies. When there is no other option, any interstate system highway can be named a preferred route.

Private carrier A company that maintains its own trucks to ship its own freight.

Progressive shifting Shifting technique whereby the vehicle is taken up to highway speed by using the least horsepower needed to get the job done, thereby saving fuel.

Proper shipping name The name of the hazardous material shown in Roman print (not italics) in Part 172.101.

psi (or p.s.i.) Pounds per square inch.

psia (or p.s.i.a.) Pounds per square inch absolute. Psia uses vacuum (or zero pressure) as a reference point.

psig (or p.s.i.g.) Pounds per square inch gauge. The difference between two pressures.

Pyrophoric liquid A liquid that will ignite by itself in dry or moist air at or below 130 degrees Fahrenheit.

R

Radial Tires constructed with the cords running directly across the top to better mold and flex their way around objects on the road, thereby offering less rolling resistance and increased mileage.

Radiator Part of the cooling system that removes engine heat from the coolant passing through it.

Radiator cap A cap that protects the radiator from impurities, allows pressure to build up, and releases excess pressure from the engine.

Radioactive material Material that gives off rays, usually harmful.

Reaction time The time that elapses between the point that a driver recognizes the need for action and the time the driver takes that action.

Rearward amplification The "crack-the-whip" effect found on tractors with trailers.

Recap Restore a tire by putting new tread on the carcass (body).

Reciprocity Mutual action; exchange of privileges such as between two states.

Reefer Commonly used industry term for refrigerated truck or trailer that hauls perishables.

Reflectors Light reflectors that, when placed along the road, warn motorists of an emergency.

Regrooved When new tread grooves have been cut into the crown of the tire after the original tread is worn down.

Regulation A law designed to control behavior. A governmental order with the force of law.

Relay emergency valve A combination valve in an air brake system that controls brake application and that also provides for automatic trailer brake application should the trailer become disconnected from the towing vehicle.

Reportable Quantity (RQ) The quantity of a material, fixed by the Environmental Protection Agency, that must be reported if spilled or released into the environment.

Representative vehicle   One that stands for the type of motor vehicle that a CDL applicant plans to drive.

Restricted articles   Commodities that can be handled only under certain specific conditions.

Rev   To operate an engine at a high speed.

Right-of-way   The right of one vehicle or pedestrian to proceed before or instead of another vehicle.

Road   The traveled part of a roadway and any shoulder alongside it.

Roadside   The part of a highway or street that is not used by vehicles and is not a sidewalk.

Roadway   A general term that refers to any surface on which vehicles travel.

Rocky Mountain double   Combination consisting of a tractor and two trailers.

RQ   Reportable Quantity.

Runaway ramp   A ramp on a steep downgrade that can be used by a truck driver to stop a runaway truck when brakes have failed. Sometimes called an escape ramp.

Rural   The countryside.

S

Saddle tanks   Barrel-type fuel tanks that hang from the sides of the tractor's frame.

SCF   Standard Cubic Foot. One cubic foot of gas measured at 60°F and 14.7 psia.

Seal   In the shipping industry, a security device to assure that truck doors have not been opened in transit.

Semitrailer   Truck trailer equipped with one or more axles and constructed so that the front end rests upon a truck tractor.

Series parallel switch   Used to connect two parallel batteries into a series circuit for a 24-volt starter motor.

Serious traffic violations   These are going more than 15 miles over the posted speed limit; reckless driving; improper or erratic lane changes; following the vehicle ahead too closely; traffic violations that result in fatal accidents.

Shipper's certification   A written statement signed by the shipper stating a shipment was prepared properly, according to law.

Shipping paper   A shipping order, bill of lading, manifest, or other shipping document serving the same purpose. These documents must have the information required by Parts 172.202, 172.203, and 172.204.

Shoulder   The part of a roadway that runs along next to it and can be used by stopped vehicles.

Sidewalk   The part of a roadway that is to be used by people who are walking (pedestrians).

Sight glass or gauge   Glass window for determining fluid levels, as in a radiator.

Sight-side   The side of the tractor visible to the driver, opposite the blind side.

Sixty-hour/seven-day or seventy-hour/eight-day rule   Department of Transportation regulation in which no driver is allowed to be on-duty for more than 60 hours in any seven-day period; or, if carriers are operating the vehicles every day of the week, they may permit drivers to be on-duty not more than 70 hours in any eight-day period.

Skid   Failure of the tires to grip the road due to loss of traction.

Slack adjuster   An adjustable device located on the brake chamber push rod that is used to compensate for brake shoe wear.

Sleeper   Truck cab with a sleeping compartment.

Sleeper berth   Area in a tractor where a driver can sleep.

Slider   Nickname for a sliding fifth wheel.

Sliding tandem   A semitrailer tandem suspension that can be adjusted forward and backward.

Sludge   Thickened oil and sediment.

Smooth bore tanker   A trailer shaped like a long hollow tube with no baffles and used for one liquid material at a time.

Snub braking   A moderate brake-and-release technique to reduce speed by five or six miles per hour, using a brake application of about 20 to 30 psi.

Solution   Any uniform liquid mixture of two or more chemical compounds or elements that won't separate while being transported under normal conditions.

Spontaneously combustible material (solid)   A solid substance that may heat up or ignite by itself while being transported under normal conditions. This includes sludges and pastes. It also includes solids that will heat up and ignite when exposed to air.

Spotter mirror Small convex mirror mounted on the mirror frame that gives a different view.

Spring brake Conventional brake chamber and emergency or parking brake mechanism for use on vehicles equipped with air brakes.

Spring brake control Controls spring-loaded parking brakes.

Starter motor The motor that cranks the engine to get it started.

Starting system Made up of the starter battery, cables, switch, and controls.

State of domicile Your home state. This is the place you plan to return to when you are absent. It's where you have your main, permanent home.

Steering axle An axle that steers the vehicle. Can be powered or non-powered.

Stopping distance The distance the vehicle travels between the time the driver recognizes the need to stop and the time the vehicle comes to a complete stop.

Straight truck A truck with the body and engine mounted on the same chassis (rather than a combination unit, such as a tractor-semitrailer).

Street The entire width of a roadway that is maintained with public funds, including the shoulder, sidewalk, and roadway.

Suspension The system of springs and other supports that holds a vehicle on its axles.

T

Tailgating Following another vehicle too closely so that, if the need arose, there is not enough room to come to a safe stop.

Tandem Semitrailer or tractor with two rear axles.

Tandem axle An assembly of two axles, either of which may be powered.

Tank vehicle (or tanker) One that transports gas or liquids in bulk. Portable tanks that carry less than 1,000 gallons are not included in this definition.

Tarp Tarpaulin cover for an open-top trailer.

Technical name A recognized chemical name currently used in scientific and technical handbooks, journals, and text. Some generic descriptions are used as technical names.

Throttle Controls the engine speed.

Tire load The most weight a tire can carry safely.

TOFC Trailer-on-flat-car.

Traffic Anything using the roadway for the purpose of travel, including vehicles, pedestrians, herded animals, or streetcars.

Trailer Freight-hauling part of a rig, designed to be pulled by a tractor and that is generally divided into three groups: semitrailer, full trailer, and pole trailer.

Trailer brake Hand-operated remote control brake located on the steering column or dashboard for the trailer only.

Trailer spotting The job of moving a trailer from one place, or spot, to another in a terminal yard.

Transmission Mechanical device that uses gearing or torque conversion to change the ratio between engine RPM and driving RPM.

Transport vehicle A cargo-carrying vehicle, such as an automobile, van, tractor, truck, semitrailer, tank car, or rail car, used to transport cargo by any mode. Each cargo-carrying body is a separate transport vehicle.

Triple Combination rig consisting of tractor, semitrailer, and two full trailers coupled together. Also called triple headers or triple bottoms.

Trolley brake A hand valve used to operate the trailer brakes independently of tractor brakes. Also called trailer brake.

Trucking industry The business of carrying goods by truck, including carrier, drivers, warehouse and terminal employees, and all others involved in any phase of trucking.

U

Ullage The amount by which a packaging falls short of being full of liquid. It's usually stated in percent by volume.

Uncontrolled intersection An intersection without any form of traffic control, such as stop signs or stop signals.

Uncouple To disconnect sections of a trailer combination.

United States The fifty states, the District of Columbia, the Commonwealth of Puerto Rico, the Virgin Islands, American Samoa, and Guam.

Universal joint Joint or coupling that permits a swing of limited angle in any direction.

Unstable cargo Cattle, swinging meat, liquids, and the like that cannot be completely secured against movement.

Urban The city or referring to the city (the opposite of rural).

V

Vacuum brake system A brake system in which the brake mechanism is activated by a vacuum.

Valve A device that opens and closes openings in a pipe, tube, or cylinder.

Viscosity A measurement of a liquid's tendency to resist flowing. A liquid with a high viscosity doesn't flow as easily as one with a low viscosity.

Viscous liquid A liquid material that has a measured viscosity in excess of 2,500 centistokes at 25 degrees Centigrade (77 degrees Fahrenheit). Viscosity is determined using ASTM procedures.

Volatility How fast materials evaporate and become vapor.

Volt A unit of measurement of electrical potential.

Voltage regulator Part of the generator regulator used to control the voltage going from the generator to the battery.

W

Water pump Part of the cooling system that circulates coolant between the engine's water jackets and the radiator.

Water reactive material (solid) Any solid substance that will ignite when it comes in contact with water. Water reactive materials are also those that will give off flammable or toxic gases in dangerous amounts when they come in contact with water.

Wedge In a truck, a commonly used type of brake.

Wet tank Part of the air brake system with a tank that must be drained at least once a day.

Winterfront A protective device to shield the front of the vehicle from cold. Usually of canvas or other sturdy material, the two-piece construction can be fully open, fully closed, or cover just the bumper or the grille.

Wrecker Truck designed for hoisting and towing disabled vehicles.

Y

Yaw instability The side-to-side motion at the rear of a vehicle combination.

Yield Give way to a driver with the right-of-way.

Index